IN SEARCH OF THE GENESIS WORLD

IN SEARCH OF THE GENESIS WORLD

DEBUNKING THE EVOLUTION MYTH

ERICH A. VON FANGE

CONCORDIA PUBLISHING HOUSE · SAINT LOUIS

I will tell you of new things, of hidden things unknown to you.
Isaiah 48:6b

Published by Concordia Publishing House
3558 S. Jefferson Ave.
St. Louis, MO 63118-3968
1-800-325-3040 • www.cph.org

Manufactured in the United State of America

Library of Congress Cataloging-in-Publication Data

Von Fange, Erich A.
 In search of the Genesis world : debunking the evolution myth / Erich A. von Fange.
    p. cm.
 Includes bibliographical references.
 ISBN 0-7586-1100-5
 1. Creationism. 2. Creation. 3. Evolution (Biology)—Religious aspects—Christianity.   I. Title.
BS651.V665 2006
231.7'652—dc22

2006009042

1 2 3 4 5 6 7 8 9 10   15 13 12 11 10 09 08 07 06

CONTENTS

PART ONE

A WORLD VERY DIFFERENT

PREFACE

It would be difficult to find a better opening statement for this book than the one John Lightfoot made in 1682: "A few and new observations upon the book of Genesis. The most of them certain, the rest probable, all harmless, strange, and rarely heard of before."[1]

Over the past several decades some very special persons influenced my thinking and my search for insights on the ancient world. This book could not have been written without them, though I take full responsibility for what appears on these pages. For different reasons I salute and thank the following for what they consciously or unconsciously did for me: Bill Rusch, Charles C. Anderson, Albert Reiner, George Rode, Arthur W. von Fange, Arthur Custance, Alfred Rehwinkel, J. A. O. Preus, Shirley Weilnau, Erich H. Kiehl, Norman Utech, Jim Sauer, Immanuel Velikovsky, David Noel Freedman, Abdullah Alireza and his sons Teymour and Fahd, Joe Oppenheim, Norman Macbeth, John Whitcomb, Henry Morris, Chuck Wagner, Edgar and Margaret Riep, William Corliss, Doug Sharp, Paul Anderson, Lorella Rouster, Ralph Lohrengel.

Above all I thank my wife, Esther, and my children, Paul, Ruth, Lois, Jana, and Judi for their patience and encouragement.

DISCLAIMER: Please note that the dates in this book are almost always based on the assumptions and beliefs of evolutionists about time. As this book makes clear, I have never found reasons to accept such millions and billions as valid. I fully accept the teachings of the Bible that we live on a young, created earth.

Erich A. von Fange, PhD

FOREWORD

Readers of this book will encounter a no-nonsense, comprehensive search for the Genesis World. In chapter 1 Erich von Fange explores the fallacy that evolutionary science is objective, unbiased, and accurate. His quotes by James Lovelock, Søren Løvtrup, Michael Wilson, and David Childress are critical to the belief that the evolutionary model of origins is a kind of mantra that covers all "scientific" efforts to discover the real truth. Chapter 2 presents an exhaustive review of the place of metals in the possible development of the ancient past. The author proposes that the Old Testament view shows a better, more accurate framework of early history.

Dr. von Fange presents an exhaustive review of the framework for understanding the belief that man and dinosaurs existed together on a young earth. With a broad list of references, he points out that dinosaurs were created during creation week; they may have lived at the time of Job; some human footprints and dinosaur tracks together may be genuine; dragons are dinosaurs; and dinosaur species not already extinct were included on Noah's ark and died later in extreme environments.

Citing the limitations of archaeological research, von Fange covers the three areas of scholarly archeology: (1) field archeology, (2) linguistic studies, and (3) biblical studies, showing how they can fit neatly with what the Bible describes and teaches.

There is an elaborate explanation and clarification on two attempts (Charles Totten and Harold Hill) to verify "Joshua's long day" as recorded in the Bible, even though these apparent scientific theories were proven to be false. The author concluded the long day really happened. "The Bible does not need to be proved. It is believed by faith."

There is an interesting coverage of "the art of misquoting Ussher." The author exposes all the sloppy and dishonest statements ridiculing Archbishop James Ussher's dating of the beginning of the earth at 4004 BC. After an exhaustive review of evolutionary statements about what Ussher said and what Ussher really published in his original volume, the author concludes: "Ussher is roughly correct and the pronouncements of evolutionists are precisely wrong." He points out that many evolutionary scientists see "only what they expect to see."

Readers will enjoy the "incredible Piltdown hoax" where the author concludes "a skull made monkeys out of anthropologists." All believers of evolution should read this chapter as a warning to those scientists who have an emotional religious belief in proving a theory that has "wishful thinking" as its motivation for carelessly interpreting the facts. Evidently the many doctoral dissertations and other scholarly studies on Piltdown Man are acts of worship of the false religion of evolution rather than critical, scholarly efforts.

In reviewing the large amount of data about plants and trees in the ancient world, it is concluded that the Christian who accepts the biblical record has nothing to fear from observing the data.

The chapter on astronomy shows that it is the most speculative of all fields of science. Examples of scientists being professionally punished for discovering data that contradicted the pre-conceived ideas of evolutionary astronomers point out that what Einstein said is true today: "A theory informs you of what you are permitted to see."

Readers will find that when it comes to the evolutionary explanation of extinctions, the denial of catastrophes in the past, and the supposed history of animal domestication, it is "the blind leading the blind." The biblical framework of history may not tell us everything, but it can make a good fit for the limited contemporary data we have. The author shows how eight fictions on the supposed evolution of the horse demonstrate how wishful thinking of evolutionists has replaced the bare facts in order to support the questionable theory of evolution.

After a comprehensive review of the opinions of scholars and cranks about prehistory, Erich von Fange recommends the Bible as the best framework for the past because any human attempt at the reconstruction of the past is full of assumptions. No one was there to record the infinite details of the drama.

I especially enjoyed certain clever, telling comments inserted into the texts, such as:

- "Interpretations begin with mountains of speculation based on molehills of evidence."

- "What does one do when irresistible bones are found in immovable strata?"

- "Vast amounts of scholarly energy are devoted to the art of explaining away inconvenient evidence."

- "Many experts are only interested in 'proving' their own opinions."

- "Evolution's 'big lie': Creation is factless faith and evolution is faithless fact."

In conclusion, the reader is led to a choice between faith in the Triune God of the Bible or faith in the triune god of evolution, that is, father time (unlimited billions of years), mother nature (only natural processes), and lady luck (chance probabilities). Dr. von Fange has presented a strong case for the former.

David A. Kaufmann, PhD, Secretary, Creation Research Society

CHAPTER ONE

What Can We Really Know about the Ancient World?

A Search for History

Ancient history holds special attraction for those who like to tackle great mysteries. For them, searching out the past is an intriguing adventure. It is no exaggeration to say that millions around the world are deeply engrossed in genealogy, searching for their ancestors. Some years ago the television special and book *Roots* captivated a nationwide audience.[1] Libraries stock sections of books that attempt to track the history of humankind, of the earth, and of the universe.

How can we uncover the real story of ancient times? Is the Bible the key for unlocking these mysteries? The first impulse for some people might be to run to the nearest encyclopedia or textbook to learn the basics of ancient history. Is the best way to learn about the past by taking courses at a university? Is it possible that the textbooks have it all wrong? After a lifetime of studying the literature dealing with the past, William Corliss concluded, "The entire picture of human exploration and colonization of our planet is probably radically different from what we have been led to believe."[2]

Similarly, British scholar Richard Rudgley summed up his thorough study of the ancient world by declaring that "the widely accepted view of the human story is wildly inaccurate," and "preconceived opinions have repeatedly led to the rejection of evidence that does not fit with present archaeological dogmas."[3]

WHERE IS THE TRUTH?

As I shall point out and document, things are not simple. We soon discover, for example, that lectures, television specials, textbooks, and articles are always written within the bonds and the boundaries of a prevailing theory in the mind of the writer. There is no secret about this. Albert Einstein said it well: "A theory informs you of what you are permitted to see!"[4] Thus what fits is used; what does not fit is discarded or attacked or ignored. This is why a bad theory can look very good when all the evidence against it is not allowed to be considered.

In this book I discuss two sharply contrasting beliefs about the ancient world: (1) that there existed a young, created world; and (2) an old world evolved through time and chance. The Bible presents origins and history within the framework of an all-powerful God who directs the course of history. The last place some would look for truth is the first place many others seek it. No matter what one believes about the ancient world, there are difficulties. Let us examine more closely the Bible as a framework for exploring the ancient past.

- "Whatever wisdom may be, is far off and most profound—who can discover it?" (Ecclesiastes 7:24)
- "Great are the works of the LORD; they are pondered by all who delight in them." (Psalm 111:2)
- "Search the past, the time before you were born, all the way back to the time when God created man on the earth. Search the entire earth." (Deuteronomy 4:32 TEV)
- "Who is this that darkens my counsel with words without knowledge?" (Job 38:2)
- "They say that what is right is wrong, and what is wrong is right; that black is white and white is black; bitter is sweet and sweet is bitter." (Isaiah 5:20 TLB)
- "A friendly discussion is as stimulating as the sparks that fly when iron strikes iron." (Proverbs 27:17 TLB)

These passages reflect some of the excitement as well as some of the problems and mysteries in exploring the past.

IS SCIENTIFIC METHOD THE WAY TO GO?

The lifeblood of science is the scientific method. This method includes the following elements: State a problem that can be tested; state your theories; gather and analyze all the available relevant data; draw tentative conclusions; repeat the study, preferably by other scientists. Fundamental to the method is healthy criticism of what is currently believed to be true about a hunch or a theory.[5] Is evolution a good example of how scientific method is used? An ardent evolutionist included the following in a letter to the *Kansas City Star*:

> [T]he evidence to support organic evolution relies on reason and common sense, not mythology, as does the Bible. The various sciences used as evidence in favor of organic evolution are in themselves quite provable: classification of species, homology, analogy, vestigial organs, physiology, embryology, paleontology, anthropology, geology and astronomy.[6]

We disagree, of course. This quote is an example of what many students today are taught at our universities. As this book unfolds I shall examine other similar statements. For the present let us consider the following statements made by scholars who point out that anyone who conducts a study soon runs into problems of bias. We find this very illuminating statement by an evolutionist who speaks of the pecking order among the sciences, as stated by a Nobel Laureate at Columbia University:

> [A]n intellectual hierarchy exists in science, with mathematics and theoretical physics on top, experimental physics just beneath, and then further down, chemistry and perhaps astronomy. Geology and paleontology, which deal with dirty objects like rocks are considerably lower on the list, and biology—at least the parts that deal with soft squishy things like entire organisms—is at the bottom. Anthropology, psychology, and the "soft sciences" like sociology are not on the physicist's list at all.[7]

As you go lower and lower on the list, there is less and less real science happening. This does not sound encouraging because many of the books that tell us about the past are from the "soft" sciences at the lower end of the pecking order. Potential error and bias lurk everywhere. A good illustration comes from Joseph Alsop, a scholar of ancient history, who discovered many examples of experts confounded with remarkable regularity. For example, in a study of Phoenicians and Greeks, he concluded that experts frequently place paralyzing straitjackets onto hard facts, for the sole purpose of justifying their own preconceived notions.[8] Truth seldom emerges in his view. In another, more general view of history, one authority, Paul Johnson, stated:

> The study of history is . . . humbling to discover how many of our glib assumptions, which seem to us novel and plausible, have been tested before, not once but many times and in innumerable guises; and discovered to be, at great human cost, wholly false.[9]

James Lovelock assesses much of the scientific establishment in this way: He stated that nearly all scientists have traded freedom of thought for good working conditions, a steady income, tenure, and a pension.[10] They are also constrained by bureaucratic forces, from the funding agencies who make clear in advance what they want "discovered," and by the tribal rules of the discipline to which they belong. He calls the peer review a self-imposed inquisition. It has degenerated into a well-meaning but narrow-minded nanny of an institution to ensure that scientists work according to the current party line and not as curiosity or inspiration might move them. They are thus entrapped in rigid dogma.

The following two comments were made by people who have experienced and evaluated society and the educational scene:

> The universities today are not places of mental adventure, but dull workhouses of conformity.[11]
> It is better to be roughly right than precisely wrong.[12]

A noted scientist, Søren Løvtrup (not a creationist), made the following remarks:

> Among the unwritten laws of scientific hierarchies, one is that you must respect the scholarship and authority of all your colleagues. Anyone who violates this law is certain to suffer ostracism.
>
> Biologists themselves are aware of the fact that the significance of Darwinism is a myth, but for reasons of piety they do not divulge the truth.
>
> Micromutations do occur, but the theory that these alone can account for evolutionary change is either falsified, or else it is an unfalsifiable, hence metaphysical, theory.
>
> I suppose that nobody will deny that it is a great misfortune if an entire branch of science becomes addicted to a false theory. But this is what has happened to biology: for a long time now people discuss evolutionary problems in a peculiar Darwinist vocabulary—adaptation, selection pressure, natural selection, etc.—thereby believing that they contribute to the explanation of natural events. They do not.
>
> I made a very remarkable and unsuspected discovery: nobody, not even Darwin and his closest friends, ever believed in Darwin's theory of natural selection: Darwinism was refuted from the moment it was conceived—a very peculiar situation in the history of biology.[13]

It is worth repeating: The above comments were stated by a prominent scientist who has no ties with religion.

Listen to a Veteran Scientist

A geologist for the state of Wyoming came up with some surprising conclusions about advances in science:

> Major breakthroughs at scientific frontiers are not usually the result of dogged application of the scientific method. The discoverers usually are using methods and theories not in the mainstream of their disciplines. Thus it is that their discoveries—in vindication of their unorthodox pro-

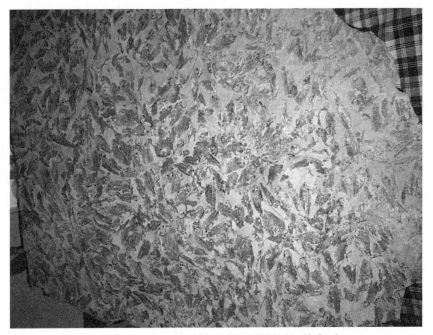

A vivid example of the trillions of fossils left by Noah's flood.

cedures—result in true scientific revolutions. Mainstream scientists, when studying clusters of data, cling to their past interpretations and ignore what does not fit. Unorthodox scientists may ignore these same clusters of data in favor of the one or two bits that do not fit the cluster. Such puzzling bits could lead to the discovery of a previously undetected law or rule.... Amateur researchers working without affiliation to learning institutions are very often scorned by official anthropologists, who doubt that such amateurs possess adequate resources with which to make meaningful contributions.... Those who write books directed to the public at large instead of to their colleagues are often the subject of derision, especially if the book is controversial, or worse yet, if it makes money.[14]

David Childress comments in a similar fashion:

Contrary to popular opinion, geology is not a very exact science, nor are there any real geological "facts." Geology is a matter of opinion and theory, and many scientific theories taught as fact in schools may never really be proven ... ever changing and evolving as old theories become replaced by newer, more "reasonable" theories.[15]

The above are some heavy hits on those who abuse scientific method. We cannot assume that textbook writers are free from the abuses stated above. Note carefully that nothing above criticizes the method itself. It is only when faulty beliefs, biases, and assumptions get in the way that problems arise. It is very plain to see that one cannot simply swallow whatever the printed page presents. We must know something of the author's belief system, which in turn sets the framework for what he or she writes. A wise man said: "The first to present his case seems right, till another comes forward and questions him" (Proverbs 18:17).

The message seems very obvious. We must do as scientific method demands. Be skeptical, question, and test what is said. Truth then has a chance to emerge. Everyone, Christian and atheist alike, operates under a belief system or framework. In contrast to the faith of the evolutionist, who depends on time and chance for explanation, creationists believe the Bible teaches the only way to eternal salvation in Christ. But it is also a true, valid framework for the study of the ancient world.

The Purpose of this Book

Above all, this book discusses fascinating mysteries about the past. This is a book about so-called prehistory. As summarized below, I will examine many of the sciences that endeavor to deal with some aspect of the ancient world. In part I treat areas of the past where other written records either do not exist or where they do not yet furnish clear answers to crucial questions. I have a special interest in those aspects of ancient history where the biblical record and other attempted explanations of the past are in serious conflict. Is the difficulty caused by the evidence or by speculation about it?

Thus I want to explore and evaluate two approaches for studying the past. First, how does the Bible fare as a framework for the ancient world in the light of scientific discoveries? That is a very important question for those who hold that the Bible is God's truth. Second, is evolution "fact" as many claim, or is it a type of mantra smothering all efforts to discover real truth? In terms of evolution, could the mantra invoked be the sacred formula believed to embody the new divinity of time and chance that possesses magical powers to explain the mysteries of the past?

Explorations in this Book

Attempting to treat the enormous complexity of the past in any comprehensive way is something like the child who decides to empty the ocean with his spoon and little tin bucket. Nevertheless, the mysteries and other challenges are much too fascinating to set aside. The following is a brief outline of what I am attempting to do in this book.

Under biblical studies, I first take a fresh look at the ancient use of metals and stone that govern much of the work in archaeology. What light does biblical archaeology shed on the past, and how do we separate fact from interpretation? One interesting story in the Bible describes Joshua's long day. This tale has received much attention in recent years, and this material cries out for evaluation. The age of the world has long been a troubling controversy. What may we say about the age of the world in light of Archbishop James Ussher's conclusions in contrast to the beliefs of evolutionists? Dinosaurs continue to make news. How do Christians, evolutionists, and others treat the fascinating mystery of the dinosaur?

I then explore the uneasy alliance between science and evolution. Did man really evolve? How could a large number of supportive doctoral dissertations and other formal scientific documents be written about the obviously fake Piltdown skull?[16] I will explore what is known about the beginning of animal domestication. Did the horse evolve as pictured in countless textbooks? The wonderful tale of so-called "false" horses in South America furnishes a window into the unusual problems of classifying the animal world. The plant world also sheds interesting light on ancient times and on how the evidence is interpreted.

What is truth and what is fiction about Charles Darwin's beliefs? After 150 years of Darwinism how is it possible for one evolutionist to publicly criticize another for incorrectly showing how gradual change occurs?[17] After all, evolution is presented as fact—not theory. What kind of answer about the past does ancient astronomy furnish us? It is hard to resist evaluating some of those who have explored the past, so I will examine the works of scholars as well as some that are called the "lunatic fringe." Finally, acknowledging the many wonders and accomplishments of science, we also need to be aware of the dark side of deception all too common in some aspects of science. The final chapter sums up what we have learned in our search within the two radically opposite ways of explaining the past.

Special Motivation

Several weeks after I had shared truths of Genesis with a Bible class in Florida, I received a very special letter. The writer was moved to say this:

> I am a very good example of someone who has been lost in a sea of unanswered questions. I am now 30 years old—was raised as a Christian, but when I was about thirteen I began wondering and asking about the beginning of the earth. I strayed from the church and chose to believe the scientists' way of thinking. I recently have been drawn back to the church . . .
> I am so desperately searching for sense to be made of my questions.

This book was written to inform and assure the reader that science was never the problem. There is a vast difference between science and speculation posing as science.

For Reflection and Discussion

1. St. Paul warned Timothy (2 Timothy 4:4) that the time would come when a great number of teachers "will turn their ears away from the truth and turn aside to myths." Is this happening today? In the coming chapters we will see many examples of plain truth wrongly interpreted. Christians need to distinguish between truth and interpretation in fields of science. Why? Is truth determined by majority opinion?

2. We are offered two frameworks in which to fit all of ancient history. The Bible tells of a young, created earth. The followers of Charles Darwin insist on a very old earth where time and chance are the foundations of everything that has happened. Can these opposite views be reconciled? When Darwin's famous book *The Origin of Species* appeared, many of the clergy in England were strong supporters, and scientists of that time were generally very critical. What does this tell us?

3. A prominent view in philosophy is that everything that exists, exists in some amount, and therefore it can be measured. Can you think of anything invisible that cannot be measured in the laboratory?

4. Einstein stated that a theory tells what one is *permitted* to see. What good thing may happen when evidence does not fit the theory? What else may happen to such evidence?

5. Is twisting plain evidence something new in human history? What is Isaiah saying in 5:20? "Woe to those who call evil good and good evil, who put darkness for light and light for darkness, who put bitter for sweet and sweet for bitter." We believe evolution is one very dangerous example. Agree or disagree? Can you think of examples of evil called good today?

6. "Scientific method" has long been recognized as vital to progress in any civilization. This includes healthy criticism of any theory to see if it actually represents the real world. May a theory be called a fact in order to avoid criticism? Is evolution theory or fact?

7. Many argue today that we must not challenge anything called science. Agree or disagree? Why or why not?

8. We sometimes hear that there is an unwritten law among scientists that they must respect the scholarship and teaching of fellow scientists. It is not rare to read that a scientist will say that there is no proof of evolution in his/her field of expertise, but there is convincing proof in other fields where he has no background to judge. Defend or oppose.

9. How did St. Paul in his witness approach the blasphemy of false gods he observed in Athens? See Acts 17:22–31. How would his witness have changed if he had exploded in anger and ridicule over their blindness? Is there a lesson

here for us? What should be our approach in speaking with those who honestly believe that evolution is really science?

10. Is it possible to strongly defend science and scientific method and yet oppose just as strongly the beliefs in evolution? Hundreds of highly qualified scientists state that evolution is not the answer to earth's history. They are attacked, not on the basis of evidence, but they are accused of dragging religion into the science classroom.

11. A noted scientist, an evolutionist, stated in the preface to a new edition of Darwin's *The Origin of Species* that evolution and creation are exactly parallel. Neither can be proved to the satisfaction of the other. Agree or disagree? Why? Why doesn't this statement appear in today's textbooks? As no lab can prove either belief, would a good solution be to take both evolution and creation out of the science classroom and teach the two opposing views in a class or other presentation on origins?

PART TWO

BIBLICAL STUDIES AND ARCHAEOLOGY

CHAPTER TWO

THE ANCIENTS AND THEIR USE OF METALS

INTRODUCTION

What the Textbooks Tell Us

When we research the ancient past, nothing seems more deeply ingrained in textbooks than the belief that over a long period of time man passed through various stages from apehood to manhood. This observation is followed almost universally in texts and other books on ancient history, anthropology, archaeology, and related fields. This view of the past is so firmly established that there appears to be little purpose in questioning it. Yet here and there are the curious anomalies, the mysteries, and the startling exceptions. Beyond all this there are the assumptions, that is, the beliefs, on which the whole pattern is built. A closer look at the whole supposed sequence, therefore, does seem to be in order.

The Drive for Metals

Because this chapter examines the use of metals in ancient times, it is interesting to find that there was a real obsession with metals by the ancient people, as shown by the names of key metal producing areas. Nubia was the Egyptian word for *gold*. Hatus, the Hittite capital, means "Silver City." Cyprus is the same as our word *copper*. Tarshish means "smelter," and the word *Barzel* or *Brazil*, which means "iron," spread into the Atlantic.[1]

A Look Ahead

In this chapter we shall look more closely at the origin of stone and metal ages, their validity, and a number of curious things reported around the world having to do in some way with metals. I will describe and show the roots of the conventional pattern of stone and metal ages. Following this, I will examine some significant statements about metals from ancient literature, including the Bible. Along the way a few of the many curious finds of metals in regions around the world will be revealed.

Does the idea of evolution or the Bible as a framework for the ancient world better describe what we know?[2] Basic emphases in this chapter are:

- How far back can we trace the use of metals?

- Did technology seem to spread from one source or were there many independent discoveries? What are the implications of our findings on the two belief systems above?

- How well does the commonly accepted time sequence of stone to bronze to iron reflect the true nature of the past?

- The world today vividly demonstrates that technology is cumulative. Is this the same as evolution? How sophisticated were the ancients in metalworking and what are the implications of what we find?

- When did deep-sea travel develop? What was its purpose and how did it originate?

THE CONVENTIONAL CHRONOLOGY

The Idea of Stone and Metal Ages

Archaeologists use their digs to make painstaking reconstructions of sequences of artifacts and events in the past. What are the assumptions of this craft? Can errors creep into the conclusions drawn?

In 1819 a Danish archaeologist, Christian Thomsen, proposed an idea so logical that no respectable author of a book on the ancient past would dream of not using it. Thomsen suggested that man first passed through a Stone Age, then a Bronze Age, and finally an Iron Age. The Iron Age ended about 333 BC at the time of Alexander the Great, after which the world entered into more modern times.

These terms had been lying around for a much longer time. Certainly Christian Thomsen popularized the concepts by means of exhibits he prepared for the National Museum in Copenhagen in 1836. Earlier, Chinese historians had suggested Stone, Bronze, and Iron ages in AD 542. In 1758 a French magistrate suggested the same periods but inserted a Copper Age. In 1813 a Dutch historian argued for Stone, Bronze, and Iron periods in Scandinavian prehistory.[3]

Each of the original three ages was divided up rather neatly into early, middle, and late periods. With a bit of a flair for sophistication, the Old Stone Age was called the Paleolithic (Old Stone), the Middle Stone Age became the Mesolithic, and the New Stone Age became the Neolithic. Some found this beautiful symmetry of the pattern lacking and inserted a Copper Age or Chalcolithic (meaning "copper-stone") Age between the Stone Age and the Bronze Age. Others inserted an Eolithic Age (dawn stone) before the Paleolithic Age. It is plain to see that the whole pattern was founded on the belief that man went from primitive to modern stages, a basic teaching of evolutionary theory.

The Bronze Age in Palestine is said to have run from 3150 BC to 1200 BC. The early Bronze Age (EB) is given as 3150–2150 BC. The Middle Bronze Age (MB) lasted from 2150–1550 BC, and the Late Bronze Age (LB) was from 1550–1200 BC. Similarly the Iron Age was divided into early (1200–1000 BC), middle (1000–587 BC), and late (587–333 BC). In other parts of the world and with different authorities, dates for the Stone and Metal ages vary.

Problems Arise

These nine periods still proved to be inadequate in many ways, so a marvelous assortment of Roman numerals, capital and small letters of the alphabet, and Arabic numerals were used to subdivide the periods even further. This method has infinite possibilities for expansion. No speck of dust, no pottery sherd, no stone flake or chip need be unassigned. For the archaeologist there is a place for everything in the neat scheme, and everything is put into its proper place.

The scheme is a triumph for man's penchant for classifying. Only one little ripple mars the placid surface of the plan. It does not work very well in many parts of the world, because it is heavy with assumptions that do not relate to the real world, and almost nothing relates to actual history.

The Stone Ages

Gabriel de Mortillet, a French archaeologist, followed up the classification system of Christian Thomsen with a refinement of the stone ages. Based on studies of what was assumed about the ice ages and of artifacts found in France, he arranged a system of time periods from oldest to youngest: Chellean, Acheulian, Mousterian, Aurignacian, Solutrean, and Magdalenian. Later the term Abbevillian replaced Chellean, because the latter turned out to be identical with Acheulian. These formidable names of ages, which sound so scholarly and which plagued generations of budding archaeologists, were simply derived from the little nearby French villages where the typical artifacts were found for each time period, or so it was thought.

The system was supposed to fit the whole world, but it was based on limited studies along with huge assumptions in a small part of Europe. Although half-hearted attempts are sometimes made to apply the system elsewhere in the world, the plan, as archaeologist Kenneth Macgowan says, is no more than a fifty-year-old straitjacket patched here and there.[4] The Mortillet system is tied closely to an assumption of four glacial periods with long intervals between each period, but scholars come up with wildly different guesses on the length of these ages and even on how many ice ages there had been.

A generation ago Will Durant wrote a monumental series of books on the history of civilization.[5] Having no better scheme to follow, he slogs through all the stone ages from Pre-Chellean (Eolithic) down through the Magdalenian, showing the developments that distinguish one from the other. At one age people chipped stones this way, and 50,000 years later they chipped them that way. Only the date for the first age is qualified—"about" 125,000 BC. More daring than most, Durant was willing to see remains of the Chellean Culture, 100,000 BC, in Europe, Greenland, the United States, Canada, Mexico, Africa, the Near East, India, and China. Still more daring, he found the Mousterian Culture, featuring Neanderthal Man at 40,000 BC on all continents. Durant allowed for pottery in the Magdalenian Culture, 16,000 BC, but Nelson Glueck placed the Pottery Stage at about 5000 BC.[6] The Mesolithic Age, also called the Natufian Age (before 7000 BC), was divided into Lower and Upper. This age saw the "sudden" development of all the cereal grains and the domestication of animals—a spectacular scientific achievement. Others denied that there ever was such an age and credited these achievements to the Neolithic.

The Neolithic Age followed, characterized by the development of villages and pottery as well as other great inventions. The Pre-pottery stage was thought to be about 7000–5000 BC, followed by the Pottery Stage, about 5000 BC. Jericho is given as the classic example of both Neolithic stages, the Chalcolithic Age, and of all the ages that followed.[7]

The fun begins when a Neolithic arrow point, as reported by J. William Dawson, is found in a skeleton from the Paleolithic Age. Evidently the arrow flew over a time gap of many, many thousands of years.[8] Of many other problems that soon followed with the whole scheme, one came out of Africa. Similar stone tools to those of Europe were found, but it was impossible to date them in the same way. They obviously came from a much different time. Kenneth Macgowan concluded the Paleolithic, Neolithic, Bronze Age, and the like, mean less than nothing as a dating mechanism for more than a single locality. Even the term *postglacial* means different things in two such neighboring countries as Sweden and Germany.[9] Clearly something is radically wrong with the whole scheme of dating. Unfortunately for the scientist imprisoned by evolution, there is no palatable alternative.

Louis Leakey muddied everything by finding sherds and marks of basketry in Paleolithic strata in Africa. No assertion about the Stone Age is safe. Gordon Childe concluded that there was no such thing as a Neolithic civilization, but for lack of any alternative, the textbooks continue to carry the whole scheme with no hint of any problem.[10]

In the midst of the incredible confusion in dating the past, Willard Libby came to the rescue in the 1950s with C14 dating that at last placed dating on a scientific basis, or so it was thought. It did not take long for a whole new set of problems to arise. Archaeologist James Griffin, a scholar's scholar, neatly summed up the problems by saying that C14 tests on the same specimen gave wildly different results far outside allowable statistical error.[11] This understandably tends to shake the faith of the social scientist or humanist in the magic of atomic science. It is no secret that the "correct" answer often must be pulled out of a hat. Like any other dating system in scientific guise or not, it is only as good as the assumptions on which it is based.

The Bronze Ages

The Bronze Age in Palestine rests on an elaborate set of assumptions, generally accepted synchronisms with ancient Egypt, and many decades of painstaking archaeological study. Already in the Early Bronze Age, the use of bronze, a mixture usually of copper and tin, was common in Mesopotamia. Users of bronze soon after were swarming into Europe and the British Isles. Swords, shields, buckles, pins, and many other artifacts of this ancient time show great artistic skill.[12] As we shall see, the worldwide search for tin was triggered by the insatiable demand for articles made of bronze.

The Iron Ages

We know that iron, usually attributed to a meteoric origin, was used in predynastic (before 3000 BC) Egypt for making beads. Smelting of iron ore for tools, weapons, and other artifacts is generally dated back to about 1300 or 1400 BC. Iron became common in western Europe about 500 BC.[13]

It is only natural that artifacts found in isolation of stone, of copper, of bronze, or of iron would tend to be placed by most students of ancient history into an appropriate age in the conventional system. For example, an iron artifact would normally not be dated earlier than about 1200 BC no matter what the circumstances were of its discovery.

Bronze and Iron Age Surprises

One of the initial surprises in reading the literature about the Bronze Age and the Iron Age is that metals are very incidental to the geologic periods. The many subdivisions are based almost entirely on pottery remains, which are plentiful and normally about the only artifacts that have the potential for being dated. Much of the confusion about time rests on several factors: First, Middle East dates are calibrated with the chronology of Egypt, but Egyptian chronology has grave problems of its own and may well be disastrously in error. Second, the long life attributed to the mud brick assumes factors about weather and rainfall in ancient times that simply are not true. In addition, other factors, such as fire and destructive invasions, played important roles in creating the many levels found in the mounds or tells (ruins) of ancient cities and towns.

Even more important, we must not assume that past climate was the same as present climatic conditions. Hundreds of large cities thrived in the Middle East in the distant past where nothing will grow today. There are two reasons why early Jericho ruins may be much younger than one might think. John Garstang found that the early floors of the city were built in part on the channel of the great spring there and the water seeped upward. This required frequent rebuilding. Further, there is good evidence that early Jericho was even wetter than the Dead Sea area during the time of Lot when it was described as watered like the garden of the Lord.[14] The conventional dating of ancient Jericho is therefore based on assumptions that simply do not hold water, to borrow a phrase.

Kenneth Macgowan has harsh words for the so-called Bronze Age, which he calls a misnomer and a phantasm, a minor phase in the use of copper. Much of the Bronze Age is limited to the area of southern Europe, Asia Minor, and (much later) the Incas in Peru. He notes that most of the world used iron before bronze.[15] We must remember, too, that iron rusts rapidly, while bronze does not. With so much of the world poorly investigated, for example, vast stretches of Siberia and most Third World countries, there is still much to be learned about early man and his use of metals. The conclusion is obvious. One cannot fix a chronological period by the percentage of bronze and iron objects found in tombs.

Other Problems in Dating Archaeological Materials

One may have the highest regard for the archaeologist and his work without becoming too much in awe when this scientist speaks with great confidence about time. Extensive reading of the literature makes plain some very simple and basic facts about the problems in fixing time periods for many archaeological sites:

- Most ancient remains have been ravaged by nature and by the destructive hand of man;
- Most of what little has survived has been ransacked, robbed, and trampled before scientists got to them;
- Most of what little has survived this devastation has never been excavated—only the tiniest fraction of one percent of sites have been scientifically examined and analyzed;

- Most of what has been excavated in the past is a matter of blunder and plunder as sophisticated scientific methods of excavation scarcely existed before the 1950s, when the work of Kathleen Kenyon began to open up a more scientific approach;

- Most of the conclusions drawn about the ancient past are based on those early crude excavations;

- Most plans for excavating important sites cannot be carried out because of location or lack of time, staff, and funding; important sites lie underneath modern cities and towns;

- Vast and important areas of the earth have not yet been surveyed in even the most casual manner for their archaeological potential;

- So-called scientific methods of dating archaeological remains are often more a scandal than a help, and scientists are reduced to using such data only if they happen to "fit";

- Assumptions and theories rule supreme—inconvenient finds are ignored or declared to be fraudulent. Ancient chronologies in the Middle East are based on assumptions that are certainly open to honest question, on a calibration with Egyptian chronology that is open to challenge, and on uniformitarian assumptions about the strata in which pottery sherds are found.

Perhaps some will think this rather dismal outline is overstated and anti-science. All ought to agree, however, that there is little room for confidence about dating the ancient world, regardless of the position taken. Of course, dates are correct if all the assumptions underlying the conclusions are correct, but the scientist ought to write down all the assumptions that went into the conclusions. The archaeologist may be astonished.

Ancient Literature on Metals

Early References to Metals

We would expect references in ancient literature to support in a general way the scheme of Stone, Bronze, and Iron ages accepted today. If this is not the case, two possible explanations would suggest themselves: (1) The ancient literature cited is in error; or (2) The dating system may contain some problems.[16]

The Bible

In the pre-flood world Tubal-Cain forged all kinds of tools out of bronze and iron (Genesis 4:22). According to Byron Nelson,[17] the Pentateuch contains thirty references to bronze (or brass or copper) and two references to iron. If we take the Bible seriously, mining, metallurgy, and smelting were highly developed long before the universal flood. We assume that Noah and his sons were the Old

Ones spoken of in many cultures, who passed on this technology after the flood. Some of Noah's descendants learned the technology before they broke away or were driven away from the first tribe of them all. The knowledge of metals or lack of it and the characteristics of the environment for a given band of ancient people determined whether or not a metal age followed for them. Thus there have been metal ages from the very beginning, even though people still live in a Stone Age culture today in remote parts of the world.

Moses spoke familiarly of iron centuries before the assumed beginning of the Iron Age.[18] It is curious indeed that Moses, unaware how modern scholars would separate Copper, Bronze, and Iron ages, spoke of copper, bronze, and iron in one breath in Genesis 4:22. Long before the Iron Age, the Promised Land for the Hebrews was described as a land whose stones are iron and out of the hills of which the people could dig copper (Deuteronomy 8:9). The name of the first teacher of metallurgy, Tubal-Cain, became the Hebrew word for "smith or worker in metals." "Ken" is another form of "Cain," and thus the Kenites were a tribe of smiths noted for their skill in metal working.[19]

The Fabled Land of Ophir

In the Bible we read of the land of Ophir, famous for its gold, and most scholars identify the name with some location on the Arabian peninsula. (See 1 Kings 9:28.) Yet how could Solomon's fleet take three years for each round-trip to Ophir? In 1571 a Spanish scholar, Arias Montanus, published a world map of biblical lands. It is surprising and interesting that he indicated Ophir in two places on the map: on the Andean coast of South America and in the coastal mountains of the western United States. Gold in the United States was not known until 1848, almost 300 years after the map was printed. Cyrus Gordon suggests a very ancient map tradition carried down over many centuries, an intriguing possibility.[20]

Two Lost Civilizations

From a variety of fragments of ancient literature, we are able to piece together other information on the use of metals in the distant past. We know from clay tablets found in Mesopotamia that during the height of the Bronze Age, two "lost" civilizations were involved in supplying Mesopotamia with huge stores of copper. Makan, the Copper Kingdom, lay to the south and east of Bahrain in Arabia. Bahrain was part of the ancient kingdom of Dilmun. The sands of Arabia still cover most of the story of Makan and Dilmun, but Geoffrey Bibby has made a very interesting beginning. The copper mines and old cities of Makan have been explored since the 1990s. Dilmun supplied the great seaports that served the merchants of Ur and of Makan.[21]

Surprising Metal Words

Some of the literature of the ancient Mesopotamian kingdom of Akkad has been recovered on clay tablets. Akkadian was an obsolete language by 1500 BC, about the time of Moses' birth. Texts in this old language use words for tin, copper, lead, gold, silver, iron, and bronze. One hymn speaks of bronze as a mixture of copper and tin.[22] The big surprise is the mention of iron. The Iron Age was centuries in the future!

Another View of Historic Ages

Hesiod, the noted eighth-century BC Greek poet, had a catastrophic view of ancient times. He described world ages, each of which ended in a great destruction. He spoke of the third world age of bronze, followed by those who used both bronze and iron, which was the time of the Trojan War. After another destruction the next generation used iron.[23]

Ancient Literature on Iron Taboos

Iron seems to be in a class by itself in ancient literature. An immensely complicating factor in the consideration of the role of iron in ancient life is the shroud of mystery and superstition that covers this metal unlike any other. Undoubtedly, the fall of iron meteorites contributed to the feelings of the ancients about iron, but the problem seems much deeper than this one factor.

Sacred Meteorites

Iron meteorites were venerated in many ancient temples, for example, in the temple of Astarte at Tyre; in the temple of Amon at Thebes; at Delphi; in Mexico; and of course the sacred black stone of Mecca enshrined in the Ka'ba.[24] In Ephesus the "image that fell from heaven" was greatly venerated (Acts 19:35). Similarly, an ancient relic was passed down among the Malay kings. This was the Sacred Lump of Iron that became part of the regalia of the rulers. The lump was regarded with the most extraordinary reverence mingled with superstitious terror. In the Celebes iron was much used in magic rites.[25]

Ancient Fears About Iron

James Frazer reviews taboos about iron. To many people, objects made of iron were especially noxious and dangerous. Roman and Sabine priests could not be shaved with an iron implement, only with bronze razors or shears. Whenever an iron graving tool was brought into the sacred grove in Rome for cutting inscriptions in stone, a sacrifice of a lamb and a pig had to be offered. The same rite was repeated when the tool was again taken from the grove.

As a general rule iron could not be brought into Greek sanctuaries. On Crete, sacrifices could not involve the use of iron in any way. The Hottentots today never use an iron knife for sacrifices or circumcision. In southwest Africa, iron could be used for the operation, but only if the iron knife was then safely buried. In Yap of the Caroline Islands, wood for the fire-drill could not be cut with iron or steel. The men who made the need-fire in Scotland and elsewhere first had to remove all metal from their persons. This was an ancient Celtic rite that persisted in many lands into the twentieth century.

Iron was forbidden in Scottish divination ceremonies. In India an old taboo affected a highly educated Hindu Rajah who would not allow iron to be used in any building on his land. His belief was that great evil would follow if the taboo was violated. The Baduwis of Java today will use no iron tools in tilling. Iron there is used as a charm for banning ghosts and spirits. The Toradyas of Central Celebes are very careful not to put iron in a coffin. They think the dead may throw it out, and they also think that iron falling on the fields might blast the rice crops.[26]

Was Iron Related to a Great Catastrophe?

To the ancients iron was evil, but the poets forgot why. The recital of the origin of such iron deposits may have been incomprehensible to the poet, yet he knew he had to recite the "deep origin" to control the deadly powers of cold iron. Magic and divination were always present in the grim business of the smith in ancient literature. Aristotle refers respectfully to the grave testimony of the ancient poets. The earliest writers behave like worried and doubting commentators in trying to explain a dimly understood tradition. They use old words whose meaning is half lost to them.[27]

Some early event or events having to do with iron left a searing, horrifying memory in mankind that is dimly recorded in ancient writings and in customs that have come down to the present day. Some writers hint at what this event may have been.

The moon and the planets were bombarded by a great swarm of huge meteors, which we can plainly see from space photos. It is only reasonable to believe that the earth also underwent the same catastrophic bombardment at the same time. The evidence that this really happened is all around, and researchers are only beginning to recognize it for what it really is. The asteroid belt that orbits about where another planet ought to exist has led to the theory of a planet that had been torn apart. The jagged fragments were torn away violently and some of them now rotate on their longer axis.

These evidences indicate a violent separation from a parent body. Satellite photos show mountain-size lumps embedded in the earth up to several hundred miles in depth. Perhaps many of our iron and nickel deposits fell to the earth in a great catastrophic event.[28] There are wildly different estimates of the time when enormous meteors struck the earth. The testimony of the poets and the customs around the world support the idea that this event happened early in man's history.

Ancient literature cannot be appealed to for any significant support of the conventional pattern of Stone, Bronze, and Iron ages. We do find support for very early sophistication in metal technology, support for the idea of worldwide travel in search of metals, and possible support for one or more catastrophic events dimly remembered.

Curious Finds Around the World

Why Intriguing Reports are Questioned

There are odd finds of metal here and there that may take us back to pre-flood days, and other discoveries appear to conflict with the usual view of how metal technology developed. Reports that differ from accepted views about the past are usually ignored or denounced, so the ones we can locate are sometimes poorly documented. The possibility of fraud or error in some cases cannot be ignored. Where metal objects from such reports still exist, the artifacts deserve careful scientific evaluation. The following brief reports of metal use in ancient times are only a tiny sample of hundreds of similar reports buried in sometimes offbeat literature.

Metals in the Ancient Middle East

Many authorities would agree that the Iranian Highlands and adjacent lands is where metallurgy probably began. From this ancient land, where future discoveries will still be made, metallurgy spread throughout the world. Following are locations and natures of a few interesting and surprising discoveries by archaeologists.

Iran: Earliest sites reveal the oldest copper tools (at Tepe Sialk), gold and silver objects, and copper casting. At Tepe Yahya bronze containing arsenic used for tools dated to 3500 BC. The Great Sand Desert, Dasht-e-Lut, was once the site of an ancient flourishing town with many highly artistic copper artifacts, but experienced sudden catastrophic destruction. In remote northwest Iran are ancient sites with mounds of iron slag and artifacts of bronze, silver, gold, and an iron ring so unexpected it was called "alleged." Susa: Hatchets, mirrors, and needles of bronze.[29]

Iraq: A working electric battery was excavated, dated to about AD 300, used for electroplating. Sumer: From the very beginning highly sophisticated copper, gold, and silver work, but none of those minerals came from that area; filigree, mold and hollow casting, intaglio, wire-drawing, beading, granulation, welding, inlaying repoussee, gilding on wood, electroplating. At ancient Ur a gold container held a set of tweezers, lancet, and pencil.[30]

Arabia: Metal mining abandoned suddenly about 3,000 years ago, and 400,000 tons of slag found at just one of many mines. One must think of great forests once thriving there to supply the fuel for incredible amounts of smelting. No trees can grow there now.[31]

Israel: A unique and extraordinary hoard of 400 copper pieces was discovered in a Judean cave near the Dead Sea that dated back to the Chalcolithic period, more than a thousand years before Abraham. It is the earliest and most important collection of metal artifacts found anywhere in the world. At Elath is the largest copper smelting location in the Middle East. Solomon only revived a very ancient copper industry here.[32]

Syria: At Ugarit, a lead sewer nine feet underground was located, with completely modern-looking manhole covers and drain holes.[33]

Turkey: Ancient Hattians were highly skilled in creating animal life bronzes, perforated metal disks, and elegant gold cups. At Troy, Heinrich Schliemann found Bronze Age artifacts below Stone Age levels. Catal Huyuk, called the world's oldest city, practiced metallurgy and used a full range of pigments derived from metal ores.[34]

Metals in Ancient Asia

A discovery was made in 1968 by Dr. Koriun Meguerchian at Medzamor, Soviet Armenia. We might say this location is just a stone's throw from Mount Ararat, that is, 15 miles. This site was termed the oldest metallurgical factory in the world and was dated at approximately 3000 BC. Objects made of all the common metals were found at the site. Rocks were hollowed where ore was crushed and treated. Many furnaces were excavated, and at least some of the ores were imported. Among the metals and alloys found there were copper, bronze, lead, zinc, iron, gold, tin, arsenic, antimony, manganese, and steel. Slender steel tweezers, still shiny, were found. Scientists from many countries have verified the find.[35] The concept of the Bronze Age—Iron Age sequence suffers a severe blow here.

On the shores of Lake Sevan in the mountains of Armenia, northeast of Mount Ararat, some exceptionally rich finds were reported in 1973. Dating back also to about 3000 BC were ceramics, utensils, jewelry, wagons, and a unique chariot with spoked wheels. Extensive gold mining was important there at that time. Gold and bronze buttons, studded with gems, were uncovered. A possible tie to the Hattians, or at least some common thread, was the find of many types of exquisitely formed miniatures in gold. Ceramics were decorated with pictures of ostriches. Oxen, deer, lions, goats, birds, and frogs were depicted in bronze. Some

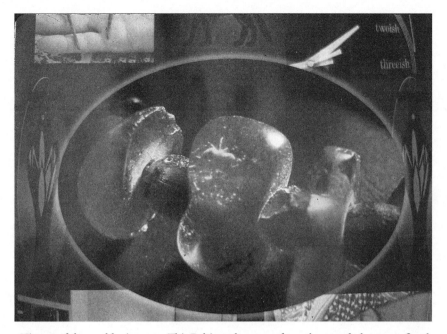

History of the world—in a way. This Baltic amber came from the sap of a huge pre-flood tree, and the sap trapped tiny insects as it rolled down the bark. During and after the flood the sap became fossilized, that is, the sap became amber. Some centuries later this piece was found on the beach of the Baltic Seas and became a part of the ancient and legendary amber trade, traveling over the Alps to the Bible lands where it was highly prized for jewelry. Along with Middle Bronze Age beads, this amber was made into a necklace and worn for generations. We can see the wear marks. Then buried for 3,500 years or more, it was found by a Bedouin east of the Jordan and offered for sale in a shop in Amman, Jordan.

of the decorations involve graining and lines so fine they can hardly be seen by the naked eye.[36] To gain some perspective of time, 3000 BC is about 1,000 years before Abraham was born!

The following shows the location and nature of other interesting discoveries related to metals in these vast regions.

India: Central India is full of dolmens and other large stone monuments, and is riddled with ancient mine workings for gold, copper, iron, and diamonds. Extraordinary skill is shown with shafts at a depth of 640 feet. Rich iron workings in the Bellary District revealed iron slag and Neolithic stone tools together. In other areas smelting places and iron slag were found with copper vessels and tools, bricks, and copper celts. This whole culture ended abruptly.[37]

Pakistan: At Harappa the earliest period was mature and sophisticated with exquisite gold work, copper and bronze rods, and all kinds of beautifully worked artifacts. Indus Valley: Bronze hatchets, mirrors, and needles.[38]

China: Ancient bronze culture began in the Yellow River valley. More recently another was discovered in South China dated a thousand years earlier. A site on the edge of the Gobi desert was a center of metallurgy, both copper and iron, at a very ancient time.[39]

Thailand: In northeast Thailand two-piece socketed bronze tools were dated at 2500 BC.[40]

Malaysia: In the Tui Valley immense gold mines were worked by the ancients. Hills for many miles are honeycombed with circular shafts radically different from more recent Chinese or Malay mines. This mining ended abruptly and was never resumed.[41]

Russia: Extensive ancient mining took place along the Yenisei River, 56°N, and enormous stone works cover the area. Wedges, mattocks, mallets, and hammers were found in place, some of copper. Gold ornaments, arrow points of copper, and fine animal figurines of molten copper were discovered here. Neolithic races spread to the Arctic coast in order to mine metals. These activities also came to an abrupt end. On the Lena, inside the Arctic circle, stone monuments were discovered. Outstanding finds have been made near Lake Yolba, and in Yakutsk a workshop of metallurgists was found for making bronze axes similar to those of the Near East and in Europe. The ancients of northeastern Siberia could extract copper from ore, melt it, and pour it into forms. They made axes, beautiful bronze spear tips, knives, and swords. A golden buckle of Scythian origin shows the image of a saber-toothed tiger on it, but these animals are thought to have been extinct many thousands of years earlier.[42]

Ancient Metalworking in Africa

In his overview of the history of man, Michael Day concluded that the earliest metal artifacts are beads and pins of beaten copper that date to about 4000 BC, and which were found in Egypt. As early as the First Dynasty plaques of gold and precious stones, necklaces of gold beads engraved with geometric patterns or shaped like snail shells, gold amulets of the bull and oryx, and gold capsules in the form of a cockroach inlaid with emblems were known. Also from the Old Kingdom vases, knives, razors, and a manicure set, all of gold, were found.[43]

Cyril Aldred marvels at the sophistication of the Archaic Period of Egypt, Dynasties I and II.[44] He described a remarkable carved dish that he dates at 3100 BC. It imitates a form originally made in metal. Eyeholes in needles and teeth of copper saw blades were punched out by an unknown harder metal, but bronze and iron are thought to have been unknown at that early time. The copper saws were used with some kind of abrasive. Exquisite workmanship with gold, silver, and precious jewels was common, but few pieces escaped thieves over the centuries.

Flinders Petrie has shown that the use of hollow metallic drills, armed with gems, was common in ancient Egypt for cutting and carving the hardest of stones. The diamond drill was used among Babylonians at a very early period. It is a curious fact that the Hebrew word for diamond is expressed by two signs that mean a boring stone.[45] At any rate, during the Old Kingdom, long before the Iron Age, sharp, finely cut lines were cut into granite, basalt, and into diorite, the steely stone, incredibly hard.

The main purpose of Egyptian sea expeditions was to secure minerals, and it seemed that no place was too far to secure these treasures. In a similar vein, Gudea, the Sumerian ruler of Lagash about 2000 BC, tells about getting minerals from different parts of the world.[46] There are discoveries about the ancient use of metals that jar the mind. Metal was used to make lightning rods in Egypt in the fourth century BC, yet credit is given to Benjamin Franklin for this invention in 1752, about 2,000 years later.

At Gerzah, 50 miles south of Cairo, iron beads were excavated at a predynastic site (about 3000 BC). An iron chisel was found between the stones of the Great Pyramid of the Fourth Dynasty (about 2500 BC), and a number of chisels and other iron tools were unearthed in Saqqara, the oldest of all the pyramids. Other iron tools were discovered in sites of the Sixth Dynasty (about 2200 BC). Some of the iron objects contained no nickel. The conclusion is that iron smelting was known in the Old Kingdom, many centuries before the so-called Iron Age began.[47]

Tanzania: The first high-grade carbon steel was produced in the nineteenth century AD. An astonished anthropologist reported that the Haya people on the west shore of Lake Victoria produced carbon steel 2,000 years ago with mud and slag furnaces and goatskin bellows. They achieved temperatures of 3,275°F![48]

South Africa: David Livingstone found survivals of ancient smelters using hematite ore, which bears no resemblance to iron. Iron tools were forged from this ore by people who never knew bronze, nor had they gone through a Stone Age. Malachite, a semi-precious green stone, gives no hint that it is rich in copper. Yet natives in central Africa smelted this ore extensively from ancient times.[49]

Northern Rhodesia: An iron foundry, slag, and ashes were found with Paleolithic materials in a cave at Mumbwa, predating the Iron Age by several thousand years according to conventional chronology.[50]

Swaziland in southern Africa: A mountain peak here is honeycombed with pits and tens of thousands of stone implements for mining in ancient times.[51]

Ancient Use of Metals in Europe

Europe and surrounding islands have their share of fascinating metal discoveries. Scientists have identified a few of these sites.

England: In Victoria Cavern, among bones of extinct animals and thousands of tools, archaeologists discovered the carving of a horse's head on bone. Strangely, the horse is hog-maned, which means the mane had been cut with metal shears![52]

Austria: Copper mines and smelters were found in southeast Austria, along with earthen vessels that were obviously parts of bellows similar to those still in use by African tribes today.[53]

Switzerland: Lake dwellers here and in Scandinavia, Germany, and northern Italy left thousands of bronze artifacts when a sudden disaster destroyed this culture.[54]

Denmark: People settled here from the sea at a very early time. Artifacts recovered include copper axes, spiral armlets, daggers, saws, and objects of gold.[55]

Greece: The earliest metal discovered in Greece was excavated at a site in Macedonia in 1969, dating back to 4000–5000 BC. Carbonized grape pits, a painted ceramic camel, pins, fishhooks, and beads of copper were noted. Slag at the site showed that the copper was extracted there from ore, demonstrating astonishing sophistication for that time.[56]

Elba: The oldest iron mining known in the Mediterranean area is on this island. There are huge subterranean excavations for iron ore that were then destroyed by some catastrophic force, several thousand years before Christ.[57]

METALS AND ANCIENT SEA TRAVEL

Old World—New World

The story of sea power and ocean trading during the copper and bronze ages has been distasteful to historians who cannot rid themselves of the belief that man went gradually from primitive to modern, a fundamental tenet of evolution. In part the absence of this dramatic story in history books is due to the fact that ancient trade routes were jealously guarded. Few details were passed on, but other evidence is convincing and fascinating.[58]

For generations students have been taught to believe that no one crossed an ocean to reach America until Columbus. All natives in the Americas were thought to be descendants of those who came here by the Bering Land Bridge. Further, as compared with Asia, Africa, and Europe, man was believed to be very much a newcomer to the Americas. Any evidence of apparent age, that is, more than a few thousand years, was ignored or attacked. While Viking landings in America are now grudgingly accepted in a very limited way, not until the 1970s has there been a new breed of scholar willing to take a serious look at the many evidences of other widespread travel across both oceans to and from the Americas. James Bailey, Henriette Mertz, Carroll Riley, Barry Fell, and Thor Heyerdahl, to name a few, pointed to compelling evidence for ancient extensive sea travel.[59] These and others paint an exciting picture of early cultural contacts and of the clear evidences left behind. When we look at peculiar finds of ancient mining and metal in the Americas, we find strong support for new ways of thinking about the ancient history of people in the Americas. Ocean travel elsewhere in the world was also common. The search for metals was a compelling reason for all of this.

The Old Copper Culture of northern Michigan is an interesting example.[60] A thousand years before Abraham there were over 5,000 prehistoric copper pits on Isle Royale alone, plus many thousands of others on the Upper Peninsula of Michigan. Hundreds of thousands of stone mauls and hammers were uncovered in the pits, some weighing up to 40 pounds. Mining engineers estimate that up to 1.5 billion pounds of copper were extracted. Excavations along the waterways show that the copper was chiefly transported eastward toward the Atlantic, rather than north or south on this continent. Among the thousands of copper artifacts, toggle pins were discovered identical to many found in Mesopotamia. It is a curious fact that copper and other metals were sometimes shipped in the shape of oxhydes as a form of money. Ancient art shows this, and cargo recovered from a sunken ship in the eastern Mediterranean contained 200 of them. Similar oxhydes were found in upper Michigan.

Based on almost identical culture patterns of northwest Europe and northeast America as early as 7000 BC in conventional dating, Carroll Riley suggested ancient two-way travel across the Atlantic.[61]

Along with much other evidence, Pierre Honore noted the influence of the Far East in America in the manner of gold working, the use of copper, core casting, granulation of metals, alloys, use of bronze, and metal dyeing.[62] In a similar way, he supported a Mediterranean influence in America, for example, in the way copper rivets were used. The lost wax method of casting, long known in the Old World, was expertly used in the New World long before Columbus.[63]

Constance Irwin marvels that the metal techniques of the Old and New Worlds were alike to the point where independent invention can safely be ruled out.[64] The similarity of techniques extends to such elaborate processes as casting an object from an alloy of gold and copper, then treating the surface with acid to dissolve the copper, leaving a pure gold surface. In both worlds the same methods were used in hammering, embossing, annealing, welding, soldering, strap joining, incising, champleve, cutout designs, and making objects of two metals.[65]

Kenneth Macgowan and others have noted that bronze and copper weapons and other tools show great similarity in South America to those of the Old

World.[66] Further, metal techniques were very similar in the New and the Old Worlds. Particularly with regard to the invention or discovery of bronze, it is difficult to conceive of independent invention in the two worlds. It is much easier to argue for transoceanic contacts in the distant past.

When the tin-bearing gravels of Cornwall, England, were reworked for tin and gold, relics of the Old Ones were found, including many stone bowls, mortars, and dishes, mostly of granite. The great similarity between these vessels and those found deeply buried in the gold-bearing gravels of California was noted already in 1881.[67] One culture spread over the world at a very ancient time.

The trade in tin from Cornwall across France to Marseilles seems to have followed a track coinciding with the line of the ancient dolmen builders. The worldwide extent of ancient mining associated with megalithic monuments is shown by the fact that erecting huge stones was still practiced by hill tribes in India, in Japan, and in North Africa until quite recently.[68]

Evidences for ancient mining were found in California. A mortar was found under volcanic matter at a depth of 150 feet at San Andreas. A triangular hatchet of stone with a drilled hole for the handle was found 75 feet below the surface in gold-bearing gravel and under basalt 300 feet from a tunnel mouth at Table Mountain in Tuolumne County. A human skeleton was found in a clay deposit at a depth of 38 feet at Placerville. A large number of stone mortars, pestles, and stone dishes were found at Murphy's in Tuolumne County. In this high barren mountain district of California the site of an old mine was discovered. A shaft 210 feet deep was found with human bones at its bottom. Here also was found an altar for worship and other artifacts. In other gold diggings numerous stone relics, mortars, grooved disks, and other artifacts were reported at various depths. Granite mortars and dishes were uncovered of perfect form that weighed up to forty pounds. A lava vessel, hard as iron, was circular in form and had three legs and a beautifully formed spout. Almost without exception, these ancient artifacts were in or near gold-bearing rock or gravel. The Calaveras Skull, found here in gold-bearing gravel, was and is highly controversial, because the gravels are dated tens of millions of years before man is supposed to have evolved.[69] Edward Lane and Robert Gentet completed a thorough restudy of the skull controversy (see chapter 7).[70] They concluded that the skull was genuine and that the discovery pointed to mining during the early centuries after the flood, before the post-flood ice age, and before the area was covered by a massive volcanic flow.

Ancient voyages to America from the Eastern Mediterranean are accepted by more and more scholars. Some suggest the same from India, Indochina, and China at an early date, and gold was one of the chief attractions.[71]

Metal Discoveries in the Ancient Americas

Many centuries before the first Europeans came to America, ancient miners left tell-tale evidence of their exploration and mining of copper, precious metals, and precious stones. Here is a sample of curious discoveries.

United States: *Michigan:*[72] The mysterious Old Copper Culture dug 10,000 pits and mined over a billion pounds of copper, then disappeared about 1000 BC. Very little of the copper has been found, supporting the view that it was taken overseas. *Ohio:*[73] The mound builders of the Mississippi-Ohio valleys built 100,000 mounds of all sizes and shapes. They worked iron meteorites, turned copper into tubing, ear-spools, and crafted raised designs in copper overlaid with silver and iron. Copper axes weighed as much as 23 pounds. *Louisiana:*[74] The enormous complex of six concentric octagons dates back to about 1500 BC with a volume of material 35 times that of the Great Pyramid in Egypt. Their trade network went in all directions up to 3,000 miles and included copper and iron ore from the Great Lakes region and galena (lead ore) from Missouri. *Arizona:*[75] An ancient tribe known as the Hohokam people cast bells that required a melting temperature of 2,066°F, similar to bells excavated in China. Ancient copper smelters have been noted along the Salt River.

Canada: At Thunder Bay, Ontario, a copper spear point was excavated in association with the bones of extinct bison and horses lying on a blue clay deposit covered with 40 feet of cross-bedded sand; human bones there were modern.[76]

Mexico: Ancient Mexicans melted metal in crucibles, cast metal in molds, and used the lost art of casting parts of the same object of different metals, for example, a fish with alternate scales of gold and silver. They made birds and animals with movable limbs and wings in a most curious fashion. Near Merida in Yucatan strange buildings, erected long before the arrival of the Spanish, had stone roofs held in place by iron rods.[77]

Peru: People here 2,000 years before the Inca smelted ores, annealed, welded, and soldered metals. With bronze scalpels they performed trepanning and other forms of surgery. Gold and silver electroplating was practiced at Chan Chan.[78]

Ecuador: Ornaments of molten platinum found here required a temperature of 3,216°F.[79]

Conclusions

There are hundreds of similar surprising discoveries in the Americas and in other parts of the world. We leave the possibility open for error in some of the above accounts. It seems very clear, however, that transoceanic travel to and from the Americas occurred for thousands of years before Columbus. All periods of time seem to be quite well represented. There is more than a faint suggestion that the Bering Land Bridge was largely built by archaeologists who were afraid of water.

Afterword

Seeing What One Wants to See

We have looked at a great deal of evidence that runs contrary to conventional modes of thinking. The present scheme was painstakingly built on an evolutionary model. Scholars and novices alike tend to see pretty much what they want to see in the evidence. Students of great and not so great scholars are understandably reluctant to deviate from the positions of their elders, regardless of how arbitrary and conjectural these positions may be.

Scholars are very slow to discard a theory simply because there is evidence that contradicts the position taken. Although one can find many examples of extreme conservatism when it comes to discarding theories and models in science, just one is given here. We can ponder the statement of a brilliant archaeologist, Franz Boas, whose life work was dedicated to establishing the late arrival of man into the New World. Boas stated that it would be a pity to have new evidence come to light that would overthrow all the admirable scientific work of the past.[80] That seems to say it all.

The Rare Exceptions

We need to cherish those scholars today who are willing to defy tradition and work directly with the evidence, wherever it may lead them. After all, that is what scientific method is all about.

Compelling Conclusions

In this small sample of fascinating glimpses of man's use of metals in the ancient past, we have repeatedly found unexpected signs of sophistication far beyond what one might expect for such a time in the past. It seems too much to believe this all gradually developed independently in numerous places by primitive Stone Age people.

It seems better to take the biblical account seriously. The pre-flood metal technology was passed on after the flood by the sons of Noah, called the Old Ones, who became the teachers of all cultures. Some of Noah's descendants learned the technology before they moved away, broke away, or were driven away from the first tribe of them all. The knowledge of metals or lack of it and the characteristics of the environment for a given group of people determined in large part whether or not a Metal Age followed for them.

Another inescapable conclusion is extensive worldwide travel on the seas in very ancient times. This seems to be the best explanation for much high technology appearing almost simultaneously in widely separated parts of the world. Ancient times emerge as filled with enormous energy, enterprise, and sophistication.

We saw again and again evidence for catastrophic events during historical times. Early civilizations were very sophisticated in technology, then suddenly disappeared, and civilizations that followed often showed evidence of degeneration. Yet here and there, despite all the catastrophes of history, finds from the past show some astonishing continuity over the centuries of skills, techniques, and patterns.

The observations of field anthropologists are directly opposite to what other scientists assume about human development. It is a simple fact that most so-called primitive peoples are extremely conservative and simply do not experiment to evolve new and better ways. This is an ironic parallel to party-line science described above. We are blind to the ways of many primitive societies where any suggestion of doing something different from the past is an insult to one's ancestors and therefore becomes an outrage to the community. The striking fact is that natives never seem to discover a new idea for themselves, nor do they modify anything in the slightest. When change has come to a community, it came from the outside. Thus the whole rationale behind the concept of stone and metal ages falters. There are Stone Age people living on top of large ore beds in various parts of the world today who continue to live as a hunting and food gathering culture. The use of metals by other peoples came long before the periods assigned to them. It seems almost unnecessary to point out that there is no support at all of any kind of sequence of stone and metal ages in the Americas.

The pattern and chronology one finds in much literature of ancient history, anthropology, archaeology, and other works are based on questionable assumptions. Hence only a very superficial and erroneous view is given to us of man's use of metals in the ancient past. Certainly there is no need to be impressed with the truth of a good many conventional explanations.

In addition to standard sources, unusual sources were consulted in this chapter, because they contain facts or alleged facts that are not found in conventional sources. With the story of the past so confused in much literature today, no potential source should be overlooked in the search to unravel at least some of the puzzles. Further investigation may help to verify or expose assertions found in unusual references.

Most interesting of all is the fact that if one takes the Old Testament seriously, that person finds a framework of history presented that is supported and illuminated by a broad variety of non-biblical sources.

For Reflection and Discussion

1. Read Job 28:1–11. Why do you suppose ancient people were obsessed with metals, especially precious metals?

2. Some scholars of the Bible believe that the Book of Job may be the oldest book in the Bible, that is, it may have been written before Moses wrote the first five

books of the Old Testament. Evolutionists believe there were vast stretches of time when people only used stones for tools and only very gradually began to use metals. Thus we have the many Stone Ages followed by various metal ages. Does the Bible tell a different story? See Genesis 4:22 about Tubal-Cain who worked with bronze and iron many centuries before the so-called Bronze Ages and Iron Ages.

3. Why do you suppose God told the Israelites about places they could dig for iron and copper in the Promised Land? See Deuteronomy 8:9.

4. Evolutionists present stone and metal ages to illustrate how humans gradually became more and more sophisticated over long periods of time. Note in the chapter that this is in sharp contrast to many actual discoveries of very sophisticated work with metals at a very early time. Why do texts on ancient history fail to discuss amazing early sophistication?

5. There are so-called Stone Age people living today who feed themselves by hunting-gathering. There are good examples of the very crude stone artifacts and tools they still use. Why haven't these people evolved, like the rest of us? Can one date an artifact on the basis of how crude or sophisticated it appears? Why or why not?

6. Why does the "correct" date for a sample often have to be pulled out of a hat?

7. When did somewhat sophisticated scientific methods really begin in archaeological excavations? What caution is in order about earlier digs (excavations)?

8. In Acts 19:35 St. Paul speaks of the "sacred stone" that fell from the sky. Why was it venerated? Why do you think the "sacred black stone" is venerated today in Mecca?

9. Technology is cumulative. Examples: auto, planes, electronics. We do not have to begin all over again to produce a new model every year. Is improved technology year by year the same as evolution? Can you give an example of some living creature that is gradually changing into another species, and, if so, could this then support the theory of evolution? Do you think human nature is gradually improving? Why or why not?

10. Discuss the statement in this chapter by Franz Boas, that it would be a pity to have new evidence come to light that would overthrow all the admirable work of the past. Does his view follow "scientific method" here?

CHAPTER THREE

WRESTLING WITH
THE DINOSAUR MYSTERIES

INTRODUCTION

Scanning Dinosaur Literature

It is no secret that almost all books, films, and television specials featuring dinosaurs promote the theory of evolution. Many Christians have not known what to say about dinosaurs when asked by their children or other curious individuals. It is not uncommon to meet Christians today who were taught by their well-meaning pastors or teachers years ago that dinosaurs never existed, that God placed fossils in rocks and dirt to test our faith, or that dinosaur stories were simply wild inventions based on a few bones and teeth. We know only too well how saturated we are with evolutionary beliefs about dinosaurs. I present here a biblical framework for understanding man and the dinosaur on a young, created earth. In this brief survey I also address the speculations of those who support neither creation nor evolution.

The new generation has been brought up on a rich but faulty fare of dinosaur lore, such as the *Jurassic Park* book and movies. There is now much new information available about these strange creatures. Looking at the fossil record, it is apparent that the ancient world really was far different in many ways from the world we live in today. We believe that most of the fossil record comes from the pre-flood cursed world. What would we do today if we had to live among dinosaurs 150 feet long, snakes 100 feet long, and centipedes 25 feet long? The fossil record clearly shows that these creatures once existed. That is only the beginning of the fossil record. Christians who accept the entire Bible as truth also accept the fact that our earliest ancestors lived at the same time as creatures such as these.

Where Were the Dinosaurs?

Hundreds of species of dinosaurs of all kinds and sizes have been identified, and fossils, footprints, or both have now been discovered on every continent, including the Antarctic, New Zealand, unfossilized bones and stomach contents on the Arctic coast of Alaska, and on such remote islands as Spitzbergen far to the north of Norway in the Arctic. A French team of scientists, in a region of the south Indian Ocean hundreds of miles from land, sampled cores from the ocean bottom. They discovered forest remains, a dinosaur tooth, and other evidence that this was a drowned land of the ancient world. Dinosaurs once lived in large numbers in every present-day climatic zone of the earth. Fossil track experts report dinosaur stomping grounds with billions of tracks preserved in stone in just one Rocky Mountain area, and there are other such sites.

Creationists Discover the Dinosaur

Why isn't ancient literature full of dinosaur stories? Clearly the dinosaur was an awesome factor during the early history of the earth. Perhaps there is more in ancient writings than we tend to think. It is only within the past several decades that creationist writers and lecturers, with just a few exceptions, discovered the dinosaur. Creationist writers began to probe the question of fitting dinosaurs into early Genesis in such exploratory articles as the following:

- Lorella Rouster: In three brief articles Rouster sheds light on seeing how the dinosaur was part of early human history.[1] It seems fair to say that Rouster, more than anyone, helped the Christian community begin to understand how the dinosaur fit into the young created earth of early Genesis. She discusses topics such as: Adam and his descendants knew the dinosaurs; dinosaurs are written about in the Bible and in ancient literature; pictures of them are found in caves and in dragon art around the world. There is no reason why we should be timid about exploring both the evidence for and our beliefs about man and the dinosaur together on a young, created earth.

- In *The Dinosaur Dilemma*, Gordon Lindsay[2] discusses the following points: Dinosaurs are an enigma for many Christians; belief that dinosaurs existed before Adam is in error; there was no death before the fall and every creature was good; dinosaurs were created during creation week; they may have lived at the time of Job; some human prints with dinosaur tracks are genuine; dragons are dinosaurs; there is the fire-breathing problem; dinosaurs were taken on the ark and died later in harsh environments; there have been modern dinosaur sightings.

- John Morris, in "How do Dinosaurs Fit In?"[3] claims dinosaurs were created on

Day 5 (marine) and Day 6 (land) of creation week; all were created to be plant eaters; the carnivore problem; plants and animals were genetically changed by the fall; taken on the ark and gradually died out later in harsh conditions; dragons are dinosaurs, and they may have been hunted to extinction.

- Anonymous, "What Happened to the Dinosaurs?"[4] Abrupt appearance and end of dinosaurs; author supports an old earth, but stresses God the Creator.
- In "What About Dinosaurs,"[5] Norman Hafley discusses whether dinosaurs are an embarrassment to the Christian; extinctions, the ark, dinosaurs and humans together, dinosaurs in the Bible.

Other authors are cited elsewhere in this chapter. I have written extensively on many aspects of the dinosaur mystery within the biblical framework of our young, created earth.[6] This chapter focuses on some interesting developments during the past several years.

Dragon's Graveyard

Northeastern Wyoming is one of many so-called dinosaur graveyards found around the world. Here in the Lance Formation is a broad band of fossils stretching for well over 100 miles, dominated by remains of the *Triceratops*. The Hanson Ranch is located here and the dinosaur fossils are so numerous, even on the

A dinosaur bone excavated near Ely in England showing cut marks. Evolutionists cannot believe that this must have been a butchering site long ago. (Photo from the *Victoria Institute*.)

surface, that the area is known as the Dragon's Graveyard. After some years of excavations by evolutionists, the ranch owners arranged for creationist scientists and volunteers to excavate in the summers of 1995 and 1996. In the tiny area dug, these fossil bones and teeth emerged: *Triceratops*, *Tyrannosaurus rex*, *Hadrosaurus* (duckbill), and *Dromeosaurus* (raptor-like dinosaur). Mingled among those remains were turtle shells, alligator scut, clams, snails, gar scales, pine seeds, and other seeds. Catastrophic water action was evident.

It immediately became all too clear that massive funding was needed for equipping and staffing a laboratory and for further excavation. Such funding is simply not available at the present time. Desirable as it is for creationists to do their own excavations, the problem is interpretation of the evidence, not the evidence itself. At present, at least, creationists cannot compete with the museums and universities who are all fully committed to evolution. It is important, however, to provide interpretations within the framework of the Bible, and to show how evolutionists have abandoned scientific method in desperate attempts to shore up their beliefs.

Dinosaurs in the News

Discoveries of large numbers of eggs, many in undisturbed mud nests, continue to make the news. Yards of fossil dinosaur skin made other headlines. The biggest push presently is to try to show that dinosaurs evolved into birds. When researchers from the University of Kansas found modern bird fossils in strata older than dinosaur layers, they were savagely attacked as being "outside the mainstream."[7] Other controversies deal with whether or not a feathered dinosaur was actually found, and how dinosaur arm bones differ from the bone structure found in bird wings. As usual, the experts differ loudly. An extraordinary embarrassment occurred in late 1999 when *National Geographic* proudly announced the perfect missing link, which leading experts in paleontology agreed on: a fossil that was clearly half-dinosaur and half-bird. Within days, however, a fraud was uncovered. Apparently a Chinese peasant had glued the two halves together and the result was exactly what excited evolutionists had hoped to find in the fossil record. This fraud is now referred to as the Piltdown Chicken, after an earlier gigantic fraud.[8]

As in previous years, much of the present focus is on mighty carnivores, with new and sensational discoveries reported in South Dakota, Montana, Oklahoma, New Mexico, Mongolia, Thailand, Morocco, Argentina, and the Antarctic. The list keeps growing.

High Tech Dinosaur Exploration

As we scan the treatment of the dinosaur since the 1990s and into the present century, we observe that a new era in sophistication has arrived. The University of Utah announced the use of a device that can pinpoint a fossil through rock up to a meter thick.[9] Several new species have already been discovered in Utah with this device.

Evolutionists Disagree about Dinosaurs

We are all quite aware of the stories of the rise and fall of the dinosaur written by evolutionists. Much of this material is purely descriptive and thus is irrelevant to the beliefs of creationists or evolutionists. There are strong disagreements among evolutionists on the interpretation of fossils and other remains. Again, almost without exception, these clashes are irrelevant to the creation/evolution debate. They are good illustrations of the limitations of what bones and imprints can tell us today.

Evolutionists, however, speak as one on the timetable of when dinosaurs appeared and when they finally became extinct. This is merely a badge of loyalty to evolution rather than something established by evidence. We have noted several areas of controversy above. Other areas of current disagreement are the following:[10]

- Classification, ancestry, cradle of origin: There is little agreement on these fundamental issues. One current argument is whether dinosaurs could have "originated" in South America and then spread to all the world.

- Metabolism: Were they mammal, reptile, or bird?

- Diet: Did the so-called carnivores such as *Tyrannosaurus rex* attack other creatures or were they mostly scavengers? Were their teeth weapons of terror or so fragile they could never be used for attack? In 1996 the controversial discovery was made of fifty-eight *Tyrannosaurus rex* bite marks on the pelvis of a six-ton *Triceratops*. Tooth force was estimated at 3,011 pounds per tooth. The conclusion was that this had been an attack on a living beast.[11]

- Anatomy: Current views on dinosaur gait, posture, and head structure are mostly wishful thinking. Other serious disagreements include tentative evidence that dinosaurs were becoming birds, the probable location of plates on the back of the *Stegosaurus*, where noses are positioned on the dinosaur face, and the function and strength of the *Tyrannosaurus rex* forelimbs.

- Reproduction and parenting: how they reproduced; were blind helpless baby dinosaurs nurtured in their nests by loving parents or did they hit the ground running when they were born or hatched?

- Extinction: Did this occur at one time or over a long time period? Was it the result of a comet striking the earth or multiple events?

The unanswered questions above are important ones, and the answers are full of speculation, but they emphasize that fossil bones and teeth tell us much less than what we would like to know about these mysteries.

Carl Sagan made an astounding statement in *The Dragons of Eden*, which apparently no evolutionist has discussed since. He concluded that the dragons in ancient art and literature were really dinosaurs, and that "man-like creatures encountered dinosaurs," such as *Tyrannosaurus rex*.[12] While he was careful not to say *humans*, he jumped over many millions of years from the last dinosaur to the first humans, according to the rigid timetable of evolutionists. Needless to say, he has violated everything evolutionists teach about dinosaurs and man. But because of his high standing, there is no outcry.

Evolutionist Vs. Creationist Writings

Evolutionists, as we might expect, do not take creationist writing on dinosaurs seriously. William Stokes,[13] as just one example, reviewed Paul Taylor, *The Great Dinosaur Mystery and the Bible*.[14] Stokes called it a contrived fairy tale. Stokes' strongest objections, as we might well imagine, are the belief in a young earth, and the belief that humans and dinosaurs lived at the same time. It is instructive to read reviews such as this to see ourselves through the eyes of the evolutionist. There can be no compromise between old and young earth, between created earth and evolved earth. In each case, it is one or the other, not the average of the two. Our choice is between the framework of the Bible or the one embracing the beliefs of evolutionists, whose gods are time and chance.

Purpose of this Survey

The following is a survey of the topics, themes, issues, and beliefs of creationists, evolutionists, and others. We include "anti-evolutionists" to consist of writers opposed to evolution but who do not necessarily believe in creation as we accept it. The reader will recognize such persons by the nature of their beliefs, and of course some of the beliefs must be immediately rejected. We shall evaluate our findings below as to which ideas seem to hold the most promise for further investigation within a biblical framework by creationists into the strange and fascinating time of the dinosaur.

ANTI-EVOLUTIONIST TREATMENT OF THE DINOSAUR

Recent Creation Research Society Activity

Michael Oard evaluated books on the possible role of catastrophic meteor bombardment and volcanic activity as related to dinosaur extinction.[15] He concluded that these events during (and possibly before) Noah's flood, along with the flood waters, are plausible explanations for the extinction. Eugene Chaffin believes there is ample evidence that wading or swimming dinosaurs could have produced the fossilized tracks found in different strata layers today.[16] Bolton Davidheiser raised insurmountable obstacles to the belief among evolutionists that dinosaurs evolved into birds.[17] His argument focused on lizard-hip and collarbone structures. Evolutionists, however, continue to push this notion.

Oard also grappled with the problem of much warmer climate vegetation and dinosaur material in the polar regions.[18] Among other problems are the estimated billions of dinosaur tracks in stone and the undisturbed nests atop thousands of meters of flood sediments. He evaluated the possibilities that the dinosaurs lived there, walked there, or floated there. No one has very satisfying solutions at present. British creationists explored the possibility that the tracks and nests date to the post-flood era, but few see any support for this view.[19]

Todd Wood did a thorough review of the supposed plesiosaur carcass found in the south Pacific by Japanese fishermen in 1977.[20] He used photos and analyses that the remains were those of a decayed basking shark and suggested that creationists no longer use this find as an argument for the plesiosaur existing into modern times. But other qualified scholars differ, and the last word is still to be said.

Carl Froede[21] reviewed a book on dinosaur extinction[22] and agreed with those authors that much of the impact theory evidence for dinosaur extinction is nothing more than pseudo-science, in other words, humbug. Certainly there are many gigantic impact sites caused by outer-space bombardment, vast volcanic activity, and cataclysmic floodwaters. No one aspect can account for all that happened to end the dinosaur age, which we hold to have occurred chiefly at the time

One of the great dinosaur fossil areas of the world—the badlands along the Red Deer River northeast of Calgary, Alberta.

of Noah's flood. Further research and impressive photos were featured in 1997 of human-like tracks among dinosaur prints in Arizona. Other tracks were discovered in New Mexico that seem several hundred million years out of place according to the evolutionist time table: simian-like (ape) tracks, probable bird tracks, wolf, and very convincing bear tracks.[23] The tracks were made in strata thought to be hundreds of millions of years before the creatures that made them had evolved.

THE McIVER SURVEY

Unless otherwise indicated, the following themes are noted in Tom McIver's book *Anti-Evolution: An Annotated Bibliography*.[24] McIver offers a comprehensive listing of resources dealing with the topics of creation and evolution. After examining the references, the following themes appeared in multiple sources. Please note that anti-evolution does not always mean pro-Bible. I agree with many of the themes noted below, but I also indicate where I strongly disagree. The following provide a good sampling of the literature, but the list should not be thought of as complete. See McIver's book for full citation listings.

Themes Based Directly or Indirectly on the Bible

1. Various writers have examined creation week, particularly Day 5 (the great creatures of the sea) and Day 6 (the wild animals, including the dinosaur).

2. The fall resulted in a "stupendous transformation" of God's creation, including monstrous dinosaurs changed from original huge, "good" creatures.

3. The pre-flood world was hostile and filled with dinosaurs.

4. The age between the fall of man and the flood was one of comparative quiet geologically. The earth was warm and probably no earthquakes or volcanic activity; abundant and animal life.

5. The earth underwent thousands of years of post-flood adjustment leading to the reinundation of vast areas and causing most of the fossil record.

6. Biblical references to pre-flood giants are related to finds of giant man-tracks associated with dinosaur tracks.

7. Noah, the ark, and the dinosaur. Dinosaurs, or their young, were taken on the ark.

8. Dinosaur extinction was entirely or primarily a result of the flood. Thus most dinosaur fossils are a product of the flood.

9. References to dinosaur-like creatures in the Bible are descriptions or memories of the dinosaur. Based on passages in Job, at least some dinosaurs may have lived for some centuries in the post-flood era.

Comment: The nine themes above in my opinion contain much truth. At our present level of understanding, however, these points should be made. I believe there is strong support for item 3 above and that 4 is only partially true. There may have been ecological zones on earth as described by item 4, and I believe the earth was warm with abundant plant and animal life, at least in some of the zones. See my book *Noah to Abram* for my treatment of the turbulent post-flood world, which may have been the cause of some of the fossil record.[25] Item 6 has not been convincingly demonstrated. Regarding items 7 and 8, I believe there were many extinctions before the flood, but there are good reasons to suppose that at least some dinosaurs were on the ark.

Themes Based on Observation and/or Scientific Study

1. Drawings, paintings, and carvings of dinosaur-like creatures stem from observation or memories of actual dinosaurs.

2. More space seems to have been devoted to the topic of human and dinosaur footprints supposedly found together than all other topics combined. Because there is so much literature on this topic, we shall treat it very briefly here. As of the late 1980s creationists are treating this subject with much caution because it has been shown that some prints, formerly thought to be human, seem clearly to belong to dinosaurs. Yet this claim is far from encompassing all the footprint finds at the Paluxy River in Texas, some of which seem quite convincing. There are other such finds in the United States and elsewhere that also deserve attention and further analysis.

3. Reports of dinosaurs in the twentieth century fit Bible passages that appear to refer to post-flood dinosaurs.

4. C14 tests of wood associated with dinosaur remains support the concept that dinosaurs lived in recent times.

5. Similarly, examination of claims that dinosaurs lived millions of years ago, long before man, shows that this is merely a belief, a time illusion, not supported by actual scientific evidence.

6. In addition to the flood of Noah, at least some dinosaur extinctions are attributed to sin in general, to hunters, to catastrophic causes, or to climatic changes caused by the destruction of the pre-flood canopy over the earth.

7. Chemical fire produced by the bombardier beetle is a parallel to references to fire-breathing dragons, which were actually dinosaurs.

Comment: The seven themes above are thought provoking. Some of the dinosaur artwork seems convincing, and the same is true of at least some of the footprints of humans among those of dinosaurs, especially some of the most recent work noted elsewhere in this chapter. I am skeptical about dinosaur sightings in the twentieth century except for some possible marine sightings. I understand 6 to be a possible description of life in the pre-flood cursed world. Number 7 is interesting, but hardly convincing in generalizing from insect to dragon-dinosaur.

Other Themes and Beliefs about the Dinosaur

1. Myths, folklore, and other ancient literature contain memories, sometimes very vivid, of the dinosaur.

2. As one example, dragon tales from all over the world are actual memories of the dinosaur.

3. Dinosaurs never existed. Fossils are examples of God's wit.

4. Dinosaurs lived and became extinct before the six-day creation described in Genesis.

5. Dinosaurs were a result of a vitachemical process that operates throughout the universe. Thus dinosaurs may be in existence now on other planets, propelled by the same vitachemical process.

Comment: I accept 1 and 2 as fruitful areas of further study. I of course reject completely the ideas expressed in 3, 4, and 5.

Brief Evaluation

It is at once apparent that the above themes are not of equal merit and that some ideas expressed conflict with others. I have highlighted those themes that for the creationist seem to fit well into a biblical framework. Others must be rejected as speculations foreign to the Bible.

Future Directions for the Study of Dinosaurs in a Young, Created Earth

The above listings may give the impression that anti-evolutionists have thoroughly explored all aspects of the problem. This, however, is far from the actual situation. As noted above, the bulk of the discussion thus far has been on dinosaur fossils and human footprints, a topic of much controversy. Future study ought to focus on such important problems and issues as the following:

- Textual and word studies in the first eleven chapters of Genesis, and in significant passages of Job and other portions of the Bible. Surprisingly, little has been done thus far in such studies. For example, the "great whales" in Genesis 1:21 (KJV) and the behemoth and leviathan in Job described as crocodiles and hippos merely reflect the opinions of medieval biblical scholars who had no knowledge of the fossil record.

- Careful surveys and studies of ancient literature, including legend and myth, for memories of or actual encounters with, dinosaurs.

- Additional studies of rock and cave art, ancient symbols, and particularly dragon art.

- Study and analysis of the stratigraphical record of the earth without the trappings of evolutionary thought. In what strata are dinosaurs found and in which are they NOT found? What strata are found below dinosaur layers? What sequences of strata are found in dinosaur fossil areas? What implications may be drawn from the soil, volcanic ash, sandstone, limestone, and other substances in which dinosaur fossils are discovered? Is the material mostly water-laid or wind deposited? What strata lie beneath undisturbed dinosaur nests? What evidence of catastrophic events are evident?

- Study of fossil footprints and how they relate to flood events; develop explanations for the vast number of prints preserved at so-called dinosaur "stomping grounds," keeping in mind the strata that lie below the prints. We can understand how vegetation has turned to coal, but explain how dinosaur footprints could have turned to coal; develop explanations for fossil prints found at many different levels in the strata, for example, in Grand Canyon formations; do print patterns indicate one or more than one past catastrophic event? Do track sequences indicate panic or a normal pattern?

What is the message of the fossil record? What can we learn from fossil sequences in the strata of the earth? What fossils are found together and which have never been discovered together? When "wrong" fossils are found together, how is this explained?

It is easy to point out errors and false beliefs of evolutionists about dinosaurs. What errors and false trails of creationists can be identified? How competent and zealous are we to disavow humbugs, regardless of source? How careful are we to distinguish between our beliefs derived from Scripture, fossil, and stratigraphical evidence, and speculation with our own framework of beliefs?

Evolutionists are interested in dinosaur questions such as these: How can the enormous mass graves be explained? How can dinosaur remains in the polar regions be accounted for? Can DNA tell scientists anything at all about the dinosaur? Creationists will be very much interested in any new information along these lines.

Summarizing Statement

Remains from hundreds of species of dinosaurs have now been discovered. Every continent, major islands, and the Arctic and Antarctic regions are represented. Clearly, the dinosaur was not just a minor blip during the early history of the earth. Evolutionary explanations of the rise and extinction of the dinosaur are far from satisfactory. This chapter surveys how creationists and other anti-evolutionists treat this great mystery.

Dinosaur themes based directly on the Bible include creation week events, stupendous changes brought about by the fall, the hostile pre-flood world, giants, young dinosaurs taken on the ark, possible references to dinosaurs in Job and elsewhere in the Bible, and the role of the flood in dinosaur extinction.

Other topics include evaluation of old carvings of dinosaurs, footprints, evidence that dinosaurs were recent, and dragon tales and illustrations from around the world.

Some of these themes hold considerable promise for creationists to shed new light on the early world of man and dinosaur. I note greatly increased interest among creationists in the dinosaur mystery and much more sophisticated efforts to relate dinosaur remains to a biblical framework of a young earth.

For Reflection and Discussion

1. There are two largely "silent" periods in early Genesis, the period from the fall into sin by Adam and Eve up to the flood, and the period of many centuries after the flood until the birth of Abram. The Bible has more important priorities than to satisfy our curiosity about ancient times. What is the purpose of the Bible? What is *not* the purpose of the Bible?

2. At the end of creation week, God saw that everything He had created was very good. How does that fit into fossils discovered of snakes 100 feet long, dinosaurs 150 feet long, and centipedes 25 feet long? Could this be related to God's curse of the earth, the animal and plant world after Adam and Eve fell into sin?

3. Read the descriptions of fearsome monsters in Job 40–41 (TEV). How well do you think these descriptions fit the dinosaur?

4. In Romans 8:20–22 we learn that all creation is groaning from the cursed world. May one relate this passage to thorns, poisonous snakes, mosquitoes, gigantism, and violence of a cursed earth?

5. Read other descriptions in different versions: Genesis 1:21—sea monsters or great creatures; Psalm 89:10—Rahab or the monster Rahab. Commentaries of the past could only think of whales or crocodiles or hippos because dinosaurs and other fossils had not yet been discovered. Could these references in the Bible also be memories of the dinosaur?

6. Carl Sagan, the late defender of evolution, studied ancient, worldwide art and literature about the dragon. He believed that memories of the danger and horror of the dinosaur passed on through 65 million years of evolution of mammals up to modern humans. This is why, in his mind, art and literature give us many vivid memories and descriptions of dinosaurs. How do you respond to his analysis?

7. Is it possible to compromise and harmonize beliefs about the old/young and created/evolved world? Why or why not?

8. One evolutionist stated that dating the rocks or strata by the fossils contained in them and then dating the fossils by the strata they are found in is an impossible method. However, they have no alternative. Is this circular reasoning not science? The age of fossils and ancient chronology are very basic problems for

discussion of creation/evolution issues. How well has the case for a young earth been presented in this book?

9. It is important to remember that the cursed ground is noted not once but three times in Genesis: 3:17, 4:11–12, 8:21. One geologist spoke of the lately tortured earth. What do you suppose are some implications of these passages that may relate to this chapter?

10. Do you suppose you could find a bison (buffalo) footprint from 150 years ago somewhere on an uncultivated prairie in the West? There are billions of dinosaur footprints preserved in stone, and we are told that these prints were gradually covered and then fossilized over millions of years. Defend or oppose this view of dinosaur footprints.

11. May one expect to find the answers to any and all questions about the ancient past in the Bible?

CHAPTER FOUR

BIBLICAL ARCHAEOLOGY:
ILLUMINATING OR UNDERMINING?

WHAT IS BIBLICAL ARCHAEOLOGY?

Early one morning during the stifling hot summer of 1989 I was sitting on the curb of a dusty street in a remote town of northeastern Syria. I was waiting for a ride back to our mud brick compound, headquarters for an all-summer dig at Tell Tuneinir, where I served as one of a dozen volunteers for this adventure. Unexpectedly, a metal door noisily rolled up across the street, and a man appeared dressed just as I had seen in illustrations of life a thousand years ago in the Middle East. His shop, perhaps a dozen feet wide, was filled with sacks of grain. Shortly after opening, he got out a scruffy old broom and swept up the spilled grain mixed with dirt, dead insects, and other debris in the narrow pathways between the piles of sacks. My first thought after seeing the filthy floor was, "It's about time!"

He continued sweeping in front of his shop. I assumed he would then sweep the whole mess into the litter-strewn street where it would never be noticed. Instead, he stooped and picked out and discarded the larger pieces of debris. With a makeshift dustpan he then swept up the unspeakable remainder, including the trampled grain.

What happened next seared itself forever into my mind. He opened up a sack of grain, dumped the filthy sweepings into it, and stirred it a bit with his hand to blend it with the other contents. I was horrified to think of a poor widow buying grain from that sack later in the day, not knowing the origin of its contents.

Two weeks after my return home I was jolted when I heard the Old Testament reading in our church. What I had witnessed to my shock in Syria was already a disgusting practice about 2,800 years ago in Israel!

> Hear this, you who trample the needy and do away with the poor of the land, saying, "When will the New Moon be over that we may sell grain, and the Sabbath be ended that we may market wheat?"—skimping the measure, boosting the price and cheating with dishonest scales, buying the poor with silver and the needy for a pair of sandals, *selling even the sweepings with the wheat.* (Amos 8:4–6, my emphasis)

Incidentally, the location of this town is along the Habor River (also spelled Habur) about 400 miles northeast of Damascus. Here is where the Assyrians on two occasions sent Israelites into permanent exile. Could this merchant have been a descendant of one of the lost tribes of Israel?

Persecution Fear Contrasted With a Childlike Faith

In the rare book section at the University of Michigan is a papyrus document dating back to the first century AD when the persecution of Christians was at its peak in the Roman Empire. The papyrus was discovered in the grave of a mother and her daughter. It was an affidavit of a kind of passport that reads as follows:

> To those in charge of the sacrifices of the village Theadelphia, from Aurelia Bellias, daughter of Peteres, and her daughter, Kapinis. We have always been constant in sacrificing to the gods, and now too, in your presence, in accordance with the regulations, I have poured libations and sacrificed and tasted the offerings, and I ask you to certify this for us below. May you continue to prosper.
>
> We, Aurelius Serenus and Aurelias Hermas, saw you sacrificing.
>
> I, Hermas, certify. The first year of the Emperor Caesar Gaius Messius Quintus Traianus Decius Pius Felix Augustus, Pauni 27.[1]

They were so fearful that they might be accused of being Christians that the document went to the grave with them. And here is a vivid contrast. Not long ago the world's oldest book was discovered in Egypt in a grave under the head of a little girl, about 12 years old. It is the complete, well-worn Book of Psalms on 490 delicately handwritten pages, bound between wooden covers, and the burial was dated about AD 400. We can see the faith and love of her parents when they buried this precious book with the body of their precious daughter.[2]

When we say that biblical archaeology at its best illuminates the Bible, the above examples show us why. These are but three of many hundreds of vivid, dramatic examples that relate places, persons, and events to Old and New Testament times.

Exciting Discoveries

Every year volunteers from all over the world pay their own expenses to experience the thrill of an actual archaeological dig. The heat, the drudgery, the hard work, the insect bites, the scorpions, and the often primitive living conditions seem a small price to pay for the anticipation of discovering something significant. Perhaps the nearest parallel is the gold prospector whose hope springs eternal about what lies hidden in the next shovel full of soil.

There are a number of lists drawn up by reputable scholars of what has been uncovered in recent years: Peter's house in Capernaum and the foundation of the synagogue there where Jesus taught; the original site of the temple in Jerusalem and what is believed to be the exact location for the ark of the covenant; a broken carved pomegranate scepter believed to have graced Solomon's temple; the first inscription to refer to the House of David; and many other artifacts and inscriptions that relate to real people and real events in the Bible.[3]

In the past few years, however, much publicity has been given to charges that some dramatic recent discoveries in the Bible lands may be expertly produced forgeries. Of course, we could never defend a fraud that attempts to relate an inscription or other artifact to the Bible. Investigations are still underway to attempt to determine which discoveries are genuine and which are fraudulent. This is a very difficult task because qualified experts disagree in many cases. It is well to remember that the truth of the Bible does not depend on archaeological discoveries, but genuine discoveries can and do much to illuminate life in biblical times.

Going Back in Time

Suppose by a whole combination of miracles you were able to determine that one of your ancestors lived in a certain village a thousand years ago and that the site of that village was accurately located, and here you are, spade in hand, to learn all you can about your ancestor. What would you find? Perhaps a few foundation stones of homes, bits of broken ceramics, buttons, metal fragments, maybe even a coin or two, but no writing of any kind is likely. Anything found beyond this would be amazing indeed. Now try to put together your ancestor's life with only a fraction of 1 percent of the original data and well over 99 percent speculation. This illustrates in a sense the impossible task laid before archaeologists of the Bible lands and why we should never be caught in the trap of thinking that somehow archaeology must prove the Bible.

What is Biblical Archaeology?

Biblical archaeology at its essence is the careful, systematic recovery and study of remains of the past. These scholars tell us much of the cultural background for the events of the Bible. In some cases such research can clarify and help to explain some biblical texts. It adds much to the history of the Bible lands. Separating evidence from interpretation, it defends the truth of the Bible against the many attacks of liberals who attempt to undermine it under the guise of pretending to apply pure science.

Biblical archaeology means different things to people, and it is wise to make some distinctions in our evaluation of this field. Here are some of the varieties called biblical archaeology:

Humbug Biblical Archaeology

Tourists to the Bible lands may be confronted with some of the more outrageous fictions and abuses: Here are the jars where Jesus made the wine at Cana. Visit the ancient tree today in northeastern Cairo under which Mary, Joseph, and the baby Jesus rested. In this basement room behind the altar of this church in old Cairo, the Holy Family hid from Herod. In some cases the statement is well meant but unprovable; in others it is a shameless pursuit of the tourist dollar. In 1998 it was reported that a rock venerated by early Christians as the place where the Virgin Mary rested on her way to Bethlehem had been rediscovered.[4]

Biblical Archaeology as Wishful Thinking

Those who accept the Bible as true history sometimes expect more of biblical archaeology than it can deliver, pursuing the field as though it can be used to prove that the Bible is true. At times it is simply a matter of drawing a wrong conclusion. Many conclusions, as we might expect, are controversial. The truth of the Scriptures is not waiting around for well-intentioned people to prove that it is true. We find such unfortunate statements in Bible studies and Bible handbooks—again well meant and some still taught as truth today—as the following: Here are pre-flood tablets found at Kish under the flood layer. Here is a list of pre-flood cities that have been excavated so far. Here is an actual layer of mud of Noah's flood, found in three places. Here is a piece of wood from the ark. Here is the actual face of the Pharaoh whom Moses confronted. This jar contains some of the darkness from the plague of darkness in Egypt. Here is the cup (the Holy Grail) Jesus used at the Last Supper. The intentions may be good, but they are pure speculation or embarrassing wishful thinking.

Michael Drosnin attempted to show in his book *The Bible Code* how the Hebrew text contains numerous coded names, messages, and prophecies about future events.[5] Using the Hebrew text without vowels, searching in any and all directions, and applying strange computer gymnastics, the author was able to come up with ancient and modern names and events by helpfully inserted appropriate vowels as needed. The most astonishing thing about the book is that some have taken it seriously.

Another form of wishful thinking, of course, is the denial of plain and compelling evidence. We shall illustrate this view below with the evidence found at Jericho.

Humanist and Minimalist Anti-biblical Archaeology

It is no secret that most of what is termed biblical archaeology is in the hands of those described by G. Ernst Wright as having "broad humanistic interests."[6] Their own preferred term currently is "mainstream" in contrast to those who are identified as "conservative." Those are the polite terms. Name calling is still common. Humanists believe in all sincerity that their speculations and interpretations are scholarship, while anyone differing from such views is demeaned and ridiculed as a fundamentalist or a literalist. Remember that most of the work done in the Bible lands is done by those with humanist beliefs. Yet these are the scholars who want to remove the words *biblical* and *Israel* and substitute something like Syro-Palestinian. They avoid speaking of BC and AD because these terms refer to God. The ones who most loudly oppose any possible connection of the Bible with any history are called minimalists, for want of a better term. I shall describe more of this bizarre humanist approach to biblical archaeology.

Biblical Archaeology As It Should Be

There is something special about biblical archaeology in contrast to all other archaeology. The Bible is the only sacred book that can be examined archaeologically. Biblical archaeology at its best carefully separates evidence from speculation about it. Biblical archaeology ideally shows the highest regard for unbiased scholarly work, for scientific method, and for the latest technology. Much archaeological work wonderfully illuminates the life, cultural setting, and times of biblical characters and events, and thus helps to explain or illuminate portions of the Bible. That is why we can make good use of the work of archaeologists even when they have no regard for the truth of the Bible. In true biblical archaeology, Christians see and accept an ever lengthening list of discoveries that fits beautifully with what the Bible describes and teaches.

Approaches to Biblical Archaeology

This important field of study includes three areas of scholarly treatment, each unfortunately living in "splendid" isolation from the other two for the most part: field or dirt archaeology; linguistic studies—the study of inscriptions by persons called epigraphers; and biblical studies—the study of the biblical text. Each has the potential of benefiting greatly from the other two.

Here for the present is just one illustration: Kenneth Kitchen observed the strange and remarkable fact that modern literary criticism of the Old Testament thrives completely separate from actual literary finds of Orientalists who study the ancient manuscripts and other inscriptions.[7] For example, Kitchen noted that a certain literary style known to be common 1,500 years before Solomon is tagged by critics as fourth century BC Greek, and therefore the wisdom literature is moved up 500 years to fit the notion. Words labeled by critics as postexilic have been found to date a thousand years or more earlier, yet such finds have no effect at all on Old Testament mainstream (that is, humanist or liberal) scholarship.

Limitations of Biblical Archaeology

How Scientific is Biblical Archaeology?

Professional archaeologists, regardless of how they interpret evidence, do use sound methods and the best technology available to them. The method is not the problem. Anyone who believes that most biblical archaeology is purely an objective science is naive and uninformed. J. Maxwell Miller stated that archaeologists tend to be overconfident regarding the possibility of reconstructing the details of biblical history.[8] We have already begun to see something of the limitations of biblical archaeology in attempting to recreate the past. Consider these points:

The fragmentary nature of the evidence from the ancient Near East complicates the drawing of inferences. One reasonable estimate is that we have at hand less than 1/1000 of the potential evidence from antiquity based on the admittedly optimistic projection that 1/4 of the available sites have been surveyed, that 1/4 of these sites have been excavated, that 1/4 of the excavated sites have been adequately examined, and that 1/4 of the materials and inscriptions have been published. In reality, hardly more than an estimated 200 of over 5,000 sites in all of Israel and Jordan have been excavated, and less than 50 of these can be considered major excavations. According to Paul Lapp less than 2 percent of the best sites in Palestine have been touched by archaeologists.[9] While much work has been done in the past several decades, we can hardly say that this situation has really changed. In Mesopotamia less than 1 percent of the total sites have been excavated. No sites in Israel except Zumrun and Masada have been totally excavated. The bulk of the potential evidence remains buried. Thus we believe that only a tiny fraction of the full picture of ancient life was excavated thus far, and these fragments are almost all mute. From there it is human inference and imagination perhaps unrelated to reality.

By its very nature, biblical archaeology is a very, very slow, careful tedious process. At the rate the important biblical site of Hazor is being excavated, Keith

Schoville noted that the work will continue for another 800 years.[10] Currently there is great excitement over the fact that the royal library will be discovered there. The great scandal of biblical archaeology, however, is the frequent failure to publish the results of the dig when it is over. However, heroic efforts are under way to correct this problem. More and more technology is being applied to digs currently, involving a broad variety of specialists from different fields as well as high tech equipment. This inevitably slows every operation and makes it more costly, but for the best of intentions.

As we illustrated above in the search for a hypothetical ancestor, field archaeology produces a lopsided view of an age. All the precious things that could best tell us of the past are those things that were burned, weathered away, or decayed. Even the documents discovered on very rare occasions are non-representative or are not open to study, and so we are brought to a halt.

Normally, only a tiny area of an entire site can ever be dug, as Kitchen informs us.[11] Tell Beit Mirsim, where William Albright "established" the standard archaeological chronology, had only about 1/4 of the surface excavated, but only parts of that area were excavated down to bedrock. All kinds of important features can be missed by accident. For example, a time period may be found in part of a site that was not excavated, and therefore is missed. If one digs 5 percent, one misses 95 percent, and if you are not on the site you thought it was, as happened at Heshbon, the error is 100 percent.

The Bible testifies that Moses was the inspired author of the Pentateuch—the first five books of the Bible. Critics, however, maintain that these books were developed from four sources long after Moses. The sources are named J, E, D, and P or the Documentary Hypothesis. This approach abandons any thought that the Bible is the inspired word of God. As Kenneth Kitchen observed, even the most ardent advocate of this documentary theory must admit that there is no single scrap of external, objective evidence that any alleged source-document ever existed.[12] But JEDP rises above, as it were, the need for any evidence. It is a kind of badge that publicly proclaims an attitude toward the Bible, that it is no different than any other ancient document. As often noted, Europe is where JEDP is a sacred doctrine, and this is where Americans have typically gone to get their degrees in biblical studies.

Modern archaeology is designed to gain maximum information from a stone-age culture, but this is not really what we want to find. Kathleen Kenyon, the founder of modern scientific archaeology around the mid-twentieth century, was characterized by George Mendenhall as one who gathered infinite amounts of useless detail, and who ignored the value of texts in shedding light on the past.[13] Her excavations covered too tiny a slice. While she carried out endless elaboration, she never got to any real results or relationships. She had no background in the ancient languages of the region. She was blinded by the trees and never saw the forest. This rather unkind critique stemmed from Mendenhall's work under her supervision at Jericho, the excavation that won for her top rank in scientific archaeology!

The following examples illustrate how difficult it often is to solve challenging problems in biblical archaeology.

Moab and Ammon. Another classic speculation repeated for decades as pure truth is that Moab and Ammon did not exist at the time of the Exodus/Conquest, and therefore those portions of the Bible are not true. More recently, investigations have concluded that archaeological sites in those areas are "much older" than previously thought.[14]

The problem with Heshbon. Consider the first battle of Israel at Heshbon before crossing the Jordan. For decades the excavation at Heshbon has been cited as "proof" that the Old Testament is unreliable historically, as remains from the time of Moses and Joshua were missing at these ruins. How does one deal with this matter?

First, one must observe that ancient chronology is a most inexact science and nothing conclusive can be established from a chronology known to be faulty. Second, scholars are now saying that a much better candidate for biblical Heshbon has been located and excavation has begun. This large site, located about 5 miles south of the former site, is Tell Jalul. It has all the strata to satisfy any chronological interpretation.[15] Moreover, there was no proof that the first site was actually the biblical Heshbon. Because it appeared to contradict the Bible, however, humanists were happy to accept it as such, and they tried to force this interpretation on all others.

Interpretation turmoil. If anyone doubts that archaeology has limitations when it comes to interpreting the data, he has only to consider the bitter archaeological debate between two prominent Israeli archaeologists: Looking at the same evidence on the conquest, Yohanan Aharoni said it was peaceful; Yigael Yadin said military conquest. (Others maintain that there never was a conquest.)

Yadin dated pottery to the time of the divided kingdom; Aharoni dated it to the time of King David. Aharoni said the destruction of Level III at Lachish was that by Sennacherib in 701 BC, but Yigael Yadin said it was a century later from the Babylonian attack of 597 BC.[16] For decades we have been led to believe that C14 dating could solve all such dating problems.

Joshua's altar? A passionate debate about the excavation on top of Mount Ebal appeared in 1986. Adam Zertal shows why he believes he has found Joshua's altar as described in Joshua 8.[17] Aharon Kempinski dismisses the structure as an Iron Age I watch tower.[18] And the battle appears to be more of ideology than of evidence.

Concluding Thought

When we consider the limitations discussed in this section, it seems miraculous that archaeologists have discovered so many things of value that give us a deeper understanding of the times and settings of the Bible lands. We owe a great debt to these scholars, past and present, for their discoveries, but not necessarily for their interpretations.

FINDS AND ILLUMINATIONS

We list just a few of the hundreds of inscriptions and other finds in Israel and in surrounding nations that marvelously illuminate the Scriptures.

Oldest Creation Tablet

Long before Moses was born, a tablet was written in cuneiform that begins just like the opening of Genesis 1. The ancients knew the true story of creation. Much later many other degenerated versions were written, but they still show signs of the original. By far the oldest creation tablet, it is one of thousands of pre-Hebrew tablets uncovered at Ebla in northwest Syria in the 1970s. It reads in part as follows:

> Lord of heaven and earth: The earth was not: you created it. The light of the day was not; you created it. The morning light you had not yet made exist.[19]

Bill Cooper has pointed out that there are also other long ignored accounts in ancient literature that follow closely the Genesis account of creation.[20] Remarkably, these accounts date from before the time when Moses wrote Genesis. Cooper further discovered memories of Genesis 10 names (Table of Nations) in records among people dated long before Christian missionaries came to these areas: China, Egypt, Greece, Rome.

The Gilgamesh Story

No ancient tablet in the world is more famous than Tablet XI, the Gilgamesh Epic. It is believed in all sincerity by many humanists since the 1870s to be the original version from which the Genesis story of the flood was later formed. Literally thousands of books have picked up and retold this belief. I. Rapaport, however, provided a devastating critique of that fond belief.[21] Rapaport maintains that the tablet does not tell about a flood at all, and the flood story got its start on the basis of a mistranslation. The reference to the birds in the tablet cannot have any relationship at all with the flood account. The vessel in Tablet XI is not an ark, and the date of the tablet is centuries later than even the most radical late dating of Genesis. This gross error has been known for more than a century, but the

The Ebla Creation Tablet made a headline in *Time* magazine. One may make a strong case that it is the oldest of all so-called creation tablets. It reads just like the beginning of Genesis, but was written many centuries before Moses wrote Genesis.

The long lost Kingdom of Ebla, mentioned in some ancient documents, emerged in the 1970s out of excavations south of Aleppo in Syria. This kindgom thrived for centures before the time of Abraham, and biblical scholars learn much about life in that time period from the excavations there and from the thousands of tablets discovered in the royal library, written in a Paleo-Hebrew language.

story goes on and on as a monument to careless scholarship, and it is routinely referred to today as solid evidence that Genesis is a late collection of myths. In truth it describes a catastrophic event after the flood.

The Beginning of Agriculture

The work of specialists on ancient agriculture and animal domestication is very illuminating despite their full commitment to evolution for explanation. Consider these comments by Charles Reed:

> Despite a lack of consensus we are left with the Near East as offering the clearest picture of a history of continuing cultural change culminating in the cultivation of plants and the domestication of animals. If village life is to be correlated with an increase in population as I believe we must accept, then the arc of hills from western Iran through northern Iraq, and southwestern Turkey, down through Palestine and western Jordan almost to the Red Sea was sprouting villages. In each such village a group would depart and found a new village. Whatever the factors, plant agriculture did arrive in the Near East, and with such a rush and such a rapid spread that we are amazed.[22]

To the creationist the above statements are a perfect description of the rapid spread of civilization to the south and west of the mountains of Ararat during the first centuries after the flood. The arc of villages described by Reed are just where we would expect them to be from the biblical account and from the geography of that region.

In poring over the studies of Reed and other specialists, the creationist finds nothing at all in the evidence that is out of character with the description of the early post-flood world, except of course the conventional dating. But Reed, an evolutionist, speaks of the irresponsible use of C14. Yet he strongly disapproves of creationists taking note of the conclusion he had drawn. Reed concluded:

> Of the various areas of the world considered in some detail by (leading specialists), the emerging pattern of agricultural beginnings for the Near East seems to me to be clearer than is that for any other area.[23]

Genesis 9:20 notes that Noah, a man of the soil, proceeded to plant a vine-yard. This was the true beginning of agriculture and of village life after the flood. A short distance southeast of the mountains of Ararat, on the Zagros plateau at Godin Tepe in present day Iran, archaeologists discovered the earliest documented evidence of wine making. Red stains in a jar dated about 3500 BC were analyzed and found to be derived from wine.[24] Creationists are not surprised.

Could this be Sodom? Bab edh-Drha is located on the southeast coast of the Dead Sea in the area where some archaeologists believe Sodom and Gomorrah once stood. We leave the question as to whether or not the first of these sites was actually Sodom and focus instead on the findings of two paleobotanists.[25] In two seasons of collecting and sorting plant remains, Dr. David McCrery and his wife found remains of these crops at the site: wheat and barley, dates, wild plums, peaches, flax, grapes, figs, pistachio nuts, almonds, olives, pine nuts, lentils, chick peas, pumpkin, watermelon, and castor-oil plants. It is striking that Ezekiel (16:48) describes the people of Sodom as overfed!

While walking along the desolate east shore of the Dead Sea we found a stone cast that once had been mud or muck encasing two stalks of papyrus. This identification was made for me by a paleontologist at the University of Michigan. The Dead Sea then was the Live Sea!

That is the evidence. Interpretation heads in two opposite directions. The conservative Christian points to Genesis 13:10: "The land well-watered everywhere, before the Lord destroyed Sodom and Gomorrah, even as the garden of the Lord." The humanist, believing that Sodom and Gomorrah never existed, looks for signs of irrigation. One needs to see the site of Bab edh-Drah today to fully appreciate the dreadful barrenness of the region.

In another interesting but controversial development, William Shea analyzed two Ebla tablets that listed a large number of ancient place names including

The terrible destruction and desolation on the east shore of the Dead Sea, which must be a consequence of the destruction of Sodom. Nearby on the seashore the author found the fossil cast of papyrus stalks from the time when the Dead Sea was the Live Sea. Josephus speaks of the sea where the valley used to be.

many in Israel and Jordan.[26] The tablets are dated at least several centuries before the destruction of Sodom and show the routes taken by a traveler of that time. Shea notes that Sodom, before its destruction, was one of the places visited, and that linguistically it is identical to the biblical city of that name. Of special interest is the fact that the place names listed on the tablet just before and after Sodom are now geographically impossible for a modern traveler. The route passes through some of the most devastated land in all the world, and the route taken to the north of Sodom is now completely blocked by a rugged range of sawtooth mountains. When the land was "like the garden of the Lord," that route would have been easy for the traveler. The tablet supports the conclusion that the land in that region changed radically at the time of the destruction. This should not be surprising, however, because the earliest Genesis passage mentioning the Dead Sea may be paraphrased as the "sea where the plain used to be." Similarly, Josephus noted in a matter of fact way that the Dead Sea did not exist before the destruction of Sodom. Humanists believe the Dead Sea was there for millions of years.

Other Discoveries

It is not the purpose here to list even the most important discoveries that illuminate the Bible. I will close this section with additional interesting discoveries:

- A receipt written on a pottery sherd was unearthed recording the gift of three silver shekels for the temple of the Lord. The date of the pottery places this event at the temple of Solomon.[27]

- *Biblical Archaeology Review*[28] announced the discovery of a clay impression of a royal seal that reads: Ahaz (son of) Jotham, King of Judah. The evaluation is that it is genuine and therefore is a most spectacular find. Ahaz ruled from 732–715 BC.

- A silver scroll dating to about 650 BC was buried with a devout Jew.[29] Inscribed on the scroll is the oldest direct quotation from the Bible, the words of Aaron's blessing that we find in Numbers 6:24–26.

The following is a small sample of other significant and dramatic finds that tie in directly with Bible characters and events:

- Villa ruins and a defaced seated statue in Goshen believed to be that of Joseph (Genesis 50:22)

- Baalam, son of Beor—inscription found at Succoth site in Jordan (Numbers 22–24)

- Nahash, King of the Ammonites, text found that expands the account of 1 Samuel 11

- House of David inscription unearthed at Dan (1 Samuel 16:13)

- Moabite Stone inscription regarding Omri and Ahab events, and refers to the House of David (2 Kings 1:1)
- Inscription: King Jehu of Israel gives tribute to Assyria (2 Kings 10:32)
- Inscription: Assyria takes captives; Hoshea replaced Pekah as king of Israel (2 Kings 15:29)
- King Hoshea's royal seal found (2 Kings 17:1)
- Inscription of Sennacherib regarding King Hezekiah (2 Chronicles 32)
- Inscription about Shebna, the disgraced royal steward of Hezekiah (Isaiah 22:15)
- Inscription of King Hezekiah and the tunnel he ordered beneath Jerusalem (2 Kings 20:20)
- Hilkiah's ring found: "He found the Book and brought it to King Josiah" (2 Kings 22:8)
- Seal of Gemariah, the secretary of King Jehoiakim (Jeremiah 36:10)
- Seal and fingerprint of Baruch, Jeremiah's secretary (Jeremiah 45:1)
- Inscription: capture of Jerusalem by Nebuchadnezzar (2 Kings 24:10)
- Tablet: King Jehoiachin's rations as prisoner in Babylon (2 Kings 24:10)
- Inscription: Zedekiah appointed king (2 Kings 24:17)
- Inscription: Jerusalem destroyed by Babylon (2 Chronicles 36:17)
- Cyrus cylinder: authorized return of the Jews from captivity as prophesied (Daniel 5)
- Inscription of Tobiah, Ammonite official, at the entrance to his banquet hall (Nehemiah 2:10)
- High priest Caiaphas' tomb inscription and ossuary (Matthew 26:3)
- Inscription naming Pontius Pilate (Matthew 27:2)
- Home site of Peter believed located in Capernaum (Luke 4:38)
- Inscription of Erastus of Corinth, friend of St. Paul (Acts 19:22)

Conclusion

We cannot fail to be impressed with even this small sample of discoveries that relate biblical archaeology with living, breathing characters and events of the Bible. Yet it is necessary to call attention to the fact that there are endless arguments in the professional journals and gatherings between experts on whether or not several of the above items are genuine or sophisticated frauds. Where there may be frauds, we suspect two possible motivations: greed, or the belief that nothing in the Bible is historically true.

We learn from the first-century historian Josephus that this beautiful valley was the estate of the Tobiah family, and the name Tobiah in Aramaic is carved next to a cave that served as a banquet hall. From twelve references in the Book of Nehemiah, we find that Tobiah was a bitter enemy of the Jews.

PROBLEMS OF CHRONOLOGY

Chronology is an Inexact Science

Isn't it interesting to know that the core of the most heated controversies in the field of biblical archaeology has always been chronological? Yet in almost all we read there appears to be no uncertainty.[30]

In addition to examples earlier in this chapter, there are other problems that have never been satisfactorily solved: the date of the sojourn in Egypt and the Exodus from Egypt; the name of the Pharaoh of the Exodus; the archaeological placement of the Conquest, that is, which destruction layer of cities destroyed during the conquest of Canaan is the one the Israelites inflicted; the *habiru* problem (Does this word in ancient writings mean the Hebrews?); Solomon's use of the copper workings in the Arabah; and was the sacking of Lachish, Level III, by Sennacherib or Nebuchadnezzar? And who can sort out the dating of the world before Abraham? Humanists are not interested in such questions because they believe all those matters are myth and other fiction. They believe Abram, Moses, Joshua, and David are fictional characters, and that there never was an Exodus or a Conquest.

Perhaps readers are familiar with conventional tables of archaeological time periods that flow out of and are based on the same assumptions as the geological time table, especially the assumption of uniformitarianism—long, slow, uniform, gradual change. A thick layer of sediment must have taken twice as long as one half as thick. The periods and ages of archaeology are universally accepted as accurate within a handful of years in almost all the literature.

The chronology of Palestinian pottery was established by William F. Albright from his work at the site of Tell Beit Mirsim in the 1920s. Albright believed that just about everything had been solved:

> Since Egyptian chronology is now fixed within a decade or two [for the Middle Bronze and Late Bronze Ages] our dates are approximately certain wherever we can establish a good correlation with Egyptian cultural history. Thanks to scarabs and inscriptional evidence, this is quite possible.[31]

Further refinements of course have been made, and the precision of ceramic dating for the biblical period is thought by some to be within a century, others say within 50 years, and one scholar in a recent lecture said that ancient pottery could be dated within 10 years. The whole structure, as we will observe, hangs on the validity of Egyptian chronology—a fragile thread indeed. It did not take long for critics to question Albright's work. Israeli scholar Anson Rainey speaks of Albright's subjective impressions about Tell Beit Mirsim, that the Albright school was in practice based more on personal opinion than on actual finds.[32] Too often the opinion of an excavator has usurped the place of true archaeological evidence.

Similarly J. Maxwell Miller observed that Albright's intuitive but faulty ideas of chronology and interpretation are so ingrained in the generation of scholars he trained at Johns Hopkins University that archaeologists no longer look at actual evidence. Yet this same chronology continues on and on.

Dr. Adnan Hadidi, the former Director of Antiquities of Jordan, made the following remarkable statement about dating accuracy:

> It is a strange anomaly that pottery of the Middle and Late Bronze Ages, can in Palestine at any rate be dated by its contexts to within 25 or 50 years with reasonable accuracy, whereas as soon as the historically far better known Roman period is reached, a couple of centuries seems to be the closest limit one can hope for.[33]

Can anyone seriously believe that accuracy improves as we move farther back in time?

Chronology is Complex

From the previous discussion we may conclude that when a site is interpreted where chronology is a key factor, for example, Jericho, the issues are hardly settled as many interpreters would have us believe. The complexities are enormous, the stones are silent, and we may never reach a point where a fully compelling solution can be offered. We say this despite the fact that interpreters such as Bruce Williams state from the humanist perspective that conservatives are overlooking "easily available evidence."[34] We have read widely, but have not seen any specifics about this. This is reason enough to encourage qualified persons to accept William's challenge to examine such "evidence" and its interpretation.

When Honest Questions are Asked

In the past several decades a few scholars who questioned the "received" chronology and who offered alternatives have been brutally attacked, including vicious personal character assassinations.[35] The profession is not ready to tolerate chronological dissidents. Lay persons who begin reading the professional journals in biblical archaeology are shocked at the bitter exchanges and attacks that appear all too commonly on journal pages. Two British scholars, David Rohl[36] and Peter James,[37] who recently pointed out some of the fallacies and absurdities of conventional Egyptian chronology, were called "archaeological charlatans" in a letter published in *Biblical Archaeology Review*.[38] Nothing was said to challenge the evidence they presented. A seminary professor once told me that if someone raised questions about the conventional treatment of biblical archaeology discoveries, no one would talk to him any more.

Is Chronology a House of Straw?

In this section I shall examine briefly why some persons, including myself, are very uncomfortable with the generally accepted chronology. The interested reader may pursue the arguments in the notes for this chapter. Is it really possible, as Jean de la Bruyere (1645–1696) said, that the exact contrary of what is generally believed is often the truth? Should we be content to receive and pass on a chronological system without question and without noting its many problems? There seems to be a great deal of lockstep thinking in mainstream biblical archaeology in matters very much open to question.

David Henige points out that those who accept chronologies based on oral or oral-derived traditions are in danger of accepting a bag full of holes.[39] Yet that is almost the only source we have. Further, scholars must be aware of these treacherous paths:

- Synchronisms, that is, an event in one nation linked with a known date in another. While this seems pure, almost no example is trouble-free of controversy, contradiction, and alternate interpretation.

- Father-son succession, such as David followed by Solomon. Even here there are many problems in establishing the dates for each ruler.

- Genealogies of rulers are surely the most suspect of all historical sources. Accuracy often gives way to propaganda purposes. There is the treacherous use of "average reigns" to try to build chronologies.

- Kings and dynasties are often shown as succeeding one another when in fact they shared available time and space. Classic examples are Sumerian King lists and Manetho's Egyptian dynasties. Too few remember that such lists were never intended to provide an accurate chronology.

- Telescoping is the accidental or intentional forgetting chunks of the past. The purpose may have been to blot out past enemies or to suppress bad news. Details are almost impossible to uncover.

The ancient Near East is a mix of secure data, uncertain evidence, assumed astronomically fixed dates, perceived gaps, and hope for new discoveries. It provides an interesting study of human nature in which world-class scholars defend shaky positions with much verbal violence.

Jericho as an Example

The Jericho excavations are an immensely complicated story, but briefly we may say that John Garstang's[40] earlier excavations were interpreted as support for the scriptural account while Kathleen Kenyon came up with no support for the biblical record. Here is what Kenyon concluded from her study of the limited excavation she supervised at Jericho:

> At just that stage when archaeology should have linked with the written record, archaeology fails us. This is regrettable. There is no question of the archaeology being needed to prove that the Bible is true, but it is needed as a help in interpretation to those older parts of the Old Testament which from the nature of their sources . . . cannot be read as a straight-forward record.[41]

More recently, the work of Bryant Wood showed the flaws of Kenyon's analyses and conclusions and presented convincing evidence of the fall of Jericho just as described in the book of Joshua.[42] The mainstream does not challenge his evidence. They simply ignore what they do not wish to hear.

Here are two comments: The argument is largely—as most are—a chronological one, and it seems safe to say that the conventional chronology is vulnerable in many ways. Second, I have personally heard one of Kenyon's students (later a world-recognized scholar in archaeology) openly scoff at Kenyon's highly subjective decisions during the Jericho excavations. Thus, Kenyon's interpretation is not as conclusive as many writers would have us believe, but it fits very well into a humanist conception of the Jericho story.

Chronological Consternation

A Chalcolithic burial cave, thought to have been sealed for 6,000 years, was discovered in northern Israel in the late 1990s full of bones and artifacts.[43] There were major shocks. The interior of the cave was covered by stalagmites, stalactites, and other geologic formations. Flowstone sometimes completely encased artifacts. Second, elegant and elaborate art styles long thought to be isolated from each other in time were found together. Expertly fashioned bronze artifacts were also found, but the burial cave had to be dated at least 500 years before the beginning of the Early Bronze Age. If the flowstone only were considered, this cave could have been dated hundreds of thousands of years old. All in all, this new discovery is a humanist chronologist's nightmare.

Toward Settling Disputes

Field or dirt archaeology is more subjective than one might gather from reading most accounts of digs. J. Maxwell Miller said that there would be many different interpretations of a 5-meter square (the normal unit for excavating at a dig), "if the director did not always have the final say in the excavation report."[44]

Consider also the comments by William Dever. Dever says the great outer wall at Gezer is Late Bronze Age; Kenyon dated the same wall to the Hellenistic period; Aharon Kempinski says Iron II. But the result, as Dever observes, is that after clearance of nearly the whole of a Palestinian site's city-wall system, together with its modern stratigraphic excavation, scholarly opinion as to the date of the outer wall differs by more than a thousand years. Dever charges Kempinski was simply ignoring overwhelming evidence. In that context, how can anyone be so positive that the Jericho destruction never happened?[45]

Willem C. van Hattem offers this comment:

> I personally cannot free myself from the suspicion that the dating of some of Bab edh-Dhra pottery [the supposed site of Sodom] was a result of wishful thinking rather than real fact finding. The "Cities of the Plain" *had* to be found in a certain era.... The weakness [of the argument] is not the biblical patriarchs, but the assumed chronology in which the archaeological facts are made to fit one way or another.... [W]e will have to be prepared to take account of new facts even when they challenge us to abandon established chronologies and widely accepted readings of history.[46]

There is interesting evidence that the long-revered sequence of Paleolithic—Mesolithic—Neolithic, supposedly covering hundreds of thousands of years, exists only in the minds of those who find it important to interpret the ancient world in uniformitarian terms. Although uniformitarianism has been discredited since the 1980s, even by leading evolutionists, the chronology based on those assumptions has achieved a life of its own. Grahame Clark cites evidence to show that hunting/gathering (that is, Paleolithic times) and settlement agriculture coexisted, that the Mesolithic in Europe had several Neolithic features, and that many so-called Neolithic societies in southwest Asia lacked some of the essential criteria for recognition as Neolithic.[47]

In C. Ernest Wright's pioneer study of the period between the Chalcolithic and the Early Bronze Age, he distinguished three groups of pottery—each supposedly dominant for a lengthy period of time in turn. Since that time, it has been shown that these groups of pottery, at least in part, are contemporary. This concurrence of the pottery groups in Palestine, and the attempts to use them also as chronological indicators have caused great confusion because each of five authorities has employed a different method of interpreting the archaeological data.[48] Referring to the analysis of finds in Mycenaean shaft tombs, John Dayton commented: "The important point is that all styles exist side by side in time. Archaeologists have been too ready to give time spans to different varieties of pottery and assume that one must succeed the other."[49] While Dayton's point is an overstatement, a great deal more attention ought to be given to the assumptions underlying pottery dating, to the validity of its dating, and to the supposed anchor points on which it rests. It is not unusual to find statements such as the following:

In 1973 Beno Rothenberg surveyed a settlement in the Sinai which he dated to the Proto-dynastic period (late Chalcolithic—Bronze Age I), but the site may, in fact, belong to the Middle Bronze I period, a difference of a thousand years or more, according to Rudolph Cohen.[50]

The Negev wilderness south of Judah played an important role in the time of David, Solomon, and the kings of Judah. Eminent archaeologists have analyzed supposedly the same pottery as follows: Rothenberg (thirteenth century BC or earlier); Aharoni (eleventh century BC); Cohen (clearly tenth century BC); Glueck, who first surveyed the area (between tenth and seventh centuries BC). C14 tests in 1980 tended to support Glueck.[51] One prominent archaeologist privately told me (not for publication) that another noted archaeologist had planted or salted the site with pottery sherds that would support his own dating!

J. Maxwell Miller believes there are good reasons to doubt the arguments advanced by Albright and others for dating the Exodus/Conquest during the thirteenth century BC.[52] He believes that John Bimson[53] presents a thorough and fair analysis of the matter, and concludes that those who hold to a thirteenth century BC date have no monopoly on the archaeological evidence. This is an important point because Kitchen, who is universally respected for his sound scholarship in biblical archaeology, and who holds a conservative view of Scripture, has accepted the late Exodus date. Despite all Kitchen's painstaking and monumental work in biblical archaeology, Peter James suggests that his efforts are hardly above criticism.[54] Siegfried Horn stated that it is high time that another detailed and penetrating study on the date of the Exodus is written, for no serious monograph on this subject has been published since J. S. Jack's now outdated work appeared in 1925.[55]

Conclusions About Chronology

David Henige sums it all up nicely:

Paradoxical though it might seem, there is probably more argument about both the details and the broader aspects of ancient Near East chronology today than there was 50 years ago, when the paucity of data itself encouraged a more comforting degree of certitude.[56]

All the assurance in the world as we read the literature—and there is arrogance as well—does not conceal the fact that ancient chronology is a highly speculative business. No one yet has come up with a decisive system beyond argument, but we can one day expect a key discovery or reinterpretation that will resolve many of the present uncertainties. That will be a welcome day.

INTERPRETATIONS, MOTIVES, AND SPINS

Evidence versus Interpretation

In reading or writing about biblical archaeology all must do a much better job of distinguishing between evidence and its interpretation. No one writes without assumptions about the ancient world, and currently these assumptions for the most part form two groups tagged at various times as scholarly, scientific, mainstream, liberal, or humanistic on the one hand, and as biblicist, fundamentalist, literalist, or conservative on the other. More recently, the labels used in biblical archaeology—mainstream versus conservative—are a hopeful sign that strongly held differing viewpoints need not be characterized by name-calling.

Hershel Shanks had the courage to state that American archaeologists working in the Near East must clearly acknowledge two legitimate but different constituencies: (1) archaeologists with a principal focus on the Bible; and (2) those whose focus is purely archaeology and who couldn't care less about the Bible.[57] Thus the focus should be on archaeology and not on insulting and demeaning those who hold an opposite view.

Critiques

James Moyer and Victor Matthews provided an evaluation, not to everyone's taste, of the most popular Bible handbooks.[58] They offered a general evaluation of each and then commented on how each handled the key "battlegrounds" of the flood of Noah and the fall of Jericho. Because they referred to the flood "story" rather than to the event, one immediately sees their orientation and presuppositions. Please note that they are examining only the treatment of biblical archaeology in the handbooks—not any other features. The critiques do attempt to be fair and deserve careful examination. Here are several sample comments:

Halley's Bible Handbook (the original edition) mostly cited finds from 50 to 60 years ago and has now been taken off the market. (The 2000 edition is much improved. The treatment of biblical archaeology is excellent.)[59] The *Abingdon Bible Handbook* is written from a mainstream (humanist) perspective but seeks to be fair to conservative viewpoints. It frequently cites both positions, and is reasonably up-to-date.[60]

The clear winner, according to the reviewers, is *The Bible Almanac*.[61] The perspective is conservative, but there is a clear statement that archaeology does not need to prove the Bible to be true. Yet the archaeological information is more substantial and valuable than any of the other handbooks up to 1985.

It is helpful also that Moyer and Matthews published a useful evaluation of the use and abuse of biblical archaeology in current one-volume Bible dictionaries.[62] They have sorted those volumes according to mainstream (which we have termed humanistic) and conservative orientations. They have provided a much needed service by identifying those that are badly out-of-date. The winner from a balanced conservative approach is J. D. Douglas (ed), *New Bible Dictionary* (1982).[63]

In addition one must note a conservative Christian archaeologist who is considered to be one of the top scholars in the world, highly respected by all viewpoints: Kenneth A. Kitchen.[64] He is necessary reading for any conservative Christian interested in biblical archaeology.

More recently, in the 1990s, the editor of *Biblical Archaeology Review* attempted to avoid controversy by simply citing publisher comments about such publications. Thus the reader knows in advance whether the treatment of biblical archaeology matters will be that of the humanist or of the conservative viewpoints.

An Archaeological Confession

A crystal clear statement appeared in the program of the annual meeting of the Society of Biblical Literature in 1969 that illustrates the so-called humanist position:

> The basic dilemma in historical Jesus (or read biblical archaeology instead of Jesus) research is not any complexus of technical problems but rather the seeming incompatibility between intellectual honesty and traditional Christian belief.[65]

Despite the unnecessary jargon, nothing could be stated more clearly: Before beginning his research, the humanist already "knows" that the Bible is not true historically. Those who relate biblical archaeology to the Bible, in their mind, are dishonest.

Absurdity reached a new peak with a transcript of a debate between mini-malists Niels Lemche and Thomas Thompson versus humanists Dever and McCarter.[66] Only inches apart in their low view of the Bible—that it is total or near-total fiction—nothing emerged from the heated exchange. Dever was out-raged at being called that dreaded word, *fundamentalist*, for stating that possibly somewhere there might be a relationship (which he calls convergence) between a few archaeological discoveries and events told us in the Bible.

Presuppositions Equal Bias

The humanists believe they are operating without presuppositions in con-trast to those who accept the Bible as historically true. Yet the assumptions gov-erning their interpretations are very plain in their writing, and note that we are speaking of assumptions—not evidence, not truth. In their minds evolution is the great unifying principle of explanation. Modern cultures evolved in a long slow path from primitive to modern; the world is very, very old; the Bible is just another ancient book full of error and myth; evidence in support of something in the Bible can be explained away; conclusions and interpretations that discredit or undermine the Bible are seized upon as proof that the biblical record is in error; pottery often must be placed in a chronological sequence rather than assuming any possibility of coexisting forms; today's climate is the climate of the ancient past; no catastrophes occurred that shaped any ancient history; Egyptian chronol-ogy rests secure on fixed astronomical points; artifacts found that seem too young are intrusive, that is, they fell into older strata from the activity of a gopher; items that seem too old are heirlooms kept for many generations and then finally buried. A few of the above items are sometimes true, but taken as a whole, the presuppo-sitions of the humanists are highly vulnerable and therefore conclusions based on them need not inspire awe or surrender.

The Old Testament as a Scholarly Battleground

John Currid illustrates the great gulf fixed between liberal and conservative beliefs with this quotation by humanist Bruce Halpern: "The actual evidence con-cerning the Exodus resembles the evidence for the unicorn."[67]

Currid comments that there is nothing new here. The present debate is merely the tired old stuff of nineteenth century liberalism wrapped in a new pack-age. To the liberal, the Old Testament is only "children's literature." In contrast, Currid cites evidence showing the Bible as a forceful witness of ancient history involving Egyptian and biblical events. Egyptian literature contains striking par-allels with the Genesis account of creation. Recent excavations in the Nile delta region (Goshen) furnishes unmistakable evidence of the Israelite presence there, to list just two of many examples discussed.

Examples of Anti-biblical Bias

When one reads extensively in the field of biblical archaeology, one cannot help noting with what glee speculation is seized upon if it runs counter to the biblical record. For example, Martin Noth stated that the Jericho discrepancy is the best and most decisive proof of the unreliable character of the historical parts of the Old Testament.[68] We need not be surprised to find other items such as the following in the literature: The humanist typically declares as solid truth—the flood never happened or it was the memory of a local event, such as the catastrophic formation of the present Black Sea; Abraham and the patriarchs never really existed; the Sodom and Gomorrah story at best is only an embellished story based on the distant memory of a bad fire that happened once in a village many centuries ago; the Exodus never happened; the Conquest of Canaan never happened; David and Solomon are pious fictions to enhance the self-esteem of Israel. All of these are just stories developed to make a point, and archaeology "proves" that the biblical stories are fictions or myths.

This item by the editors of *Biblical Archaeology Review* appeared years ago regarding the discovery of sites identified as possibly Sodom and Gomorrah:

> One prominent scholar who was supporting a grant for their excavations threatened to withdraw his support if they were indeed identifying their sites with the Biblical Cities of the Plain.
>
> There is a segment of the scholarly community which regards it as unscholarly to focus on possible connections between archaeological evidence and the Biblical record, because the evidence is often so tentative. Yet the most far-fetched speculation is permitted in other areas of archaeological scholarship.[69]

One gets the impression that the question of evidence is not raised as long as the Bible is under attack. Yet there are exceptions. During travel to biblical sites in Israel and Jordan, I asked a leading archaeologist there this question: "If you separate evidence from interpretation in biblical archaeology, how much of archaeology conflicts with the Bible?" His one-word answer was, "Nothing."

The Geological Society of London

This is the world's oldest such society, and its fame rests on the fact that it popularized the idea of an old, old earth in the early 1800s and thus made Darwin possible a generation later. In order to achieve this goal, the competent scientists of that day were excluded from membership because they all accepted a catastrophic view of earth's history, including Noah's flood. The goal of this anti-monarchy society was to undermine Genesis and thus to get rid of both the moral code and the monarchy.[70]

In 1997 the society re-emerged into the light of day when it made Ian Plimer of Australia an honorary fellow. Plimer's claim to fame is that he sued Allen Roberts for claiming that he had found the remains of Noah's ark in eastern Turkey in 1992.[71] Roberts, to be sure, joins a long list of persons who have made similar claims about the ark, none of which has been verified to date. The absurdity here in making Plimer a fellow is the propaganda device used by the society to attack the Bible in the hope that this would influence the decision of the court.

Explaining Away

Unfortunately we must assume that otherwise intelligent persons will devote great energy to explaining away what is unacceptable to their belief system. Man has not changed from the days of Pentecost, when scoffers were confronted by the miracle of unlearned apostles speaking in foreign tongues. Their conclusion was that the apostles were filled with new wine. We must assume and will often note in the following discussion that biblical archaeology is frequently used to explain away the Scriptures without informing the reader where the evidence has left off and where speculation has begun.

The Plague of Misinterpretations

It is not at all difficult to make a collection of embarrassing misinterpretations by "mainstream" archaeologists, but the same is sometimes true of well-meaning "conservatives." This is reason enough to strongly encourage qualified persons to become active in archaeology.

William Corliss developed a widely heralded systematic collection of anomalies and errors in various branches of the sciences, most of which deal with the ancient past, such as geology, paleontology, ancient history, astronomy, and some aspects of archaeology.[72] A similar project is in the works for the systematic examination of the literature of biblical archaeology and ancient chronology. This will be an excellent means of distinguishing between evidence and interpretation.

Anson Rainey included a short-lived section on "some recent howlers" in his discussion of historical geography and archaeology.[73] He illustrated how the most eminent scholars easily go gloriously wrong in their interpretations.

Several illustrations of recent errors or questionable interpretations will suffice to illustrate the point. Inscriptions excitedly accepted as Philistine writing by some of the world's leading authorities on ancient inscriptions turned out in the early 1980s to be a most amateurish fraud.[74] In fact, one fake Philistine manuscript that fooled world-class experts was a widely known Hebrew inscription written backwards!

In 1929 C. Leonard Woolley discovered a two-meter thick layer of pure clay in the city of Ur in ancient Sumer. He immediately proclaimed to the world that he had found Noah's flood. This amazing bit of nonsense has been archaeological truth ever since and is published in just about every Bible dictionary and handbook. Other scientists have noted that this clay layer was so minor it did not even cover all of the town of Ur. Second, competent scientists who studied the layer concluded that it consisted of windblown material. No flooding was involved.[75] This is not what critics of the Bible want to hear, so this information has been ignored down to the present.

Bible as History?

One popular book, *The Bible as History*, contains much interesting information in relating the Bible to ancient world history.[76] The author, however, accepts the flawed view of an old earth of millions and billions of years, and some of the explaining away of the humanists. To his credit, however, he rejects Woolley's story of finding the flood layer. He also points to the failure of many attempts to locate the ark on Mount Ararat. It is an often overlooked fact that the Bible does not say the ark landed there. Rather it "came to rest on the mountains of Ararat," (Genesis 8:4) that is, somewhere in that mountain range.

More recently, Charles Pellegrino, a geologist and paleontologist, wrote a book along similar lines that contains much fascinating material. He adopts many humanist views of the Bible, such as a 4.6 billion year old earth, thinks an ancient flood was later dressed up into Noah's flood, and subscribes to the ape to man evolution view. He has a fixation that the explosion of the island of Thera (perhaps Atlantis) was the cause of the plagues in Egypt, followed by the Exodus. His study of the Bible shocked him:

> One of the most fascinating surprises of my life has been the dawning realization that some of the more dramatic episodes of the Bible, things that appear very strange to most of us living today, perhaps even miraculous, seem actually to have occurred . . . [77]

What a surprise, indeed! Of course, this is much less than conservative Christians would say, but this was a major step forward for him. Among other things, Pellegrino found that archaeologists were using a chronology based on a house of cards, very much open to criticism. He relates that some scholars actually threatened suicide when he showed them errors in their assumptions and bias, and others got into actual food fights over arguments about chronology. He strongly agreed with the statement of another scientist that "scientific" dating should never be taken as gospel. Much to our surprise he examined and then accepted in full the evidence Bryant Wood assembled to show that Jericho perished exactly as related in the book of Joshua. Working through this and similar books is a useful exercise in separating wheat from chaff.

CONCLUSIONS

Getting Involved

Conservative Christians ought to invest in biblical archaeology, and many more ought to qualify themselves for serious work in field archaeology and in the ancient languages of the biblical world. G. Ernest Wright's comment is as true now as a generation ago, that money from pious, conservative, or fundamentalist sources has never played a very important role in archaeology. The major excavations have been sponsored by sources dominated by a broad humanistic interest.[78] As we have noted before, humanists look at the evidence through glasses colored by humanist presuppositions.

The Trail to Nowhere

This view of the ancient biblical world has led to the kind of excesses and sterile explanations described by P. J. Wiseman long ago as

> continuing along outworn paths, explaining away new facts which have come to light; as material left to the side, uninterpreted, and by and large, ignored; as using an assumed chronology in which the archaeological facts are made to fit one way or another; as resorting to name-calling instead of the use of evidence; as a refusal to accept new facts because they challenge scholars to abandon comfortably held but false chronologies and widely accepted but erroneous readings of history.[79]

We can hardly say that anything has changed. Again we must say that the analysis of potsherds and stones has severe limitations for reconstructing ancient history. The focus must be on the illumination of the Scriptures—not proving them.

The Need to Study Chronology

A brief summary of the chronological debate may be found in an article by William Stiebing.[80] Creationists ought to encourage the responsible study of chronological problems. The final answers are not as yet in for this immensely complicated problem of dating much of the past. To a large extent we must play a waiting game and hope that in future excavations some incontrovertible synchronism will be found that will put at rest the present uncertainly about dating the Exodus and other important issues. One conclusion seems safe. No side or faction has yet come up with a satisfactory solution to dating the biblical world before 1000 BC, and the challenge lies there waiting for a person of considerable genius to put all the pieces together.

Literary Treasures

More energy ought to be devoted also to examining and analyzing the literature of biblical archaeology. One can find great treasure in the illumination of

Scripture by reading such journals as *Bible and Spade* and *Biblical Archaeology Review*. Bible handbooks and other literature on archaeological findings include excellent and helpful summaries. Authors seldom label what is fact and what is speculation. We don't read our daily newspaper expecting Christian interpretations of news items. Yet most Christians benefit from reading newspapers, and they do their own interpreting. Likewise we can benefit greatly from reading the literature of biblical archaeology while realizing that nearly all of the interpretations will be humanist ones.

The Need to Question

We must not forget how limited archaeology is in its ability to speak specifically to issues and debates. Most of the evidence is stones and broken bits of pottery. But while we recognize and accept our own past errors and weaknesses, there is no need at all to remain silent about wild speculations that appear to be designed to erode the Bible. We must keep asking, "Where is the evidence for your speculation?" Conservative scholars need to show that they are interested in gathering truth and that they do not fear truth.

Another Pressing Need

When one learns to sort fact from speculation, biblical archaeology greatly illuminates our understanding of the setting in which the mighty acts of God took place in history. Yet this literature can do great damage. A concerned Christian canceled a subscription to *Biblical Archeology Review* with these words: "Maybe I misunderstand your magazine, but I wanted the subscription in order to build my faith, not lose it."[81]

Conservative Christians need to work much harder preparing good materials to help fellow Christians in their understanding of the Scriptures. But one can read and benefit from most of the literature of biblical archaeology by simply learning to discern between fact and fiction.

In the past few years there has been much more calm reflective thought on evaluating the field and noting its many weaknesses than in earlier times. There are excellent materials that ought to be collected and published for the information of conservative Christians who sometimes have been led to believe that biblical archaeology has assembled much evidence that contradicts plain history told in the Bible. Nothing could be further from the truth, once evidence and speculation are separated.

Biblical Archaeology is a Great Treasure

Where does all this discussion leave us? Biblical archaeology, combined with the geography of the land, is a wonderful treasure house of illumination for us of the setting of the Scriptures. It enables us to picture Old and New Testament events and helps us understand many things we cannot derive from the text alone. But that is not all of it. George Mendenhall quite properly stated that "unless biblical history is to be relegated to the domain of unreality and myth, the biblical and the archaeological must be correlated."[82] If he means archaeological "evidence" rather than mere speculation, he is correct. We know that archaeological information, by its very nature, is frequently incomplete or defective information. Still we have a certain quiet expectation—quite opposite to that of the humanist. We anticipate that archaeological finds that are unambiguous will be in harmony with the scriptural account, and the list of just such finds is already a very long one. We can use biblical archaeology discoveries quietly and sanely to help us better understand the Bible in its world.[83]

For Discussion and Reflection

1. Discuss the difference between evidence and the interpretation of evidence. For example, while visiting Cairo in Egypt, our guide showed us the dried up remains of a long dead tree that would have provided welcome shade for travelers when it was alive. He explained that this was the tree Mary and Joseph rested under when they fled to Egypt. Would you doubt a paid guide?

2. Why do you think there is so much speculation about discoveries in the Bible lands? Why not deal only with hard facts?

3. Museums have been paying huge sums for artifacts and inscriptions that relate to biblical sites or persons. Some of these are now being attacked as fakes. What might be the motivation for a highly skilled expert to create a fake biblical artifact?

4. In this chapter we read about amazing discoveries by paleobotanists near the Dead Sea where Sodom is thought to have existed. See Ezekiel 16:49 (NIV) that the people of Sodom were overfed. Lot chose this area because it was like the garden of the Lord (Eden). What does all this say about accuracy of details in the Bible?

5. Why is pottery, mostly sherds, the almost exclusively used method for dating excavations in the Bible lands? Do you suppose there are possibilities for error with this method?

6. Note that liberal scholars rejoiced in their claim from pottery analysis that Jericho proved that the Bible was not history. But more recent analysis showed very convincingly that the book of Joshua describes Jericho very accurately. Does this indicate that we must be cautious about claims by humanist archaeologists attacking the Bible?

7. Can you find examples in this chapter that illustrate "the need to believe" by those attempting to undermine the Bible?

8. How does Jericho illustrate the "need to believe"?

9. When you read a story in the media about a discovery related to the Bible in some way, do you think it wise to ask yourself or another whether the story seems genuine or wishful thinking or humbug or a pronouncement from a minimalist who believes there is nothing historical in the Bible?

10. What do you make of the observation that there would be different conclusions about an archaeological dig if one person did not have the final say about it?

11. When British archaeologist C. Leonard Woolley discovered a 6-foot layer of silt at Ur more than 75 years ago, he announced that he had found Noah's flood. Later it was found that this flood layer did not even extend over this entire ancient city. Other authorities concluded the silt was wind blown, not a flood at all. What is the message?

12. Archaeologists state that they are able to pinpoint a date correct to within 25 or less years on the basis of ancient pottery fragments. See the statement by Dr. Adnan Hadidi in the chapter about the puzzle of dating far newer deposits. What is the message?

CHAPTER FIVE

A False Trail for Joshua's Long Day

Two Great Miracles

Two Old Testament miracles have received much attention. As is always the case, humanists scoff at such events or try to explain away all miracles, while those who believe that the Bible is God-breathed take the Bible as it reads. There is no way humans can explain any of God's mighty acts. In the case of the miracles discussed in this chapter, however, we shall see that there is a bizarre twist to the story.

The tale began over a century ago when a strange story was related about proof for Joshua's long day. The same narrative reappeared in the 1970s dressed up in modern clothes. The two familiar events are almost always included in Sunday School lessons: The first took place when Joshua was conquering the heathen Canaanites in the Promised Land of Israel (Joshua 10:12–14):

> Joshua said to the Lord in the presence of Israel: "O sun, stand still over Gibeon, O Moon, over the Valley of Aijalon." So the sun stood still, and the moon stopped, till the nation avenged itself on its enemies, as it is written in the Book of Jashar. The sun stopped in the middle of the sky and delayed going down about a full day. There has never been a day like it before or since, a day when the Lord listened to a man.

The second event took place hundreds of years later. King Hezekiah of Judah asked God through the prophet Isaiah for assurance that God would really heal him from his fatal illness:

> Hezekiah had asked Isaiah, "What will be the sign that the Lord will heal me and that I will go up to the temple of the Lord on the third day from now?" Isaiah answered, "This is the Lord's sign to you that the Lord will

do what he has promised: Shall the shadow go forward ten steps, or shall it go back ten steps?" "It is a simple matter for the shadow to go forward ten steps," said Hezekiah. "Rather, have it go back ten steps." Then the prophet Isaiah called upon the Lord, and the Lord made the shadow go back the ten steps it had gone down on the stairway of Ahaz. (2 Kings 20:8–11)

Are There Other Memories of These Events?

Those who believe that myth and legends are coded history like to study world myths to see if dramatic events are recorded in them. It is no surprise, therefore, that some have explored such literature for traces or accounts of the Joshua and Hezekiah miracles. What have they found?

E. Walter Maunder discovered that the great record-keeping countries of the ancient world, Greece, Egypt, China, as well as India, all have ancient records of a long day.[1] However, the records differ much in details. In China, for example, Yang of Lu raised his spear during a battle and shook it at the declining sun, which straightway went backward in the sky to the extent of three signs. This story could relate to the miracle in Hezekiah's day.

It is interesting that in the Mexican Annals of Cuauhtitlan from the fifteenth century it is related that during a cosmic event (nine time zones away) the night did not end for a long time. These annals relate events of a very ancient date. The Midrashim, the books of ancient Jewish traditions not embodied in the Bible, relate that the sun and the moon stood still for 18 hours. The Hezekiah miracle occurred very near to the time when God destroyed the Assyrian army. Thus it is interesting that the Greek historian, Herodotus, relates accounts from Egyptian priests that the sun, more than once, changed its direction. Another Greek historian, Apollodorus, speaks of a time when the sun went backward. Other folklore about the sun going backward is preserved among the Latins, Icelanders, Finnish, Japanese, Polynesians, and Native Americans.[2]

A Story About These Stories

In 1970 the "missing day" story had widespread circulation. It was reprinted in tracts and columns by various personalities. In Dallas a television newsman passed around copies of the story with his picture at the top of the page. To the man on the street, the story appeared plausible and from a reliable source.[3]

We now know how the story began. In 1890 Prof. C. A. Totten of Yale University wrote an interesting account of how an astronomer had rediscovered Joshua's missing day.[4] Harry Rimmer gives us this summary:

Sir Edwin Ball, the great British astronomer, found that 24 hours had been lost out of solar time. Where did that go, what was the cause of this

94

strange lapse, and how did it happen? The answer may be expected in vain from sources of human wisdom and learning![5]

A Yale professor and accomplished astronomer made the strange discovery that the earth was 24 hours out of schedule! That is to say, there had been 24 hours lost out of time. In discussing this point with his fellow professors, Prof. Totten challenged this man to investigate the question of the inspiration of the Bible. He said, "You do not believe the Bible to be the Word of God, and I do. Now here is a fine opportunity to prove whether or not the Bible is inspired. You begin to read at the very beginning and read as far as need be, and see if the Bible can account for your missing time." The astronomer accepted the challenge and began to read. Some time later, when the two men chanced to meet on the campus, Prof. Totten asked his friend if he had probed the question to his satisfaction. His colleague replied, "I believe I have definitely proved that the Bible is not the Word of God. In the tenth chapter of Joshua, I found the missing 24 hours accounted for. Then I went back and checked up on my figures, and found that at the time of Joshua there were only 23 hours and 20 minutes lost. If the Bible made a mistake of 40 minutes, it is not the Book of God!"

Prof. Totten replied, "You are right, in part at least. But does the Bible say that a whole day was lost at the time of Joshua?" So they looked and saw that the text said, "about the space of a whole day."

The word *about* changed the whole situation, and the astronomer took up his reading again. In 2 Kings we have the story of King Hezekiah, who was deathly sick. In response to his prayer, God promised to add 15 more years to his life.

This settles the case, for 10 degrees on the sundial is 40 minutes on the face of the clock. So the accuracy of the Book was established to the satisfaction of this exacting critic. When the astronomer found his day of missing time thus accounted for, he laid down the Book and worshiped its writer, saying, "Lord, I believe!"

The Story Reappears

Apparently few people heard of Totten's book until it reappeared almost 50 years later, much embellished, in Harry Rimmer's book, *The Harmony of Science and Scripture*. Undoubtedly Rimmer's stirring defense of Totten's book led many people to accept the story as true.

Sydney Collett is perhaps typical of those who wrote a glowing tribute to the genius of Prof. Totten. Collett also gives a brief summary of Totten's research:

Prof. Totten has studied (Joshua's long day) from an astronomical point of view and has published the result in an elaborate mathematical calculation with the following remarkable conclusion, that, by taking the equinoxes, eclipses, and transits, and working from the present time backwards to the winter solstice of Joshua's day, it is found to fall on a Wednesday; whereas, by calculating from the prime date of creation onwards to the winter solstice of Joshua's day, it is found to fall on a Tuesday; and he argues that by no possible mathematics can you avoid the conclusion that a whole day of exactly 24 hours has been inserted into the world's history.[6]

More than this, E. Walter Maunders of the Royal Observatory, Greenwich, traces not only the spot on which Joshua must have been standing at the time but also the date and the time of day when this remarkable phenomenon took place.

Collett goes on to say that Totten's calculations actually showed 23 and 1/3 hours at the time of Joshua, in remarkable accord with the words, "about a whole day." The full day that astronomy demands should be accounted for is exactly made up by the fact that in Hezekiah's time the shadow on the dial of Ahaz was made to go back 10 degrees or 40 minutes—the balance to the minute of what was needed to make up exactly 24 hours! This concert of testimony is not to be shaken by any ingenuity of man or devil!

The Story Now Appears in New Dress

A new generation grew up that never heard of Totten and Rimmer. In 1970 startling articles about high tech and the Bible appeared in many religious periodicals and in the public press.

These articles reported that Harold Hill, president of the Curtis Engine Co. in Baltimore, Maryland, engineer and consultant to the space program, had proof that scientists at the Goddard Space Center in Greenbelt, Maryland, had, through a computer study, found that at the time of Joshua there was nearly a day missing (23 hours and 20 minutes). Another 40 minutes were found "missing" at the time of Hezekiah, when he ordered the sundial to turn back ten steps, or degrees. These 40 minutes added to the 23 hours and 20 minutes constituted one whole missing day, it was claimed.

This is a part of the story that Harold Hill repeatedly affirmed was true:

> One of the stories I told often had to do with a part of the necessary ahead-of-time statistical preparation for the moon walk. The space scientists were checking the position of the sun, moon, and planets out in space, calculating where they would be 100 and 1,000 years from now. In addition, they were looking into the trajectories of known asteroids and meteors so we wouldn't send astronauts and satellites up only to have them bump into something. . . .
>
> Well, as they ran the computer measurement back and forth over the centuries, it came to a halt. The computer stopped and put up a red flag, which meant that there was something wrong either with the information fed into the computer or with the results as compared to the standards. They called in the service department to check it out.

"Nothing's wrong with the computer," the technicians said. "It's operating perfectly. What makes you think something's wrong?" "Well, the computer shows there's a day missing somewhere in elapsed time," the operators said. They rechecked their data and scratched their Educated Idiot Boxes. There was no answer, no logical explanation. They were at a baffled standstill. Then someone who remembered Sunday School lessons about Joshua and King Hezekiah saved the day by explaining where the missing 24 hours were. This is the missing day that the space scientists had to make allowance for in the logbook.[7]

Stirring Defense of the Story

As with all spectacular stories, there are always the doubters. This is especially true when such stories seek to defend the truth of the Bible. When pressed to defend the story, Hill responded to thousands of inquiries with a form letter. He had witnessed the computer feat about two years previously. He had misplaced the source information and so was unable to give names, dates, and places. He would send it as soon as his notes were located.

Harold Hill referred the skeptics to Dr. C. A. Totten's book, *Joshua's Long Day* (1890), and to chapter 13 of his own book, *How to Live Like a King's Kid*. Here they would find the whole story.

V. L. Westberg gave his full support to Hill by reviewing Totten's book as a "tremendous mathematical work," years before the computer was even a dream.[8] He said that Totten deserves the highest commendation of our newer generation not only for his tremendous mathematical genius but for the finest description of the (spiritual) purpose of his work.

The Story Questioned

We must remember that the skeptics we consider here were not doubting the biblical accounts. They questioned the astronomer's work and the computer program that could discover a missing day.

As someone has said, some stories are so good that you don't *want* to find out if they might not be true. Other illustrations of this strange need to believe the unbelievable include the huge amount of literature about the Loch Ness Monster, Bigfoot, flying saucers, aliens, and much more.

Among those who immediately denounced the story as a hoax, however, were *Eternity, Christianity Today,* and the *Journal of the American Scientific Affiliation*. In reviewing the episode, Don DeYoung stated:

Unfortunately, false and uncertain science continues to be used in an attempt to bring drama and enticement into Scripture interpretation. The false announcement that computers had detected the "long day of Joshua" is one such example.[9]

In personal correspondence with Westberg, Hill revealed how creation groups severely criticized his story.[10] He had expected favorable comment.

What Was Wrong with the Story?

Various problems about the account quickly emerged:[11]

- Readers of the story who had doubts wrote to Mr. Hill at the Curtis Engine Company in Baltimore; their letters were returned with the notation there was no such firm.

- Although Harold Hill had given permission to Ms. Hazel Brown of Baltimore, Maryland, to reprint the story, which apparently had been taped from one of his lectures, he then disavowed the article as written.

- In a form letter to thousands of inquiries, Harold Hill stated that he could not remember where he received the information on which the article was based, but that it was true.

- Dr. Bolton Davidheiser wrote the NASA office at Greenbelt, Maryland. They replied that they knew nothing of Harold Hill and could not corroborate the lost day reference. They further stated that many of their computer programs ran back long before the time of Joshua and experienced no difficulties. No astronauts and space scientists at Greenbelt were ever involved in the lost day story attributed to Mr. Hill. Another NASA official stated that the story was a complete and untrue fabrication without a shred of truth.

- Many readers noted that the computer angle was added to Totten's original story to give it a modern flavor. All the 1890 details were unchanged.

- John Read of the technical staff at Hughes Aircraft pointed out the following: Checking the positions of the sun, moon, and planets is totally irrelevant to do what Totten and Hill claimed. It was also irrelevant to use checkpoints of these bodies of 100 and 1,000 years. In current space work, orbits were studied for a projected seven years only at that time.

- Incredibly, Harold Hill stated in a 1970 letter to *Bible Science News* that Totten's 1890 book contained complete data for writing a computer program to find the missing day. This is not true.

- A key point for Totten's calculation from creation to Joshua's long day was that it fell on a Tuesday, whereas counting back from 1890 to the same event came out to a Wednesday. Totten, however, began his calculation on Day 1 of creation, four days before the sun, moon, and stars were created. This fact alone destroys his argument.

- Dr. Mulfinger found a fatal flaw in Totten's thesis. Totten assumed there was a new moon on the day of the battle, yet the Bible states clearly that it was not a new moon because Joshua could see both the sun and the moon in different directions. Aware of this problem, Totten was forced to call this reference to the moon in the Bible anomalous; in other words, the Bible was wrong.

- Computers do not operate in the manner Hill described in his story. A computer can only print out what is first fed into it. Furthermore, there is no way that a missing day could be found through mathematics, by new moon phases, or through computers.

- Astronomers assert that exact dating cannot be done on the basis of eclipses, yet Totten is very precise, claiming that Joshua's Long Day took place on the 933,285th to 933,286th day from the beginning of the world. He further claimed it was the 22,852nd eclipse that had occurred since the beginning of time. There is no rational basis for deriving any of those values.

- While 23 hours and 20 minutes is indeed about a day (unless 12 hours was meant for the daylight hours of a day), there are an infinite number of other values that could be called "about" a day. Similarly we have no way of knowing what time span ten steps meant to the ancients at the time of Hezekiah.

- According to Robert Oden, research consultant, Totten obtained his material on chronology from J. B. Dimbleby's book, *All Past Time*.[12] He was the premier chronologist of the British Chronological Society. Unfortunately, astronomers of his time discovered that his tables and method of calculating time and lunar and solar years were in error. Oden concluded that Totten relied heavily on Dimbleby's unreliable ideas and calculations. In turn, Totten's ideas are the basis for Hill's computer story, and so these also are not reliable.

- C. J. Ransom observed that scientists have to waste time refuting stories that are fabricated by someone and foisted upon the public as science.[13] One example is the missing day story. Claiming that this type of calculation is possible would be similar to claiming that a computer program proved that there is a mile missing between anywhere and Chicago. How could anyone prove or disprove that statement?

- Totten claimed to have made calculations showing that there actually was a missing day at the time of Joshua, but he said, "The mere figures are of no interest save to the verifier," so they were not included in his book! We thus have hand calculations that were never shown to anyone, a missing computer program and programmers who cannot be found.

- In order to do what Totten (and later Hill) claimed, one must have precise calendar and time points before and after each of the two events described in the Bible, but no such data exist. In addition, the chronology for this time period is in great dispute among scholars.

A Final Thought

Seven years after the story of the missing day was widely published in this country and beyond, Hill stoutly maintained that it was true and again referred people to read chapter 13 of his book, *How to Live Like a King's Kid*.[14] Sad to say, that chapter is of no help at all in verifying the story.

We cannot read the mind of Totten and Hill or know their motivations in telling their stories. We do know that the story is false, even though we have no doubt whatever about the truth of the biblical account. The events really happened. It is unfortunate that some people believe that spreading this kind of misinformation helps the spread of the Gospel. The Bible does not need to be proved. It is believed by faith.

For Discussion and Reflection

1. Read Joshua 6:20. Picture the army of Israel surrounding the city. What is the significance of the phrase "straight in"?

2. Like many ancient cities, Jericho possessed a double wall. Why did the army not enter through the city gate? Recent research shows that both the inner and the outer walls collapsed outward, thus forming rubble ramps for the invading army to enter the city from any point. Is this attention to detail in the Bible surprising to you?

3. What does this chapter do for your understanding of the long day of Joshua?

4. Why would anyone make up a story about proving Joshua's long day?

5. What is the difference in believing that Joshua's long day really happened and that a computer proved that it really happened?

6. Older persons read fairy tales when they were children. Today, children learn to believe for a time in such myths as Santa, the tooth fairy, and the Easter Bunny. From the time children enter school and on through adulthood other myths are taught as truth, as we shall see. Is evolution one of these? Why or why not?

7. In the Nicene Creed, Christians confess that God is the Maker of all things, visible and invisible. What are some invisible things we believe in? Is science able to examine invisible things?

8. How can a computer discover and prove that a day is missing from the past?

CHAPTER SIX

THE ART OF MISQUOTING ARCHBISHOP USSHER

INTRODUCTION

A Serious Question

When we study Genesis, the book of beginnings, we inevitably involve our-selves in serious questions about time. The problems are massive. Chronology, the backbone of history, according to Edwin Thiele,[1] lies at the core of many of the most heated controversies in archaeology and ancient history, simply because there is much more uncertainty about dating ancient events than we are normally led to believe.[2] The basic and most obvious question, of course, is: Do we live on a young earth or an old one? The Bible clearly teaches a young earth, and this entire book explains why I believe this truth.

James Ussher and Charles Darwin, buried incongruously under the same roof in Westminster Abbey, London, typify the debate over the past centuries. On the one hand, Ussher believed in a young, created earth. Darwin is the symbol of the radical opposition to undermine and destroy Genesis, and with it the entire Bible. Darwin had an urgent need for vast periods of time if evolution was to be defended. He and his followers since have used a veneer of "science" to support their fantasy. When Christians ask honest questions about all this and request evidence, the ridicule and intimidation begin.

Ussher's Life and Work

James Ussher (1581–1656) took office as archbishop of Armagh (Ireland) in 1626, long before publication of his famous work on biblical chronology. He was born in Dublin, January 4, 1581. In 1593, still a young boy, he was admitted into the college of Dublin, and, between the ages of 15 and 16, he had made such proficiency in chronology, that he had drawn up, in Latin, an exact chronicle of the Bible, as far as the books of Kings, not much differing from his *Annals*. His chronology rests on the assumption that the Bible presents a complete chronology of the world from Adam on. In 1650 he published the first part (three large volumes) of his *Annals of the Old Testament*. The final volume, which included New Testament chronology, was published in 1654. The original title of this massive work was *Annales Veteris et Novi Testamenti*. It is in the first volume of the *Annals of the Old Testament* that his famous estimate of the age of the world was expounded.[3]

In 1701 Bishop William Lloyd had Ussher's dates inserted in the margins of the Authorized or King James Version of the Bible. It soon was considered inseparable from the text for the past three centuries.[4]

The Ussher / Lightfoot Era

Truth By Ridicule?

According to evolutionists, scientific precision has replaced the old outmoded view of the age of the world. Anyone who does not accept the "new truth" is considered to be anti-scientific. Who can estimate how many thousands of times in books, magazines, and in university classrooms Ussher's famous date of 4004 BC for the creation of the world has been noted in order to draw belly laughs about the blindness and ignorance of a handful of uptight fundamentalists? The ultimate squelch of creationists is to cite this date for creation. Probably no other quotation in literature has been subjected to as much mockery in the past century. This choice bit of ridicule is still designed to smoke out any Christians in the classroom who might attempt to hold on to such an outmoded and simplistic belief. Have you ever wondered why evolution is the only aspect of the scientific world where ridicule, intimidation, fear, and force can be basic teaching methods in the classroom? In this chapter we shall explore just who might be roughly correct and who might be precisely wrong.

Quoting from Prime Sources

The scientific method developed over the centuries because the Renaissance had invaluable ingredients that have helped speed progress in some of the sciences and in all of the technologies. Among the essentials of scientific method are a healthy skepticism of what is thought to be true about the universe we live in and a recognition that "theory" is only an attempt to put observations into some kind of pattern as a cautious step toward scientific truth, that is, toward discovering laws of nature. To equate theory with truth or to confuse it with fact is a violation of the scientific method.

Another element of scientific method is to use the utmost care in recording observations. The history of science is full of the grotesque outcomes of the careless recording of information. Sometimes it is negligent; sometimes it is a deliberate attempt to deceive. Anyone who explores the history of science will find famous names who used fudged data to achieve planned outcomes.

It is no secret evolutionists believe that those holding to creation are very unscientific, that is, unable or unwilling to follow the scientific method. It seems fair to assume that these enlightened scientists will demonstrate the scientific method at its best in what they write.

We first examine the use of prime sources by scientists. Those who live by the scientific method often stress the importance of accuracy of observation and of going to prime sources for information. Because James Ussher is so widely quoted by scientists, let us see how accuracy fares in the hands of his detractors. Only when one has collected a number of such quotations does the picture emerge.

There is no greater sin in the scientific community than the failure to go back to prime sources for information, that is, to the original source. Thus, if Ussher is criticized, the scientist should go back to Ussher's own writings to find out exactly what Ussher said. Many scientists do quote Ussher's original works on the chronology of the world, which date back to 1650–1654. It is interesting and instructive to examine how well these scientists use the scientific method in dealing with Ussher.

How Ussher is Quoted

One may conclude at the end of this chapter which of the two greatly different chronologies is close to being correct. It is of interest to examine just what Ussher did say, especially because so many are so adept, we learned, at misquoting what seems to be a simple statement by Ussher, regarding the date and time of creation. An astonishing number of quoting errors occurred regarding the Ussher matter, despite the fact that many of these scholars cited the prime source, *Annales Veteris et Novi Testamenti*, from which they presumably drew their information.[5]

Out of many dozens of examples we find in the literature, we cite the following, underlining the errors, to illustrate our point:

In <u>1694</u>, James Ussher, <u>Episcopal</u> archbishop of Armagh and primate of Ireland, carefully tracked all the biblical begats and concluded that Creation took place at <u>exactly</u> 9:00 a.m. on October <u>26, 4004 B.C</u>. This astonishing 'fact' might well have passed unnoticed, but for an <u>unknown scribe</u> who included it as a marginal reference note in the King James version of the Bible. It didn't take long for the date to become <u>doctrine</u> and for anyone who assumed another date, especially an earlier one, to be <u>branded a heretic</u>.[6]

There are eight errors in the above quotation. A college freshman who was so careless would be severely dealt with by any instructor who held to any standards at all. But things get even worse.

The most garbled misquotation we found is the following:

<u>[I]n the face of Darwin's theories</u> and <u>mounting evidence</u>, <u>twisted scenarios</u> about the earth and its creatures were devised to make the facts fit what humans then wanted to believe. One was catastrophism, suggested by <u>Bishop</u> Ussher <u>of Lightfoot</u>(sic), who said that during the history of the earth <u>27 separate catastrophes</u> had wiped out everything, and each time God had started over. It was not until the <u>last Creation</u>, which began at 9:00 a.m., October <u>23</u>, 4004 B.C., that <u>human beings appeared</u>.[7]

This "twisted scenario" was not an easy thing for Ussher, since he died 200 years before Darwin's book was published. I count ten errors in these lines.

The following is a sample of various errors I have discovered. I have the documentation for all of the following, but indicate only a few, as there is no point to identifying all this scholarly sloppiness. In scanning numerous sources, I observed:

Name: Not only were the two spellings of his name misused—Ussher (and Usher)—but some gave him the title of bishop instead of the correct title, archbishop.[8]

When did Ussher publish his chronology? There are eight different definite answers to that question: seventeenth century, eighteenth century, 1642, 1650, 1654, 1658, 1694, the 1850s, and from a very prominent source—1751!

Which creation day? Ussher spoke of Day 1 of creation week. Others state that Ussher specified creation, the birth of Adam, world, man, creation and man, earth, human life, universe, and the re-creation and rehabilitation of the earth. These people were all "quoting" Ussher, but it is evident that they were very careless about it and obviously did not go back to the prime source. Further, there was much confusion over what Ussher said or what John Lightfoot said. Some credit Archbishop Langland with exactly the same statements as others attribute to Ussher or Lightfoot. This is further evidence of careless scholarship and laziness in checking sources. So far none of the persons who cited Ussher can be accused of having carefully checked the original source, even though they may have listed the source in their references.

The month of creation: Many are certain that Ussher never indicated a month at all; others are certain that he selected September or October to mark the beginning of some creation event; one writer specified autumn. One is rather nonplused to account for such different responses from one simple source.

The day of the month of creation: There are more surprises here. Writers affirm that Ussher did not indicate a day of the month. Others quote Ussher as saying September 17, October 22, 23, 26, or 29. Few, if any, who cited Ussher apparently looked at his work.

The day of the week: Ussher is often said to have specified the day of the week on which a creation event occurred. When we tabulated an informal sample of such quotations from texts, papers, books, reports, and the like, we found five different answers to another simple concept. Some say Ussher never indicated a day of the week; one person wins our admiration by saying that Ussher specified "the exact day," so we have no way of faulting him. Others quoted Ussher as saying Sunday, Friday, or Saturday.

The year: Because of Ussher, 4004 is one of the most familiar numbers in the world. Yet we found one writer stating the year 4044 BC instead. Actually, as we shall see below, Ussher never mentioned the year 4004 BC!

The hour of the day: Ussher is commonly reported to have indicated the hour of the day when some creation event began. Anyone who has actually examined his writing on chronology ought to be able to pinpoint any such statement. Again writers differed wildly, as follows: Ussher did not specify an hour, Ussher stated 9:00 a.m., 3:30 p.m., or 8:00 p.m. One writer noted that it all began 9:00 a.m., London time.[9]

Even those who mock Ussher would need to admit to an extraordinary degree of scholarly shoddiness above in copying down a rather simple one-line quotation—hardly the kind of performance to inspire confidence in whatever else they wish to say.

Scholarly Sarcasm

The following comments about Ussher are typical:[10]

According to marginal notations in several older versions of the Bible, the universe was allegedly created . . .

A reader will sometimes encounter, as a ludicrous target, the date proposed by Archbishop James Ussher . . .

Sir Mortimer Wheeler: That idea (4004 B.C.) we have long since chewed up . . . pulverized . . . scattered to the winds!

We no longer accept the clerical dictum that Earth was formed . . .

Biblical exegesis led Bishop Ussher to believe that the earth was created in 4004 BC, a date still accepted by the most naive of the faithful, though nothing in the Bible really so much as suggests it.

Obviously such a date is not in keeping with geological evidence regarding the age of the earth. Neither is it in keeping with geological evidence that indicates human activity on the earth goes back two million years or more.

This view (that man came recently to the New World) was almost as fixed and biased as the doctrine of Bishop Ussher, who in the eighteenth (sic) century . . .

Ussher's conclusion was trying to put cosmic events to the puny scale of human history . . .

What Did Ussher Really Say?

This chapter is not a defense of the Ussher chronology, except that we shall conclude which of two radically different chronologies is close to being correct. Nevertheless, it is of interest to examine just what James Ussher did say, especially because so many have been so skilled, we learn, at misquoting him.

We shall make several observations and cite brief quotations. It is indeed remarkable that Ussher begins the introduction to his massive work with a lengthy reference to a little book, *De Die Natali* (AD 238) by Censorinus, that likewise treated the birthday of the world. One may say that a Sothic cycle date reported by Censorinus is the keystone, rightly or wrongly, of Egyptian chronology. This takes on even greater importance when it is realized that all Middle East and Eastern Mediterranean cultures are synchronized with Egyptian chronology.

Ussher was only one of hundreds of serious scholars who attempted to solve the great chronological problem of the age of the world. Out of his intensive study of an impressive array of ancient writings and of the Bible, Ussher stated:

> I concluded that the world had been founded before the end of the year of the Julian period 710, since the beginning of the creation of things undoubtedly had been placed in autumn. Inasmuch as the first day of the age commenced with the evening of the first day of the week (that is, Saturday evening), I observed that the first day of the week, which in the year 710 approached most closely to the autumnal equinox from the astronomical tables corresponded to the 23rd day of October of the Julian year. Preceding that day of the Julian year by an evening, I concluded that first movement of time must have begun with the first day of creation. (Man then was created on Friday, October 28. As above, this day in our reckoning began the previous evening, Oct. 27.)[11]

Some Complications

While it is of passing interest to be able to say that Ussher does not say 4004 BC—the only "fact" almost all his quoters are agreed on—year 710 of the Julian period does correspond to the year 4004 BC. It comes as a surprise that Ussher specifies October 23 when John Lightfoot so often is credited with this addition to Ussher's estimate. But there is a subtle point involved here that makes the day of the month a somewhat difficult thing to specify.

In the pre-flood world and later among the Hebrews and other peoples, the new day began at sundown. Ussher specifically begins the first day of creation on the evening preceding the 23rd. For the Hebrew this is still the 23rd day of October, but for the Gentile it is the 22nd of October. Thus we have the strange phenomenon of two correct answers, depending on whether one is Jew or not. The non-Jew must then say, according to Ussher, that creation began at 6:00 p.m., Saturday, October 22, 4004 BC. The Hebrew (and anyone else whose culture begins the new day at sundown) must hold, according to Ussher, that creation began at 6:00 p.m., Sunday, October 23, 4004 BC. Since Ussher speaks of Day 1 of creation here, other October days would be correct if the writer is referring to other creation days.

There is another small problem. Great Britain passed a Calendar Act in 1750 to correct an eleven-day error in its calendar. Thus this correction for October 22/23 plus 11 days gives us November 3 (Gentile) or November 4 (Jew) for the beginning of creation.

R. Buick Knox observed that Ussher was aware of the provisional nature of his estimate, because it depended on the textual source used.[12] Ussher stated his willingness to consider fresh evidence and to revise his own figures if he found it necessary to do so.

Who Was Lightfoot?

John Lightfoot, D.D. (1602–1675), English divine and rabbinical scholar, is often quoted as adding a more specific time to Ussher's calculation. He was born at Stoke-on-Trent, in Staffordshire. After his degree at Cambridge where he developed extraordinary proficiency in languages, he entered into orders. He was chosen minister of St. Bartholomew's in London and later entered the office of Vice-Chancellor of Cambridge.

How Lightfoot is Quoted

Many critics, after mockingly referring to Bishop Ussher's famous statement on when the world began, often quote Bishop Lightfoot as well:

> To those who wished even greater precision, Dr. John Lightfoot, Vice
> Chancellor of the University of Cambridge, the great rabbinical scholar of

his time, gave his famous demonstration from our sacred books that "heaven and earth, centre and circumference, were created all together, in the same instant, and clouds full of water," and that "this work took place and man was created by the Trinity on the twenty-third of October, 4004 B.C. at nine o'clock in the morning."[13]

(Note: When author Christopher Cerf repeats the above, we are informed that this amplifies and corrects by some 15 hours Bishop Ussher's estimate made two centuries earlier.[14] Lightfoot died in 1675, about 25 years *after* Ussher's book was published.)

> John Lightfoot, Vice-Chancellor of Cambridge University, announced that the earth was created in 4004 B.C., on the 29th day of October, at 9 o'clock in the morning.

> No one with educational pretensions still subscribes to John Lightfoot's chronology, who dated the appearance of human life on this planet at 9:00 a.m. on September 12th, 3928 B.C.

To this widely quoted statement attributed to Lightfoot, we add other remarkable quotatios credited to him.

> Dr. John Lightfoot, Chancellor of Cambridge, trumpeted that Adam and not the crude half-ape, half-human creature found in the Java back country was the first ancestor of man. To heighten the contrast, Dr. Lightfoot fixed the time of Adam's birth at 4004 B.C. The claim that the outlandish Java creature could have had anything to do with man was called rank heresy.[15]

(Note: Lightfoot died in 1675, and so-called Java man was found in 1891. This is one of those extremely rare but little studied cases where an evolutionist tells us this cleric was able to "trumpet" about a fossil discovered more than 200 years after his death.)

> If you are one of those individuals who craves exactitude and who can only visualize history as a long string of dates, you can try to work out the method by which Dr. Lightfoot was able to determine, in 1654, that man was created at 9:00 a.m., October 26, 4004 B.C., and then apply the same method to man's first arrival in North America. Incidentally, if you can decipher Dr. Lightfoot's code please let us know, as we have always been curious as to how he did it.[16]

> This view (that Creation took place in September at the Equinox) was accepted by Dr. John Lightfoot, Master of St. Catherine's and Vice Chancellor of the University of Cambridge. Further consideration prompted Lightfoot to amend the month to October. He declared in 1642 that he entirely endorsed Archbishop Ussher . . .

(Note: On the surface this may have been difficult to do as Ussher's volume on the subject was published 8 years later.)

When Lightfoot is Cited

John Lightfoot is often cited, but difficulties arise quickly, as the following illustrate:

Which creation day? The writers refer to one of these beginnings: birth of Adam; creation; man; creation and man; human life.

The month: For the month when creation activity began, most quote Lightfoot as saying October, but some September.

The day of the week: The supposed day of the week is somewhat slippery also: Some say he specified the exact day but do not tell us which it is; some say Lightfoot never specified the day.

The day of the month: Lightfoot is said to have specified any of eight different days: exact day specified by him; no date specified; 12th, 13th, 18th, 23rd, 26th, 29th.

The exact hour: Some scholars state that Lightfoot indicated the exact hour of creation, but are not specific; others say he said no such thing; still others quote him as saying 9:00 a.m.; or 9:00 a.m. Greenwich time.

Another explanation of the whole matter was offered by W. Raymond Drake, who informs us that Archbishop Langland is the one who furnished the month, day, and hour of creation.[17] Things are complicated enough without following this unlikely trail.

What Did Lightfoot Actually Say?

The title of his book (1642) is irresistible: "A few, and new observations upon the book of Genesis. The most of them certain, the rest probable, all harmless, strange, and rarely heard of before." We have done what appears to be unthinkable for the evolutionists cited in this chapter. We have gone to the original source to see what Lightfoot actually said. Previously, John Klotz conducted a similar search.[18] He concluded that Andrew White must have used hearsay or he totally fabricated his statement about Lightfoot. Thus the famous quotation from Lightfoot, found so often in the writing of those who want to ridicule the idea of a young created earth, is not true, and apparently originated with White[19] and then mindlessly copied forever afterward by others.

These are the chief statements, edited, from Lightfoot:

- Genesis 1:26: It is very probable that man was created by the Trinity about the third hour of the day, or nine of the clock in the morning.

- In the month Tisri (about our month of September) the world was created.

- The world was made at equinox, the 12th of September.[20]

Lightfoot states that the year AD 1644 was 5,572 years after creation, thus suggesting the date of about 3928 BC as the date of creation. (We have not found any mention of 4004 BC in his writings.)

THE CONTRAST OF EVOLUTION TIME

Darwin's Method

A new look in dating the age of the world was fundamental to Charles Darwin who above all knew he had to have enormous blocks of time if his theory of evolution was to gain acceptance. He illustrated his style when he examined the open rolling uplands of England, known as wealds, and estimated that 306,662,400 years were required to bring about the formation.[21] He offered no evidence for his speculation, yet his followers blindly accepted it. The old date offered by Ussher was religious superstition to be forever discarded. Even Darwin became embarrassed with his own nonsense and later withdrew that estimate.

The age Darwin gave for the weald was, of course, pure guesswork, but it was this same Darwin who told his followers to avoid too many zeros at the end of an estimated age. One should not give the impression, Darwin advised, that one is guessing! This may explain the mysterious "400" at the end of the estimate.

Problem Solving by Adding Ages

As evolutionary theory unfolded over the decades since Darwin, problem after problem emerged that defied explanation. The unfailing solution was to add more ages to the earth and the universe. Time was offered to solve every problem. Thus Henry Whipple has pointed out that on the average, the "age" of the earth has been doubling every 15 years for the past three centuries.[22] Apparently almost any problem of evolution can be solved by placing supposed events farther and farther into the past.

The Orthodox Age of the World

Scientists and text writers these days have the freedom to choose between 4.5 or 4.6 billion years for the age of the earth. It would be unthinkable for any evolutionist to use another figure. One might think that such uniformity is a result of reasonably clear evidence. Cyrus Gordon unkindly suggests another reason for a related situation:

> No politically astute member of the establishment who prizes his professional reputation is likely to risk his good name for the sake of a truth that his peers (and therefore the public) may not be prepared to accept for fifty or a hundred years.[23]

There are some who suggest the date ought to be put still farther into the past, evidently to solve more problems. After all, when you pass a tree and call it your cousin, as the late evolutionist, Carl Sagan, liked to do, it takes time for trees and people to differ as much as they do today. However, courage will be required to popularize a new and older guess.

For those clutching onto radiometric dating, radiation physicist H.C. Dudley insisted that the equations describing radioactive decay rates were crudely derived long ago. "Bluntly, they are incorrect." In addition, decay rates have varied as a result of pressure, temperature, electric and magnetic fields, and stress factors.[24]

Here is a sample of what the believer in evolution has to work with. George Simpson, measuring tooth changes, declared that the average duration of a horse genus was 5.5 million years. Ammonite species had a life span of 20 million years. The lingula, a lampshell, showed no change for 550 million years, so Simpson helpfully noted that this indicates an even slower rate of evolution! The evolution of social insects are also difficult problems, as the honey bee was unchanged for at least 30 million years. Termites and ants reveal no change for tens of millions of years.[25]

The kindest thing we can say is that these are wild, unsupported guesses. When Louis Leakey gave up after 41 attempts with K/Ar dating of his most famous fossil skull, he accepted a date based on the length of fossil pig teeth in that region.[26]

Earlier Orthodoxy

Many today may not recall seeing any date offered besides 4.5 or 4.6 billion years for the age of the earth, but if we back up several generations, we see that lockstep conformity had not yet ruled over the question. One collection of dates for the age of the earth offered by top scientists of the world in the 1930s is interesting:

Playfair: infinite age	O. Lodge: 100 million
Ramsay: 10 billion	G. Darwin: 60 million
E. Dubois: 1 billion	Sollas: 55 million
Goodchild: 700 million	Kelvin: 24 million
Lyell: 400 million	Croll: 20 million
C. Darwin: 300 million	Tait: 10 million[27]

More recently the discipline has "improved." Scientists no longer dare to deviate on the age question. The question is whether the accuracy has improved any since the time of James Ussher.

Evolutionists who poke fun at Ussher forget the many revisions that anthropologists have had to make of their own calculations in the guise of discovering more and more truth. In the early part of this century William Sollas stated that early man appeared 65,000 years ago. Frederick Zeuner in the 1950s extended this to 500,000 years. Those who follow Louis Leakey give approximately two million years for this development. More recently estimates have moved back to five million.[28] These guesses are endlessly juggled.

Faulty Scientific Pronouncements

Because James Ussher's date from the mid-seventeenth century has been and is subjected to so much ridicule by those who can only parrot an old-age date in its place, it seems fair to look at several scientific pronouncements from that general era. It is difficult to believe that this was a Renaissance scientific explanation of stone axes and arrow points that a churchman had explained correctly as early as the late 1500s:

> These stones were made by an admixture of a certain exhalation of thunder and lightning with metallic matter, chiefly in dark clouds, which is coagulated by the circumfused moisture and conglutinated into a mass (like flour with water), and subsequently indurated by heat, like a brick.[29]

About 150 years after Ussher, the Academy of Sciences of France issued its official evaluation of alleged meteorites.[30] "In our enlightened age there can still be people so superstitious as to believe stones fall from the sky!" The apostle Paul correctly spoke of a meteorite about 1,700 years earlier (Acts 19:35).

Radiometric Dating Becomes the New Truth

But that was then. This is now. Because radiometric dating is often cited and discussed for scientific dating of the past, the claims for this method require evaluation. Today's scholars issue pronouncements such as these:

> . . . Earth is about 4.5 billion years old . .

> . . . light [began] perhaps 10 billion years before that . . .

> Perhaps 5 billion years after the formation of the galaxy . . .

> And perhaps 10 billion years after the birth of the universe . . .

> Life, they say, was not so quick in coming. The first bacteria reared its head about 3.1 billion years ago . . .

> . . . Ramapithecus [is dated] about 19 million years ago . . .

> And Homo sapiens [arrived] about 100,000 years ago.[31]

Such numbers as the above are supposed to be scientific precision replacing the old outmoded view of a young, created world. *Millions* and *billions* are tossed about like confetti as though that is all that is needed to be scientific. Anyone who does not accept such figures is considered anti-scientific and closed-minded. However, Robert Charroux gives a candid view of the problems encountered in dating the past:

> Groping our way, selecting probable truths and eliminating proven adulterations, we can re-establish the order of the great events that give a face to the past. But history is also a matter of dates and we have no chronological markers that give the exact times of those events. In a century when distances are measured in thousandths of a micron and times in millionths of a second, we still have only grossly empirical means of delimiting the vast periods of protohistory.[32]

Two Modern Examples of Scientific Dating

Jerry Trout stated that from 1924–1988 a visitor's sign at the entrance to Carlsbad Cavern gave its age to be at least 260 million years.[33] In 1988 this was changed to 7–10 million years; then for a little while it became 2 million years. Now the sign is gone.

In a recent visit to the amazing Florissant Fossil Beds in Colorado I noted a large sign at the entrance that told visitors to remember that two inches of sediment there equaled one million years. When you do a bit of math with that statement, you learn that a human hair with a diameter of .005 inch would require 2,500 years to be "gradually" covered.

WHAT IS THE TRUTH?

The Solution

This discussion shows why I believe James Ussher is roughly correct and the pronouncements of evolutionists are precisely wrong. Darwinism, evolution, is based on the faulty assumption of gradualism, a notion that some prominent evolutionists now reject, even though they will not budge away from the conventional age of the universe. Most of all, gradualism is believed because Darwin and his followers knew they had to have enormous spans of time for things to evolve into other things, but this, too, is a faulty belief.

Ussher is as good as his assumptions about the completeness of the biblical record in chronological matters. I differ from Ussher on chronology in a minor way. I agree fully with Ussher that we live on a young, created earth. For a good exercise in searching out the real truth, I recommend Carl Zimmer and John Woodmorappe as two up-to-date but opposite views of dating the ancient past.[34]

Another exceptional source that treats the early centuries after Noah's flood and documents much evidence in support of our young earth is John Cooper.[35]

What better way to close this discussion than to quote an evolutionist, solar physicist, John Eddy:

> I suspect that the sun is 4.5 billion years old. However, given some new and unexpected results to the contrary, and some time for frantic recalculation and theoretical readjustment, I suspect that we could live with Bishop Ussher's value for the age of the Earth and Sun. I don't think we have much in the way of observational evidence in astronomy to conflict with that.[36]

For Discussion and Reflection

1. What do you suppose is the real issue regarding attacks on Archbishop Ussher and his estimate of the age of the world?

2. What were Ussher's assumptions about finding the age of the world?

3. Discuss: What is the support for an old earth? What is the support for a young earth?

4. Did David know that he lived around 1000 BC?

5. Comment on Dr. Eddy, an evolutionist and scientist, who said there was no real evidence against Ussher's belief in a young earth, as opposed to most evolutionists insisting on an earth 4.5 billion years old.

6. Scientific method is the foundation of all the sciences. It requires careful use of facts. Why do you suppose are there so many outrageous misquotes by evolutionists regarding what Ussher actually said?

PART THREE

When **Evolutionists** **Search for Our** **Origins**

CHAPTER SEVEN

DID MAN REALLY EVOLVE?

A BRIEF OVERVIEW

Out of Africa

Each year *Time, Newsweek, U.S. News & World Report, National Geographic, Discover,* and other various science magazines and journals, seemingly without exception, publish their annual salute to human evolution. Artists, rather than forensic scientists, are employed to produce the illustrations based on one or a few newly discovered fossil bones, that will cause all previous theories to be reevaluated. Having followed these displays for more than a half century, I note that the life span of each discovery is seldom more than a year, when it gives way to the next great fossil find. In late 2004 we learned of the discovery of fossil tiny people,[1] and in early 2005 a fossil bone dated at 4 million years is believed to be from the very first walking hominid or ancestor of modern man.[2]

Johanna Rajca discusses one of the many annual magazine salutes to human evolution on how apes became human.[3] From several years of intense searching in the great rift valley in Ethiopia, a tiny handful of fossil ape bones was collected. The star exhibit, however, was a fossil toe bone from 10 miles away and in different strata that indicated a creature that may have walked upright. This collection of bone fragments was so meager that *Time* was apparently too embarrassed to show the result to its readers. Nevertheless, an artist was commissioned to show on the cover what this great new discovery looked like in the ancient past for the benefit of believers. Creationists see no reason to relate the toe bone with the ape fossil fragments. This chapter will discuss why we are not impressed with the above and with many other equally unimpressive discoveries.

A Never-ending Story

From its earliest traces among the Greeks in the third century AD until the late 1700s, many of the finest minds and countless family fortunes were deeply involved in an all-consuming tantalizing problem. This was alchemy, the quest to change lead into gold and the search for the elixir of life so that one could live forever to enjoy all the pleasures of life. Reason and logic were discarded, and the relentless search went on century after century.

In the past two centuries we have an astonishing parallel to this futile effort. This is the equally vain quest to explain how something was created out of nothing, how life came from non-life, how the enormously complex developed from the simple, and how all of this was unaided by anything but time and chance. Billions of dollars and hours are spent in teaching millions that this is fact. Enormous efforts are spent in field and laboratory studies to give this the veneer of science. In this chapter we look at the crown jewel of all this expenditure of time and money—the supposed evolution of man from the ape world.

Why Evolutionists Want You to Believe

This is the mind of the evolutionist: We believe and can prove that water consists of hydrogen and oxygen, so we should also believe in evolution. Ashley Montagu at Princeton gives this explanation: "The attack on evolution, the most thoroughly authenticated fact in the whole history of science, is an attack on science itself."[4]

But in searching through the vast literature of evolution for reasons to believe and for evidence to support such a belief, we find the going very slim indeed. It does not take long to see trouble brewing. For example, in the fervent debate about the extinct ape, Lucy, one critic observed, "many arguments turn on the subtlest features of a particular bone."[5] Paleontologists have commonly observed that those who work with ape/human fossils seem to forget everything they ever learned about variation among creatures of the same species.[6] Consider the great amount of variation in the dog family, yet they are one species and no amount of variation indicates that any breeds are changing into other species or kinds. Among humans we may observe enormous variation among the people who walk by us at a mall. Yet among the paleontologists known as the "bone peddlers," the slightest difference is hailed as a new species that will cause all the texts to be rewritten. Age, gender, normal variation between persons, and environmental conditions are all ignored as though none of this has the slightest effect on the skeletal structure. Donald Johanson, who is as guilty of this practice as any other evolutionist, nevertheless stated that it took a long time for anthropologists to get it through their own skulls that populations are extremely variable.[7]

Here are reasons evolutionists advance on why every honest, right-thinking, decent person should believe that humans evolved from apes:[8]

Man shows a remarkable resemblance in his bodily structure to the lower animals. It now seems astonishing to us that his kinship with them should ever have been seriously controverted. *Comment:* There are also radical differences. What about even more striking similarities between horses and "false" horses (See chapter 11) that are completely ignored? Similarity in structure, called homology, does not prove relationship.

Embryos of all mammals (including man) pass through a stage of development during which a foundation of gill arches is laid down in the neck region, precisely similar to that which in fishes finally leads to establishment of functional gills. This must be due to the inheritance from ancestral forms over millions of years of gradual change. *Comment:* This quaint belief, called recapitulation, is a great embarrassment to biologists today, though they are loyal to evolutionary beliefs. Like a bad penny, this flawed belief has no basis in fact but keeps reappearing in the writings of evolutionists who demonstrate ignorance of biology. William Fix can hardly believe his eyes that the late famous evolution guru, Carl Sagan, endorsed the recapitulation theory and referred to the "fish stage" of human embryology, stating that the fish stage even has gill slits.[9] Biologists have known for almost a century that at no stage of its development does the human embryo have gill slits.

Logic is that "large" is primitive, while "small" is humanlike for hominids, for example, teeth.[10] Therefore we can see evolution in action by studying fossils showing that sequence. *Comment:* This same logic does not explain just the opposite—why small is primitive and large is modern, such as in the supposed sequence of ape to man, and the supposed sequence of the evolution of the horse and numerous other mammals.

Footprints 3.6 million years old found by Mary Leakey show a major evolutionary development—footprints exactly like those of modern persons as well as upright walking.[11] *Comment:* In the creationist view the prints were actually made by persons no different from us except perhaps in size. The foot of the ape is radically different from the human foot. Truth was violated when the St. Louis Zoo reconstructed the ape Lucy with completely human hands and feet.[12] Isn't it instructive to note that the zoo's director of education said that if they had shown this ape as it really was, people might think of Lucy merely as an extinct ape that has nothing whatsoever to do with man (or with evolution). And that is exactly what other paleontologists say.[13]

Those who study communication learn the techniques used by propagandists. It is an amazing fact that in the textbooks and lectures on our topic, most of the persuasion is based on one or another basic technique of propaganda, such as:

1. Getting on the bandwagon. This technique goes something like this: All intelligent people, all scientists, believe evolution is true. It is unthinkable that you would dare challenge such an obvious truth. Join the majority.

2. Using the broad brush: If you attack evolution, you are attacking chemistry, physics, astronomy, geology, biology—all of science.

3. Use of threats, fear, punishment, ridicule, humiliation. Evolutionists hold positions of power in education and in the media, and thus use intimidation because they are unable to show actual evidence for their belief. Students have told me that if you question evolution in the classroom, you fail the course.

4. Use of the big lie, deception, and self-deception. In the past 150 years there is a long procession of fossils that finally prove the evolution of man. Each of these has had a very short life span. There are outright frauds as well as seeing many things in bones that are not there.

Two Important Critiques

The reader does not need still another detailed examination of each collection of bones in the supposed ape to human fossil record. There are many such competent evaluations. We list two devastating critiques here, one by a competent creationist[14] and one by a non-creationist.[15] Each of these examines the evidence, fossil by fossil, and finds nothing at all credible. Each of these is well worth reading and study.

Marvin Lubenow made an exhaustive study of the fossil record in the supposed sequence from ape to man. Of almost countless family trees constructed to try to show this evolution, the current dominant pattern is to begin with *Homo habilis* to *Homo erectus* to *Homo sapiens*. His well-documented conclusions are that the two forms of *Homo habilis* (*gracile* and *robustus*) consist of one 100 percent ape and one 100 percent *Homo sapiens*. By calling the two radically different fossils by the same name, evolutionists try to slip by the problem of getting from ape to man. *Homo erectus*, Lubenow concludes, is simply a variation of *Homo sapiens*. It is interesting that after Lubenow's book appeared, paleontologists from Cornell University reclassified one old *Homo erectus* as a *Homo sapiens*. Other experts also decided that some Chinese *Homo erectus* fossils should be classified as *Homo sapiens*. An article in *Geotimes* bears a very appropriate title. "*Homo erectus* never existed?"[16] Thus in Lubenow's view, strongly supported by the actual evidence, there are fossil apes and fossil *Homo sapiens*. Nothing else. Nothing in between.

If anything, the book by William Fix is the more damaging to the supposed ape-to-human sequence, because he is not a creationist. He simply investigated what was given in support of each fossil find. Although he ends up with some strange mystical conclusions, we learn some interesting things from Fix.[17] Here are a few sample gems, paraphrased:

- People with valid evidence have nothing to fear from competing ideas. The so-called Christian right is not the problem. The scientists themselves are driving people away with vacuous absurdities about the ancient world. The chief culprit in Fix's view is the late Carl Sagan.

- Fully modern human skulls supposedly hundreds of thousands of years old or much more are ignored. These finds are too embarrassing for the theory.

- Johanson has ignored plain evidence that his famous fossil Lucy is a composite of several species, not one, and neither is a human ancestor.

- The rise of man from animal is mostly a catalog of fiascoes.

- The new position is that species changed so rapidly they left no fossils.

- Creation as the proper interpretation [of the ancient past], of course, is unthinkable.

- It seems odd that while physics has abandoned its earlier mechanical basis, biology is moving closer to it. Evolutionists seem ignorant of this change and still rely on the physics, chemistry, and biology of the 1800s.

- There ought to be major professional excavations along the Paluxy River in Texas with an open mind to settle the truth about the dinosaur and alleged human footprints there.

- Like it or not, a committed creationist can still maintain that he or she is under no compulsion from logic or evidence to abandon the theory of the special creation of man.

Species Multiplying Like Rabbits

In each decade we have a seemingly endless parade of fossils and fragments of fossils, each of which is the claim to fame of a paleontologist. William Fix has unkindly noted that when it is time to apply for another grant or renew a grant, we see the newspapers and other media loaded with bone pictures and excited prophecies that the researcher is right on the cutting edge of solving the great mystery of human evolution.[18]

I still have a newspaper page from 1950 full of apish illustrations with the promise that only ten more years would be needed to solve the remaining problems about human evolution.[19] We need hardly add that none of the fossils discussed in the feature have been mentioned for decades. The situation at present is even more confused than decades ago.[20]

It is no exaggeration to say that universities operate with a publish-or-perish mentality, and what could be sweeter and draw more foundation grants than a paleontologist writing and lecturing endlessly about a revolutionary new species he himself discovered in a remote African desert area. That sort of thing is the lifeblood for fame, promotions, heading a department, and pursuing grant funds. Finding a fossil just like one that another has found and named is worthless. We cannot fault research and writing, but we must say that much research in human paleontology is akin to the earlier obsession with alchemy. To the unbiased eye, nothing is happening in human paleontology despite the excited headlines several times every year.

The Great Fossil Roster

One could fill hundreds of pages describing the outlandish and ever-changing claims for such fossils as *Neanderthal, Homo erectus, Piltdown, Hesperopithecus,* and the name *Australopithecus* followed by a whole menagerie of supposed species names, *Gigantopithecus, Zinjanthropus, Homo habilis, Ramapithecus,* and the like. This all sounds very scientific. New supposed human and pre-human fossils are breathlessly announced every year. When you scan the literature, however, you see what a short lifespan such discoveries have.

A few definitions will help to make some sense out of the names. The definitions, of course, carry an evolutionary bias. The genus/species system of classification is subjective.

Homo refers to any extinct or living species of man. The Bible clearly states that only one species of man ever existed. When *Homo* is followed by another word, such as *Homo habilis, habilis* refers to one of the species of the genus *Homo,* and *habilis* means "handy." The first word, or genus, is always capitalized and the second word, the species word, is never capitalized, thus *Homo sapiens* is the "wise human," that is, us, because sapiens means "wise." If modern human bones are found too early, that is, in strata lower than or with a supposed earlier form, then evolutionists call the bones archaic (old) *Homo sapiens.* In other known cases such bones are given a different name to conceal the fact that they are too early in the record. The fancy name for Lucy is *Australopithecus* (meaning "southern ape") *afarensis* (meaning that she was found in the Afar region of Ethiopia). We know that more often than not, fossils found in the "wrong" strata were discarded or ignored, because they did not fit the mold.

One must be careful. *Hominid* refers only to us, that is, *Homo sapiens,* but *hominoid* refers to man *and* any ape. *Homo erectus* means "erect man," and these fossils are fully human.

With all this jargon we are reminded that in order to keep the written language out of the hands of the common people in Egypt, the priests shelved a simple alphabet and developed the much more complicated hieroglyphic form of writing. We support scientific classification. We oppose imaginary species.

Some Contradictions and Frauds

How Bias Colors Our Perceptions

An observation made a half century ago shows vividly how little progress has been made. The debate centered on the 1864 display of the first Neanderthal discovered in a cave at Gibraltar. Was it man or half ape? These perceptive words apply to anyone who does or studies research today as well as in the past.

> The pat answers of Cuvier and the religio-scientific dogmas surrounding the search for man's origins are almost forgotten in our own enlightenment, but it is sometimes disconcerting to find the Cuviers, the Mayers, and the Virchows—all as equally devoted to the methods and objectives of modern science as any of today's scientists—with intellects imprisoned and imaginations shackled by hypotheses of their own making, hypotheses purportedly based on fact and uninfluenced by metaphysical considerations. It is disconcerting to realize that as their intellects were shaped and limited by the dogmas—often scientific—of their day, so may the intellect of the modern investigator be shaped by the *a priori* judgments of his time, the unproved hypotheses and overgeneralizations, the results either of the nonscientific environment in which he lives and works or of the sometimes equally nonscientific traditions he follows.[21]

How Much is Actually Known About Early Man?

A half century ago, in 1953, D. Dewar listed twelve mutually contradictory theories of man's ancestry, concluding that much of the actual evidence is against that sort of assumption.[22] That is not much to show for more than a hundred years of intensive work by evolutionists. No one has ever said it better than Stewart Easton:

> The truth is that we know very little indeed about prehistoric man. The unremitting labors of archaeologists and anthropologists . . . have only scratched the surface of our almost total ignorance. Besides, no two experts are ever in agreement on all points in their interpretation of the meager data available. . . . In this age, on principle, we are inclined to prefer even the most far-fetched of material explanations to the possibility of any kind of divine guidance or intervention, or the fulfillment of any divine purpose.[23]

Ernst Haeckel, a century ago, is the perfect example of a completely convinced evolutionist about man's descent from the ape world. He blamed Christianity, "so fatal to scientific culture," for having raised insuperable obstacles in all branches of secular knowledge.[24] Now "science" had overcome all obstacles to reveal the real truth. The evidences, in case any still doubted, were the following:

- The fossil ape-man of Java was in truth the much-sought missing link. (By 1936 the discoverer, Dubois, stated it was merely a large gibbon or ape, but this old fraud, like many others, was already in the textbooks and still is there today.)[25]

- An ape in India that sings a full octave in perfectly pure, harmonious half-tones as proof positive the our language has been slowly and gradually developed out of the imperfect speech of our simian ancestors.

- The discovery of a fair number of fossil skeletons showed the complete connecting chain of ancestors from pre-apes to humans.

We need scarcely add that each of the above and other supposed proofs were discarded as laughable long ago. The evolutionist never gives up, however, so the above pious frauds have been replaced by other equally unsupported claims.

What Helped Evolution Look Good to Its Believers?

One reason is that during the latter half of the nineteenth century every apparently early skull found, if not ape-like, was discredited, no matter how good its credentials. Even a committed evolutionist such as Robert Broom observed that practically all sapiens-like remains from early times (remains that looked just like modern persons) have shown a strange tendency to disappear. Arthur Custance also noted that practically all changes in man's physical structure were the result of dietary deficiencies and unfavorable environments.[26]

Similarly, Alvin Josephy commented on the meager number of human bones of early man in the Americas.[27] There was a special reason. The experts were not looking for them. They were looking instead for an ape man. Therefore, skulls and bones of claimed antiquity were examined, argued about, and judged on the basis of possible simian and other *pre-Homo sapiens* characteristics, and because no such bones fit that requirement, they were thrown out.

Here is just one of hundreds of similar examples. A mix of fractured bones of extinct and living species has been found in the fissures of mountains and hills worldwide. They show all the evidence of having been swept in by floodwaters. The bones include human remains and artifacts. We read, for example, that fifteen human skeletons were found in the Cattedown fissure in England, and mixed with the bones were oyster shells that show no ordinary flood had caused the event. But because the bones were those of "modern" man, they were discarded, a staggering loss to scientific study. In another fissure a human humerus bone was found. When the finder was told the bone was human, it was discarded.[28]

This fossil clam is one of trillions of bivalves in limestone and other rocks all over the world with the two halves tightly closed. Such shells separate after death and then decay or are scavenged. What sudden world-wide event caused the almost instant death and covering of this marine life? Noah's flood is a good answer.

Wonderfully preserved insect larvae from near Grand Junction, Colorado, showing rapid covering and fossilization before decay or scavenging can begin.

A Plethora of Frauds and Self-delusion

At the infamous Scopes monkey trial in Tennessee in 1925 the two main lines of evidence for evolution were the Piltdown Man and Nebraska Man in a setting of unprecedented insult and vilification at William Bryan, who held to a biblical view of creation. The Piltdown fraud, which deceived a whole generation of leading experts in the British Museum and elsewhere, consisted of a chimpanzee jaw and a modern human skull, artificially aged and shaped to fit together. Nebraska Man, consisting of one tooth, "the answer to American anthropologists' prayers," deceived many experts for five years at which time the tooth turned out to be a peccary (pig) tooth.[29]

But nothing is embarrassing for the evolutionist. Only an evolutionist could rise above these disasters and say that those bits of wishful thinking and self delusion illustrate the great "self-correcting mechanism of a science that daily draws a more precise picture of human evolution"[30] When Ernst Haeckel was confronted for faking evidence to show human evolution, his defense was that he merely filled in where the evidence was thin, and he said unblushingly that hundreds of the best observers and biologists of his day lay under the same charge.[31] This was and is the kind of science that was to replace the "myth and superstition" of the Bible.

National Geographic publishes an almost endless number of features on the supposed evolution of early man. Note how differently creationists would view the following account of 3-million-year-old fossils at Hadar, Ethiopia. The astonishing array of animal fossils included many that "resemble" modern species (can't tell the difference). The supposed not-yet human fossil hand bones bear an "uncanny resemblance to our own in size, shape, and function." They did not walk on their knuckles as African apes do. The thumb rotates, making it possible to manipulate tools with finesse. Foot bones also correspond closely to that of modern man. The skull mandible is U-shaped, like those of humans today, not V-shaped like those of *Australopithecines* (extinct apes).

One evening, for a lark, members of the research expedition made clay casts of their own teeth; one woman's jaw bore a startling resemblance to a three-million-year-old specimen.[32]

No one would have dared to suggest that they had simply found old skeletal material no different from that of people today. There never has been any evolution, but in their minds man had not yet emerged from the ape world.

Donald Johanson gives the classic uniformitarian interpretation about Lucy's death.[33] Her carcass remained untouched for countless years, slowly covered by sand or mud, buried deeper and deeper, the sand hardening into rock under the weight of subsequent depositions. No predator ever wandered by. Why then, we ask, was only 40 percent of the skeleton discovered? Other reports state that Lucy is a composite of more than one ape skeleton and of bones found widely scattered and in different strata. Competent paleontologists sharply disagree with Johanson that Lucy could walk upright in any meaningful sense. These matters and much controversy on dating the fossil show this is merely another or several meaningless extinct ape fossils.

Speaking of controversy, as of 1995 the finder of Lucy lost the support of his largest benefactor, has been sued by his former colleagues, and has alienated many of his fellow anthropologists for alleged breaches of professional ethics. Yet at present Johanson is highly esteemed in his field. He believes that all is well with the "pivotal importance" of Lucy.[34] The dating of bones and artifacts is another endless story of wild guesses and persistently seeking the right date to "prove" a point. Johanson shows the state of the art when he commented on how Dart derived the age of the famous Taung skull: "It was a guess based on a hunch derived from speculation about the age of some extinct animal fossils from the same limestone cave."[35] As we show in detail elsewhere, radiometric dating has not improved the situation even today.

Normally a block of sandstone containing fossil bones would be dated as millions of years old. Unexpectedly, however, one stone discovered in Egypt contained a well-preserved skeleton with brow ridges like the Neanderthal Man. Therefore the rock was dated to around 60–80,000 years, when Neanderthals were supposed to have lived.[36] One notes that the bones are simply slotted into the time period when that creature was supposed to have lived. It is, of course, pure speculation.

The literature on ancient man is full of analyses of artifacts, and countless remains have been dated by how crude or refined the artifacts appeared. Here Johanson is on the right track except for his chronology when he stated that you can't tell apart tools made about a thousand years ago and a million years earlier.[37] It is really impossible to date surface-found tools at Lake Turkana in Africa, because modern humans were making similar implements in profusion as recently as a thousand years ago, and there were even a few people who are making them there today!

But this plain truth is ignored by others. Soviet geologists exploring the bluffs above the Lena River in remote northeastern Siberia discovered crude stone tools.[38] To the finder they resembled and even seemed older than the tools dated in Africa's Olduvai Gorge at 2 to 2.5 million years ago. They also estimated the age by "calculating" the age of the sediment layer they were found in with the faulty assumptions of gradualism. In their excitement they then claimed that perhaps humans originated in icy northern Siberia instead of the common view among evolutionists that this happened in southeast Africa.

The examples of such self-delusion seem endless. Roger Lewin sums up the situation nicely.[39] He stated that ever since Darwin, throughout the history of the study of human fossils, the experts commit what is called Tyson's error (seeing only what one expects to see) time and time again: Neanderthal, Piltdown, Nebraska Man, *Australopithecus, Ramapithecus, Zinjanthropus*—each in its turn has been the object of the exaggeration of traits favored by evolutionists whose theories demanded them. As Matt Cartmill said, all sciences are odd in some way, but paleoanthropology (the study of human and ape fossils) is one of the oddest. Like groupies surrounding a rock star, each big name in the futile human origins search is surrounded by a fawning coterie of subordinate colleagues, and current and former graduate students. His is the only theory and explanation worth looking at. All the others are blind and incompetent. As William Fix described today's university setting rather crudely:

> Professor Braintrack, head of the department, declares his belief. Associate Professor can tell which way the wind is blowing. Assistant Professor Frogmire jumps aboard. It hardly requires the perspicacity of an Hercules Poirot to predict which way graduate student Netlebottom will run.[40]

It is no wonder that Donald Johanson says that students tend to believe what you tell them.[41] They had better, or their degrees and careers are jeopardized by vindictive professors. Nowhere is this more true than in the fields where evolution is the dominant dogma.

How Ape Man/Cave Man is Pictured

A century ago the world's foremost authority on "primitive" man was Marcellin Boule. He is the one who came up with the apish caricature of Neanderthal Man that is still often used today to show how modern people supposedly emerged out of the ape world. It is not to his credit that he believed Piltdown Man was the true ancestor of modern man, and that Neanderthal Man was merely an extinct side branch. As we have long known, Piltdown was an even more gross fraud than what was done with Neanderthal Man. It is plain to see that the illustrations of early man over the past century or more followed certain rules in order to plant the teaching that people descended from the ape world. Here are the rules:

- If the skull and other remains were obviously human, every possible apelike characteristic was added, especially in the face, jaw, bent knees, low-hanging arms, and above all, lots of shaggy hair.

- On the other hand, if the remains were plainly ape, such as Lucy, every possible human characteristic was given in the reconstruction in order to bridge the gap between ape and human.

Jack Cuozzo, orthodontist, was granted access to the world's most famous Neanderthal skeletons in European museums.[42] He used X-rays and photographs for his analyses. Among his discoveries in the exhibits was that jaws were pulled forward so the teeth no longer fit properly. This distortion gave the skeleton a marked apish appearance.

Impressionable minds were and are strongly influenced by such illustrations. It is of special interest that in the past several years competent forensic scientists were commissioned to show their skill with ancient skulls. Their expertise is normally used in criminal cases where they have demonstrated their skill in reconstructing the facial features of victims whose skulls have been discovered.

In the case of the Neanderthal Man the results are amazing. Completely gone are all the ape characteristics. The face could be that of anyone's grandfather. Should anyone be surprised at how little publicity has been given to this development? The textbooks will go right on showing hairy ape men, because that is the standard conventional theory of how man came to be. Truth doesn't matter when a false theory must be upheld. The truth is that every fossil specimen is either 100 percent ape or 100 percent human, recognizing of course that there is variation in either case.

How Little Has Been Learned

One has only to scan Brian Fagan's recent book to see how little has been learned by evolutionists in the past century about ancient man.[43] Here are just a few examples from Fagan of what might be considered a backlash about so-called human evolution, with Neanderthal Man returning more apelike and hairier than ever.

- No one doubts gradual changes in human morphology (physical form). *Comment:* This is tired classic Darwinism that leaders in evolution such as Gould abandoned decades ago in favor of abrupt change.

- There is passionate debate on how early man crossed open water. *Comment:* Fagan appears to be ignorant of a vast body of literature showing that early man was highly sophisticated, including much early transoceanic sea travel.

- Prehistorians relied heavily on stone tools to trace human cultural evolution. There are many puzzles about advanced tools in old layers. They disappear and reappear many thousands of years later again. *Comment:* Experts of greater competence have concluded that both crude and sophisticated tools are found at every level of strata, and that such tools are worthless for dating purposes. Much depends on the purpose of the tool, the material itself, the time available to make it, and the skill of the maker.

- Forest hunters today snare and trap animals in the same way as in prehistoric times. *Comment:* This is true and illustrates the fallacy of assuming that old hunting campsites with crude artifacts automatically mean thousands and millions of years in the past.

- Modern and Neanderthal skeletons found together in Israel, for example, show the moment in time when Neanderthal Man was in the process of evolving into modern humans. *Comment:* This is nonsense. Both are *Homo sapiens* and show no more variety than one can find in any gathering of people. Around the world today there are many radically different body shapes and forms, but all are *Homo sapiens.*

Who could improve on G. K. Chesterton's comment in 1923 on supposed missing links, monkey men, and bits of bone.

> Nobody dared to suggest that such evidence was rather slight. . . . Any minute bit of moldy bone was good enough for the purpose so long as the evolutionists recognized it as a good purpose. Anything proved anything, so long as it proved the proper, progressive, really evolutionary thing.[44]

EVIL SPIN-OFFS FROM THE SEARCH FOR HUMAN ORIGINS

The Decay of Scientific Method

Some of the writings of evolutionists are a chamber of horrors in what they reveal about the minds and ethics of their fellow workers. Here are just a few samples:

There are endless squabbles about dating and redating a fossil. Donald Johanson, for example, criticized a rival by saying that he can't turn a 3-million year old Hipparion (horse) into a 2-million year old horse just because he wanted to.[45]

A university we shall leave nameless reported that a graduate led a team of geologists to Java where they discovered that skulls previously thought to be 300,000 years old were actually 27,000 years old. This supposedly gives us another human-like species long after modern man supposedly evolved. The second date is just as speculative as the first.

Charles Darwin viewed the Civil War in America thus: In the long run a million horrid deaths would be amply repaid in the cause of humanity because the fittest survived. Similarly Darwin noted that the more civilized Caucasian races defeated the Turkish in the struggle for existence. Darwin looked forward to the day when an endless number of the lower races would have been eliminated by the higher civilized races throughout the world. This was plainly the motivation for Hitler's slaughter of "inferior" races.[46] The 1903 *Encyclopedia Britannica* gave the African-American the lowest position in the evolutionary scale, mirroring Huxley who said that no rational man believed the African-American was the equal of the white.[47] After 1859 there was a focus on race as illustrating the forces of evolution.

Donald Johanson made these observations about some of the dark side of human paleontology.

1. Mary Leakey once stated that certain anatomists didn't want to admit that somebody had found something different: "We'll wait until he publishes, then we'll carve him up."

2. He noted a seamy, vengeful side of anthropology. A rival in revenge accused him in writing to government officials of Johanson's incompetence, writing false reports, stealing fossils out of the host country, and other offences.

3. Johanson stated that he knew anthropologists and geologists who would cling to a point long after it had become overwhelmingly clear that they were wrong.[48]

Marvin Lubenow[49] described other struggles in this field:

In another dating struggle, one anthropologist accused another of deliberate falsification of data in order to achieve a much desired result. Moreover, when a dating expert gets the "wrong" date for some rocks, he is charged with using contaminated material and defective methodology. Such a charge need not be proved. The proof is already there because he had gotten the wrong date! Radiometric dating is a classic example of self-deception and circular reasoning.

Donald Johanson, of Lucy fame, said that only those in the inner circle get to see the fossils—only those who agree with the particular interpretation of a particular investigator. An anthropologist who asked to see Lucy and other fossils collected by Johanson was informed that permission would be given only if Johanson reviewed any article she might write on them before she sent it to a journal. In other words, censorship would be applied.

Even Leakey could view the Piltdown skull for only a few moments before it was taken away at the British Museum. We now know that the file marks of this fraud were visible on the original pieces, but not on the casts, which is all that scholars ever saw.

Theory shapes the way we think about, even perceive, data. Major museums store an immense amount of fossil individuals. They are left in limbo because they do not show what evolutionists want to see. Scientists demonstrate an incredible faith and trust in the work of colleagues. Checking or self-policing is always implied but hardly ever done.

It is a fallacy to equate simple with primitive tools. Almost every basic style of tool has been found with almost every category of human fossil material, covering supposedly several million years.

UNWELCOME, FORBIDDEN DISCOVERIES

The Danger of Finding the Wrong Things

Now and then we have heard in all seriousness that graduate students are instructed to quietly bury and forget any discovery that could upset the conventional view about early man. Long entrenched professors are not amused when some upstart comes up with something that contradicts earlier "truth." What we have discussed thus far in this chapter lends much support to that statement.

Collections of Embarrassing Discoveries

William Corliss has made a career out of gleaning the unusual out of scientific journals, including many discoveries by scientists that upset the conventional party line. As one example from dozens of his fascinating volumes, *Ancient Man: A Handbook of Puzzling Artifacts* (1978), he treats many interesting puzzles and mysteries. Corliss attempts to remain neutral, thus avoiding ire from anyone's view of the ancient world. He comments as follows:

> A few apparently human traces have been discovered in rocks millions of years old—far older than current anthropological theory allows. All such finds of course are hotly contested and frequently rejected out of hand. Such finds rarely see print, and when they do the reports are often found in fringe publications. They do not appear in "respectable" books and journals because they simply do not fit into any current theory of man. In addition to artifacts, some human or near-human skeletons have been found in coal beds or in other strata that are plainly too old. The accepted time pattern of human evolution automatically rules such discoveries out.[50]

William Corliss concludes that such discoveries are not really so much controversial as they are inadmissible to the court of "science." This, we might comment, is an interesting device to get rid of inconvenient discoveries.

Among the items included in his book are illustrations and discussions of fossil human footprints, various objects found embedded in rock many millions of years too old for such discoveries, uncanny evidences of sophisticated knowledge thousands of years too early, and curious human skeletons in strata millions of years before they were thought to have evolved, according to evolutionists.[51]

A comprehensive collection of "impossible discoveries" is found in Michael Cremo and Richard Thompson, *Forbidden Archaeology*.[52] This book reviews and evaluates all the main discoveries related to the ape-to-man debate over the past 150 years. The book shows what is reported and what is glossed over or hidden away in each case. This remarkable collection also includes numerous obscure scientific reports that never reached mainstream texts. The reason is obvious. These discoveries by competent scientists contradict what evolutionists are teaching about man's origin.

Other Reports and Evaluations

The following are good examples of discoveries described in the above sources:

Long ago Theodore Graebner wrote of a remarkable fully modern fossil skeleton from Olduvai Gorge in 1913.[53] These bones, in what today is called Bed II, should be several million years old as evolutionists have calculated the age of the beds there. However, though they were completely fossilized, they "lamentably showed few primitive characteristics." What should be the showpiece of ancient man was quietly discarded or hidden away because it violated the biases of the evolutionists there.

Science reported on new evidence that Neanderthal Man was artistic before modern man arrived. This was very unwelcome news for Prof. Klein of Stanford, a specialist on the Neanderthals, who stated: "I want the Neanderthals to be biologically incapable of modern behavior. So the evidence is a real problem."[54]

G. Frederick Wright[55] (1912) and Robert Gentet[56] describe the evidence for the Calaveras Skull found in Tertiary gravels in California. As these gravels were presumably laid down in a 65-million year period before the 1.6 million year Quaternary period, there is little wonder about why evolutionists ignored or attacked this discovery. This is just one of hundreds of such finds elsewhere, but they all fall outside the acceptable range of evolutionary theory. The fully modern skull and many artifacts were found 150 feet below the surface, 100 feet of which was lava. The skull was encased in cemented gravel and had apparently rolled along the channel of an ancient stream at an earlier time.

Bill Cooper described the fossil human skeleton stored in the basement of the Natural History Museum, London.[57] It is embedded in limestone of the Miocene Era, supposedly 25 million years old. The limestone lies below a coral reef that evolutionists date as at least one million years old. The remains are fully modern. These remains, so puzzling to the evolutionist, are either ignored or attacked as somehow very recent.

Conclusion

Science or Desperation?

Philip Johnson noted that the human brain and cells contain information equivalent to about 200 trillion sets of the Encyclopedia Britannica.[58] This echoes the Bible in saying we are fearfully and wonderfully made. Try as we might, we find nothing convincing to support what is taught as truth to our children and to adults as well about the origin of man from nothing plus time and chance.

Francis Crick, who won a Nobel Prize for helping discover the structure of DNA, brings our topic to a close. He despaired of explaining how life could have evolved from non-life on earth. His solution, called directed panspermia, is that life sprang from tiny organisms sent from a distant planet by spaceship as part of a deliberate act of seeding.[59] As we note in chapter 14, some excited evolutionists are now promoting the notion of life from Mars arriving on earth in the cracks of meteorites. Same song, different verse. If evolutionists need to "evolve" these kinds of beginnings of life from other planets, we can better understand how the supposed story of man evolving from the ape world is told with a straight face. This is not science. It is desperation.

For Discussion and Reflection

1. Read the account of the creation of man and woman (Genesis 1:26–29 and Genesis 2:1–25). The Bible says that we were made a little lower than the angels, and that one day believers will judge angels (1 Corinthians 6:3). Why is this unacceptable to the evolutionist?

2. Look at any biology textbook at any level elementary, high school, college, and graduate school. What is the story of how man came to be? Agree or disagree and state why? What can we say to our children when they ask questions about this matter?

3. An exhibit a generation ago and another in 2005 in London showed humans behind bars in a zoo as exhibits of people next to their ancestors, the apes. Is this consistent with what all the media are pushing today? How many of your friends do you think find this acceptable?

4. Every year we see illustrations in many magazines and other media featuring accounts of how humans emerged out of the ape world. This is presented as science. Agree or disagree and why?

5. Evolutionists today always show our so-called ape ancestors with completely modern human hands and feet, but the fossils themselves do not support this. Why do you suppose they do this for the public, or, we might ask, why do they do this to the public?

6. The evolutionist believes and teaches that very slight changes over immense periods of time will cause a species to gradually change into another species. This is called microevolution. Can you think of any example where one species is gradually changing into another?

7. Visit a dog show and see the fantastic amount of variation among the different breeds. Do you believe they are gradually changing into some other species? Defend or oppose.

8. We can also see remarkable variation occurring when we sit down at a mall and watch the people go by. Is it possible that we are gradually changing into some other species if we give this process enough time? Why or why not?

9. Look at the pictures of so-called ancient pre-humans. May we assume the illustrations are based on a scientific analysis of the fossils for that creature?

10. Evolutionists have a choice between having a highly skilled forensic scientist do a reconstruction from the fossil evidence or instructing an artist to illustrate the stage between ape and man that the fossil is thought to represent. Can you guess which alternative is selected about 100 percent of the time? Why do you suppose this is true?

CHAPTER EIGHT

THE INCREDIBLE PILTDOWN HOAX

THE PILTDOWN FINDS

A Nation Entranced

What kind of fossil discovery would so captivate a nation and its envious neighbors that no other topic for months on end seemed worth anyone's attention? At universities in England and elsewhere over the next several decades numerous doctoral dissertations and other learned studies were written about this handful of fossil bones and artifacts.[1] The doctoral studies were perceived as the gateway to scholarly fame and prominence. Those closest to the interpretation of the bones were knighted for their scholarly work. Leading paleontologists of the world could only view the bones from a distance for a few minutes and were not allowed to touch them. They were treated as if they were more precious than the crown jewels. Only plaster casts were made available for study.[2] Here is the story.

A Most Exciting Day

Wednesday, December 18, 1912, was hailed as a red-letter day for geology in general and for the members of the Geological Society of London in particular. Great excitement swept England as the jaw and skull of the oldest Englishman, and perhaps the oldest man of all Europe, went on display in London. As prophesied in 1871 by Charles Darwin, the ideal missing link had at last been found! In the summer of 1912, Charles Dawson of Lewes made the discovery in the dried-up bed of a pond near Uckfield in Sussex. A highlight on this special day was the unveiling of a beautiful restoration of the jaw made by Dr. Smith Woodward, keeper of the Geological Department of the British Museum. A few days later the

Illustrated London News showed Piltdown man (then called Sussex man) full-length and with a complete reconstruction of the head.

The Piltdown Discoveries

Piltdown is a village in southern England. The actual site was on Barkham Manor estate, a Sussex moor land about a mile west of the river Ouse. Dawson, who had been appointed Steward of the Manor of the Barkham estate, was walking along a road at Piltdown in 1908 when he noticed the road was being repaired with some peculiar brown flints. He then found the gravel pit, a shallow trench about four feet deep, lying between a hedge and a road leading to a farmhouse. The gravel was in strata, laid down by running water, and had been cemented together by an oxide that had stained everything a deep brown. He asked the workmen to watch out for anything unusual in the gravel. A few days later a workman handed him a piece of human skull bone. From this time Dawson was a frequent visitor to the gravel bed, but nothing else was found for years.

The Search Continues

His persistence paid off. One day in the fall of 1911 Dawson found another piece of skull, including part of the left brow ridge, lying on the rain-washed gravel. Dawson was now convinced he was onto something important, and he then asked for the assistance of his friend, Dr. Arthur Smith Woodward, the noted paleontologist of the British Museum.

Again persistence paid off. Dawson, Woodward, and several trusted friends continued searching. In the summer of 1912 a startling find was made: Part of a lower jaw with two molars, and artifacts. The human bones and the crude stone artifacts came from the lowest and darkest layer of the gravel, and some of the stones had been blunted and rounded by rolling and washing in the ancient bed of the Ouse River that once had flowed across the moor land. Here and there in the gravel were bits of other fossils: Elephant, hippopotamus, mastodon, and beaver. In all, 37 pieces of bone and stone were recovered. During a seven-year span, the principal Piltdown Man finds were as follows.[3]

1908 Left cranial bone: Dawson

1911 Part of the forehead and back of the skull: Dawson

1912 Back of the skull and right lower jaw: Smith Woodward

1913 Two nasal bones: Dawson

1913 Canine tooth: Teilhard

1915 Skull fragments and molar at second site 2 miles away: Dawson

How Old?

According to the authorities who studied the finds, these animals had lived at widely different times in the distant past. Piltdown Man was believed to be the same age as the oldest of the fossils, about half a million years old. Others thought Piltdown Man could be as much as a million years old. To the great astonishment of the scientists the lower jaw fragment and the brain case found before fit together very well despite the fact that the brain case seemed to be human while the jaw fragment seemed to be ape-like. This was far different than most scientists had thought the missing link would be like. Due in large measure to Dr. Smith Woodward's authority, Piltdown Man was almost unanimously accepted as the long-sought missing link.[4]

Piltdown Man Described

As recent as 1960, the *Rand McNally Histomap of Evolution* gave the following description: *Eoanthropus* or the Piltdown Man, probably the first real man, would have looked quite human to us except for his interlocking canine teeth and slightly protruding jaw. He shaped flints into rough scrapers of the Pre-Chellean type, examples of which were found with the skull. *Eoanthropus* used fire, but was a good gatherer, not a hunter, eating berries and fruit.[5]

A True Missing Link?

Due to the fragmentary nature of the finds one can hardly be surprised that there were some differences of opinion regarding the reconstruction of Piltdown Man. Arguments on the size of the brain pitted the British Museum against the Royal College of Surgeons. Dr. Smith Woodward for the British Museum held that Piltdown Man was a true missing link with a brain capacity midway between man and ape of 1070 cc, while Dr. Arthur Keith reconstructed a skull of 1500 cc, a capacity well suited to any intellectual task of modern man.[6]

One Miracle After Another

Another point of difference had to do with the jaw. The crucial canine tooth, pointed in apes but flatter in man, was missing. Dr. Smith Woodward was convinced that tireless searching would locate the missing tooth, and he even made a model to show just how it would look when it would be discovered. Sure enough, more searching produced the tooth shaped precisely as predicted.[7]

Honor to the Finder

As the first and principal finder of the Piltdown remains, Charles Dawson fittingly was given the high honor of having the skull named for him: Dawson's Dawn Man, *Eonthropus dawsoni*.[8] Charles Dawson was a member of a local society of science hobbyists and antiquarians in Lewes, East Sussex, England. For years before the Piltdown discovery, other members of the club took considerable pride in displaying crudely chipped flints as evidence of some kind of "dawn men" who had lived in that part of England eons ago. Dawson considered this nonsensical and devised a simple test to support his opposing view. Bringing a sack of flints to a society meeting, he proceeded to jump on the flints until many were fragmented. Then he triumphantly showed his fellow members that this action had produced artifacts identical to their finds. Instead of accepting the demonstration, his colleagues considered his test an unforgivable sin and henceforth snubbed him. Strange to say, his later amazing Piltdown discovery that made him famous showed that his colleagues seemed right after all—there must have once been a Dawn Man.[9]

Only in England

As an extra bonus in the series of finds, one of the objects found in 1914 by Dawson, shaped out of elephant bone, resembled a modern cricket bat, lending Piltdown Man a peculiarly but fitting British character—a nuance eagerly seized upon by the press.[10]

The World Comes to Piltdown

The evidence of another Piltdown Man in a field at Sheffield Park just two miles away in 1915 was a most convincing verification of the earlier finds, and at this point most remaining skeptics were won over. Visitors from all over the world flocked to Piltdown, and hundreds of casts of Piltdown Man were dispatched to the museums of the world. In 1938 a special memorial stone was erected by the gravel pit where the first Piltdown Man finds had been made. In 1948 Dr. Woodward's book, *Piltdown Man, The Earliest Englishman*, was published. In 1950 Piltdown gravel was fitted with a glass panel and explanatory notices and duly declared a National Monument in honor of the first Englishman.[11]

THE TRIUMPH FOR ENGLAND

Frenzied Fossil Searches

As the theory of evolution spread through the scientific world in the second half of the nineteenth century, it was a matter of great pride in England that Charles Darwin was an Englishman. England was at center stage. But with the passing of such world-famous figures as Darwin, Lyell, and Huxley, the attention of scientists all over the world was drawn away to other arenas. All the spectacular finds were coming out of Germany, France, and other western European countries. British scientists, while maintaining their legendary stiff upper lip, were in a state of near frenzy. A survey of late nineteenth and early twentieth century scientific literature in England showed that all types of trivial objects, such as pebbles shaped a bit out of the ordinary, old sticks containing a knothole, and almost any chipped pieces of gravel were held up as evidence that Paleolithic Man had indeed walked the fields of Britain.

France Dominates Prehistory

But all these rather pathetic offerings were as nothing compared with the sensational Neanderthal and Cro-Magnon remains and the astonishing harvest of ancient cave paintings found in Western Europe. Almost overnight French prehistoric archaeology and paleontology dominated the field with one remarkable find after another. The British were consumed with envy. Exquisite insults from the French made the pain all the more excruciating. The French called the eminent British scientists "Pebble Hunters." Several generations earlier the tables had been turned. The British then had the honor from their preeminence in the field of geology of naming the geologic strata. Now the abundance of finds in France led to French village names for the ancient cultures uncovered. By 1912 a great French scholar and expert on fossil man noted that twenty authenticated Neanderthal men had been discovered, and he pointedly noted that none were English—a bitter truth the English hardly needed to hear. Only those with a deep sense of Western European history can fully appreciate the depth of muted rage and frustration among British scientists at this wretched state of affairs.

The British Turn the Tables at Last

The discovery of Piltdown Man was first announced in *Nature* on December 5, 1912. A week later *Nature* was already able to put Piltdown Man into his proper place in the history of the world. Piltdown Man, who possessed a high forehead, lived long before the low-browed Neanderthal men spread over Western Europe. Accordingly, Dr. Woodward leaned toward the theory that the Neanderthal race was merely a degenerate offshoot of early man, while surviving modern man may have arisen directly from Piltdown Man. Thus the only genuine ancestor of modern man was represented by Piltdown Man, and Neanderthal Man was merely a freak. Best of all, Pildown was truly British. Nothing could alter this marvelous discovery. At one stroke, British human paleontology once again achieved preeminence. The tables had been deliciously turned.[12]

Piltdown Dated Again

Arthur Keith confidently dated Piltdown Man in 1914 at about a million years. He stated that from what was known and inferred of the ancestry of Piltdown Man, of Neanderthal, and of modern man, reasonable grounds were established for presuming that man was approaching the human standard in size of brain by the beginning of the Pliocene period. (The Pliocene today is conventionally dated as beginning about 13 million years ago.)

Piltdown, the Perfect "Missing Link"

In December 1912, Dr. W. P. Pycraft, an ornithologist who headed the British Museum's anthropology section, evaluated Piltdown Man thus:

> The remains thus far recovered leave no possible doubt but that they represent not merely a fossil man, but a man who must be regarded as affording us a link with our remote ancestors, the apes, and hence their surpassing interest.

> The evidence for the interpretation which has been placed on them is incontrovertible. In the first place, the lower jaw is unmistakably ape-like, while presenting other features indubitably human. It is ape-like in its massiveness, in the absence of a chin, and in the absence of a peculiar ridge along the inner surface which in the typical human jaw is extremely well-marked, and serves for the attachment of muscles concerned with the act of swallowing.

> Evidence that the remains are human is conclusive from the presence of mastoid process, which in apes is wanting.[13]

Pycraft noted that the finds were early Pleistocene and thus could be dated at several hundred thousand years, perhaps as much as a million years. One of the leading British anthropologists stated that no event in the annals of the Geological Society had created such a profound sensation among its members, and no discovery of human remains had equaled them in importance—an unkind reminder to the French that the British once more ruled the world of science.

The Ape Blends into Human Form?

Dr. Grafton Elliot Smith, another acknowledged expert on fossil man, stated that Piltdown Man was the first evidence of a hitherto unknown group of humans fundamentally distinct from all other early fossil men found in Europe. Certain features were so cleverly ape-like that they definitely confirmed the generally admitted kinship to African apes. Dawn Man differed sharply and clearly from all other human remains. Yet in other ways Dawn Man closely resembled modern man much more so than Neanderthal Man. This curious association of features—ape-like and yet man-like—was not paradoxical as some assumed. The small archaic brain, despite the thick skull, was undoubtedly human, but the jaw, despite its human molars, was more ape-like than human—not an impossible combination. On the contrary, the combination of human brain and simian features was precisely what Dr. Smith had anticipated in a published scientific paper months before he knew of the Piltdown skull. It was uncanny how the Piltdown skull looked almost exactly like the prophetic drawing Smith had made in advance of its discovery. Smith was confirmed in his belief that man's brain had led the way in human evolution.[14] In agreement with Pycraft and Smith, the *Manchester*

Guardian proclaimed that Piltdown Man was incontrovertible evidence of the ape origins of man, and that the discovery was clearly the oldest, the first English-man.[15]

Our "True" Ancestor Revealed at Last

Years later, in 1944, the famous anthropologist, Clyde Kluckhohn, had no difficulty in accepting the views of scholars on Piltdown Man, whom they described as bearing certain resemblances to modern man, and they concluded that he must be dated back to the first interglacial, that is, approaching an age of a million years.[16] Harold Gladwin examined all the options and concluded that when it comes to choosing a fossil ancestor for Caucasians or Nordics or white, in general, as of 1947 most would choose the Piltdown Man as their *Homo sapiens* ancestor.[17]

Dissent is Sternly Rebuked

In 1953 Ruth Moore could still say that Piltdown Man was one of the most important remains of early man ever found, but it was the subject of an international controversy. Controversy, of course, tends to swirl around every important find. Early after the Piltdown finds were announced, Dr. Pycraft spoke against the few who dissented:

> It has been contended that in attempting such a reconstruction, a mistake was made—that, as a matter of fact, the brain case is that of a man while the jaw is that of an ape. But no one competent to express an opinion would accept this interpretation. . . . A little reflection on the remote antiquity of these remains was sufficient to turn surprise back on itself. . . . Here was no imaginary but a real missing link.[18]

From time to time other objections were voiced. In 1940 Franz Weidenreich, recognized as the world's leading anatomist, dared to say the Piltdown Man was the artificial combination of a modern human brain case and an ape jaw. Britain held fast. It was their finest anthropological hour. As one of those most closely associated with Piltdown, Sir Arthur Keith gave the scathing reply: "This is one way of getting rid of facts which do not fit into a preconceived theory; the usual way pursued by men of science is, not to get rid of facts, but frame theory to fit them!"[19]

As the *New York Times* noted in retrospect in 1979, Darwin's monumental theory appeared to have been confirmed at one dramatic stroke, and the world recognized in Piltdown Man a discovery of the first magnitude. Dawson and his collaborator, the anthropologist Arthur Smith Woodward, achieved immediate fame. Dawson died in 1916, but Woodward and two other prominent British scientists, Grafton Elliot Smith and Arthur Keith, were knighted chiefly for their dedicated work in interpreting Piltdown Man, and for defending his validity and importance against all comers.

Who Was Teilhard?

Teilhard's Interest in Fossil Man

Occasionally it is mentioned that Pierre Teilhard de Chardin played a minor role in the discovery of Piltdown Man. As a paleontologist, Teilhard had achieved world-wide recognition by his discovery of the first of the skulls of Peking Man. This, however, turned out to be just another short-lived attempt to show the missing link between apes and man. As Marvin Lubenow, however, clearly demonstrates, Peking Man and Neanderthal Man were so-called *Homo erectus*, nothing other than *Homo sapiens*, that is, modern man.

Who Was Teilhard?

Not until his death in 1955 were his philosophic works published, and he became a popular cult figure in what may best be termed pseudo-theology.[20] From 1955 into the 1970s it seemed that everywhere one turned, one was greeted by a brief but unintelligible quotation from the works of Teilhard. As we shall see, it is no surprise that a leading evolutionist called him "the greatest philosopher and theologian the world has ever known."[21]

Pierre Teilhard de Chardin (1881–1955) was born in Auvergne, France, of a gentleman farmer who was interested in natural history. At age 10, Teilhard went to a Jesuit College where he became devoted to field geology and mineralogy. Already as a child he collected rock and fossil specimens. When he was 18, he decided to become a Jesuit and took the steps to enter this order. In 1905 he enjoyed an interlude in Jersey to study philosophy, after which he was sent to teach physics and chemistry at a Jesuit College in Cairo until 1908. From 1908–1912 he lived in Sussex, England, studying theology, and he was ordained a priest in 1912. During the period of 1905–1912 he acquired a real competence in geology and paleontology. A reading of Bergson's *Evolution Creatrice* had helped to inspire in him a profound interest in the general facts and theories of evolution. In 1912 he went to Paris to pursue his geological studies and was eagerly received by Marcellin Boule, the leading prehistorian and archaeologist of France, in his Institute of Human Paleontology at the Museum of Natural History. Teilhard earned his doctorate in 1922.[22] In 1925 Teilhard joined with Edouard Le Roy, the brilliant Bersonian philosopher, to refute Vialleton's book denouncing evolution as a fiction.

Teilhard and Boule

We begin to understand Teilhard when we see what he has to say about Neanderthal Man. He discusses Neanderthal Man—the specialty of his famous teacher, Marcellin Boule. Is he primitive, degenerate, a survival, a throw-back? Although he does not believe science was yet ready with a definitive answer, the marked characteristics of inferiority to be found in the skeletons of this epoch are not accidental, according to Teilhard. Much superior beings lived at the same time as Neanderthal Man. What more proof of evolution could one want, however, when one looks at Neanderthal Man? Teilhard pays tribute to his teacher:

> No one was better qualified than M. Boule to note the phases and to describe the present state of this important change in our views of ancient times. As a result of Boule's study, we know that in Neanderthal Man we reach the utmost human fringe of true fossil humanity. After him comes *Homo sapiens*. Before him, obviously, was the animal form.[23]

The rich documentation of Boule's book and the solid chain of his arguments are beyond refutation, according to Teilhard. The best-known of fossil men, Neanderthal, has a far less developed face than ours. His chin is more apelike than human, and his skull shape falls between those of the ape and modern man. Neanderthal Man is archaic and backward. He probably represents a survival of one of the oldest strata of humanity into the glacial age, and he apparently left no survivors.

Teilhard states that when laying before a public more or less unfamiliar with the natural sciences and with the account of the few positive facts on which human paleontology is directly based, one generally finds one's hearers astonished that prehistory dares to erect the vast structure of its conclusion on so slight a foundation. Their astonishment, according to Teilhard, is unjustified. In the absence of any absolutely sensational discovery in prehistory, there is an up-to-date and scientific manner of understanding man, which is solidly based on paleontology. All appearances suggest that man, appearing on earth at the end of the Pliocene, was a sort of final or even central objective, long pursued by nature in a repeated series of rough models and successive approximations.

Teilhard says, "We must not be taken in by words." By this he means that we cannot allow people to say that some human groups can be regarded as "regressive" which today form (or once formed) the outer edge of humanity. The present day Australian aborigines, for example, or Neanderthals, were said to be degenerate. This view of things seems unscientific. It is impossible to see by what miracle a degeneracy could have removed a chin, flattened a nose, lengthened a face, made the canine fossae disappear, and a tubercle appear on the molars, that is to say, what miracle could have succeeded in increasing man's resemblance to the apes. No, we must reject this view, Teilhard says.

Regarding a Christian belief in the creation of man, Teilhard has this to say: "Let us not play these people's game by believing as they do that, if a being is to come from heaven (that is, from God's creative act), we must be ignorant of the temporal conditions of his origin."[24]

Teilhard believed that Peking Man clearly favors, at least in a general way, that man actually did descend from animal ancestors. He then illustrates his insistence on retaining orthodox language by adding that we must state with insistence that Peking Man in no way threatens (quite the opposite) a spiritual conception of humanity. This is his approach to wedding modern science and the Bible.

Teilhard continues his review of Boule's "master work," *Les Hommes Fossiles*. Teilhard notes that here and there in Boule's remarkable chapter of conclusions, he employs expressions unacceptable to Christian thought as they stand. He therefore says that consequently Boule's book cannot be placed in just anyone's hands without explanation. When philosophers or theologians read these questionable sentences, they should not, according to Teilhard, let themselves be influenced by the words, but try to transpose into orthodox language a teaching of the broad lines which, beneath a veil still heavy with conjecture and hypothesis, seem to conform to reality. He illustrates this point as follows: The letter of the Bible shows us the Creator forming the body of man from earth. Careful observation of the world tends to make us see today that by this "earth" we must understand a substance slowly developed from the totality of things, so that man has been drawn not precisely from a little formless matter, but by a prolonged effort of "earth" as a whole. Despite the serious difficulties that still prevent us from fully reconciling the new truth of evolution with the way creation is described in Genesis, these new ideas should not upset us. Teilhard believed that not one word of the Bible, or one fact of science need be sacrificed. Gradually agreement will be reached, quite naturally, between science and dogma in the burning field, that is, the great controversy, about human origins. In the meantime, he pleaded, let us take care not to reject the least ray of light from any side. Faith has need of all the truth!

Returning once more to Neanderthal Man, Teilhard considers the question whether Neanderthal was a vanished type of primitive man (Huxley's view), or whether he was only a deformed individual, an idiot, a view he attributed to the most famous scientist of his day, Rudolf Virchow. Teilhard affirms that now we know that Huxley was right, and that in a general way one may say that Neanderthal Man represents a sort of evolved prehuman. While Neanderthal's brain size was far higher than that of *Pithecanthropus* (which Teilhard thought was an ape), the great number of animal characteristics was unmistakable in Neanderthal:

low forehead; orbital ridge; long, flat skull; no chin; no depression between the upper canine and the cheek bone, and other items that could be enumerated.

Neanderthal is so clearly distinguished from modern man, in Teilhard's view, that a skilled paleontologist cannot fail to recognize him, even from an isolated bone. Teilhard invoked a series of small directed mutations that acted on pre-man to shorten his face, bring forward his chin, fold and raise his brain pan and his brain.

For the wavering, for those not quite ready to abandon the account of man's creation as told in Genesis, Teilhard had a convincing and disarming invitation. If you don't believe me, he liked to say, read the Master to put all your concerns to rest. (We note here that for Christians, the Master was and is Jesus Christ.) For Teilhard, the Master was his famous teacher, Boule, whose monographs and reconstruction of Neanderthal Man were thought to be the last word on the question of man's ancestry around the turn of the century.

How Teilhard is Assessed Today

The secular scientists whom Teilhard had hoped to attract tended to ignore his work. British historian Hugh Trevor-Roper dismissed him as one of the "great charlatans of modern letters."[25] His influence among Protestant thinkers is minimal. His whole scheme of the evolution of early man sketched above, without exception, is now discarded as naive and erroneous. For example, his writings present Western races as developing from the long-headed, half-stooping Neanderthal Man, who in turn was supposed to have developed from something like the Piltdown Man with the apelike jaw.[26] It is devastating to observe that the famous reconstruction of Neanderthal Man by Teilhard's revered teacher, that is, the "Master," was totally erroneous if not fraudulent, and Piltdown Man, where Teilhard's role has emerged, is a complete hoax.

When C. S. Lewis read Teilhard's book[27] *The Phenomenon of Man* (1959), which was being praised to the skies, Lewis wrote that this was evolution run mad! Lewis thought Teilhard's own Jesuits were quite right in forbidding him to publish any more books on the subject, not that the truth was so dangerous, but rather that there was nothing there but nonsense.[28] The emperor wore no clothes.

THE HOAX FOUND OUT

Doubts Began to Grow

As with other fossil finds, the verdict on Piltdown's authenticity was not unanimous. After all, only a tiny fraction of a complete skeleton had been found, and many crucial elements were missing that could have resolved vital questions if those pieces had been present. The reconstructed skull resembled nothing else on earth, except, curiously, a sketch of what the real missing link ought to look like in a British scientific journal some months before Piltdown was found. No one could deny that the skull was very like a modern man in shape despite its unusual thickness. The jaw, however, resembled that of an ape far more than that of any human. Some strong opinions were expressed from time to time that the remains indeed belonged to two very different species. Then there were the fossil animal remains found in more or less association with Piltdown Man. They covered an extraordinary time span in the same layer of sediment according to evolutionists. The *Elephas planifrons* was thought to be at least 500,000 years old; some hippopotamus fossil bones were dated much later. Most recent of all were extinct fossil beaver bones dated a mere 50,000 years old. The unusual associations raised questions in the minds of some scientists. For many years, the pattern followed was an occasional attack on the validity of Piltdown, invariably followed by a stirring and scathing defense of the first Englishman by prominent British scientists.

An Outsider Dares to Question

It is a bit on the ironic side that the persistent objections of a dentist, Alvan Marston, finder of the Swanscombe skull in 1935 along the Thames River, probably had more to do with Piltdown ultimately being exposed as a fake than anything else. Yet initially Marston merely launched a very single-minded campaign to get his Swanscombe fossil recognized as older than Piltdown, on the grounds that the brain case for Piltdown was fully modern in shape and size. Marston wrote a series of papers that eventually and grudgingly led to a restudy of Piltdown Man. In 1949 Marston, for example, wrote:

> It is clear that many mistakes have been made concerning Piltdown Man—many mistakes by many highly qualified and highly placed men. To err is human and none of these men have been divine. Let the mistakes be recognized.[29]

The New Fluorine Test

Marston then called for the fluorine test under development by Dr. Kenneth P. Oakley of the British Museum to be applied to the Piltdown remains. While the fluorine test did not in itself date an object, it could determine whether objects found together were of the same age, a test ideally suited to settle some of the controversy about Piltdown Man. In 1949 the Keeper of Geology of the British Museum authorized the test. A dental drill was used to secure the tiny samples needed. A total of seventeen samples was secured from all Piltdown I and II specimens. The results were striking. All bones and teeth from the two sites showed the same approximate age. There had not been a "miracle" mix of bones as some cynics had alleged. While the test could not determine that the bones had come from the same skeleton, at least one vital question seemed to have been answered. Jaw and brain case were contemporary and could have belonged to the same being.

To the astonishment of many scientists, however, the extinct fossil beaver bones showed the same age as those of Piltdown Man, thus making Piltdown far more recent than had been suspected. Instead of being from 500,000 to a million years old, Oakley expressed the opinion in 1952 that Piltdown was only about 50,000 years old, indicating a remarkably rapid evolution in the past millennia of modern man.[30] This was horrible news for the paleontologists who needed much more time than that for this half-ape to evolve into modern man.

As Ruth Moore noted in 1953, just a few months before the final verdict was made on Piltdown Man:

> In the light of these new understandings, much that has been taught about the time of man and his development must now be changed. Books must be rewritten and courses revised. For the new timing, the new fossil finds, the new pattern of evolution are bringing about a new and major revision in the theory of the origin of man.[31]

Moore was referring to what she thought was authentic 50,000 year-old Piltdown Man.

New Honors for Piltdown Man

Apparently the prestige of the British Museum and the scientific aura around the newly developed fluorine test done in 1949 laid most of the suspicions to rest, for in 1950 steps were taken to designate the Piltdown gravels as a national monument in honor of the first Englishman. However, an inexorable chain of events and things remembered from the past combined to trigger the final horror. Because of nagging doubts, a new section of the Piltdown gravel was carefully sifted in 1950 by British Museum scientist Oakley and others. The purpose was to find additional material in support of the original finds. To the perplexity of all concerned, nothing at all was found.[32]

Old Suspicions Recalled

Various facts were now recalled. It was known that Charles Dawson had stained the Piltdown bones in the mistaken notion that the chemicals would harden the bones and thus preserve them for study. Yet this action now raised the dark suspicion that baser motives were at work, and that the staining was done to make the bones look older. It was recalled that in 1912 Woodward stated that he would have said the jaw was that of an ape if it had not been for the teeth. Another comment made back in 1916 by a dental authority, C. W. Lyne, was recalled. Contrary to opinions of the experts from the British Museum, Lyne stated that the canine was an immature tooth. The unusual amount of wear on the tooth made no sense at all.

Already back in 1925 F. H. Edmonds of the Geological Survey had reported that Charles Dawson had been in error about the height of the Piltdown gravel above sea level by about 18 feet. The corrected height greatly reduced the alleged age of the Piltdown gravel. No one noticed the report at the time. In 1950 Edmonds dropped the other shoe. The older group of Piltdown fossil animal remains could be explained as having washed down into the gravel from a higher Pliocene bed. The problem, Edmonds noted, was that there was no such bed anywhere around for such fossils to wash out from. Another interesting observation was made in 1951. The bones of the skull were extraordinarily thick, yet the jaw bone thickness was normal and ordinary. Authorities such as Ales Hrdlicka and Ashley Montagu in the United States noted that nowhere in the world had another skull case differed so from its jaw.

Another point was made. It had been frequently argued that the Piltdown jaw could not be ape, because no ape remains had ever been found in England. But an American anthropologist, Montagu, cited British literature to the British experts in which the remains of two such anthropoids found in England had been described in the past. This was embarrassing. Therefore the jaw and skull could have belonged to two different creatures after all.

Hrdlicka unkindly suggested that the molar tooth from Piltdown II, two miles away, had come from the same jaw as the molars of Piltdown I. The suggestion smacked of fraud but was so outlandish that no one attempted a rebuttal.

More Odd Discoveries

In the announcement of the results of the fluorine tests in 1949, Kenneth Oakley made an observation that should have been startling. Just below the surface stain of the teeth, the dentine was pure white—just like new teeth. The million-year-old teeth were not fossilized. Apparently no one at the time gave any thought to the devastating implications. Further, the bone of the lower jaw was so "recent" that the drill used for taking samples produced a smell like that of burning animal horn.[33] Even a total beginner hunting fossils can tell the difference immediately between a recent unfossilized bone (very light) from a mineralized fossil (heavy). How could such distinguished experts of the British Museum be so easily taken in?

The Pressure Grows

Dentist Alvan Marston again added pressure. He took to the pages of the British Medical Journal to show how he had inserted a model of the Piltdown canine into the socket of a female orangutan jaw. He demonstrated that the altered canine actually belonged to a young adult ape. In addition he discussed how the original Piltdown X-rays of the jaw showed typical anthropoid bone structure. But Marston was an outsider. What could a dentist know about teeth? It is no surprise that his crushing factual exposure was never acknowledged, but the pressure was being felt.

A Missing Record of the Missing Link

An offhand remark at a dinner led to the final stage for the exposure of Pilt-down. Oakley mentioned that due to Dawson's sudden death in 1916, the British Museum had no record of the exact spot where Piltdown II had been found. Dr. J. S. Weiner, who was present at the dinner, found it absolutely incredible that the location of Piltdown II had not been officially recorded by the finder, Charles Dawson, nor by his close friend and Keeper of the British Museum, Sir Arthur Smith Woodward, who had personally examined the new site. After all, Piltdown II made believers out of the skeptics who had been so dubious about Piltdown I.[34] With suspicion resting on suspicion, Weiner and Le Gros Clark at long last examined the Piltdown casts at Oxford University. They immediately noted that artificial grinding or filing seemed a distinct possibility. Weiner next took a chimpanzee jaw and by filing and staining was able to produce the exact appearance of the Piltdown jaw.

The Bones are Examined at Last

Weiner and Clark then arranged to meet with Oakley at the British Museum to examine the original evidence. Faking was immediately apparent. For the first time it was noted that each molar was worn at a different plane or angle—an absolute impossibility in any jaw. The crisscross scratches from filing were so visible that the scientists were appalled that the obvious had not been noted by even the most casual glance. In grinding down the canine into a more human-like shape, the hoaxer had filed one place too far. A small patch formerly thought to be secondary dentin turned out to be chewing gum! Grains of sand in the root channels were completely foreign to the Piltdown region. The black coating on the bones had been described by Dawson as iron-stain. This coating turned out to be nothing more than some kind of barn paint.

The results of more fluorine tests were then announced, as though there might still be some doubt about the whole tawdry mess. New and improved techniques supposedly gave the final verdict on the fraud. The real truth was too awful to confess. No explanation was ever given on why the earlier state-of-the-art tests on the first seventeen samples were now discovered to be completely wrong.

More Appalling Fakes

Now the other fossils and artifacts were given a closer look. Many were faked and others had been imported from other sites. The striking elephant bone shaped like a British cricket bat had been shaped with a steel tool—probably a knife—in modern times.[35] We can hardly be surprised to learn that all these faked bones and artifacts "mysteriously" disappeared from the drawers in the British Museum collection.

The Ghastly Truth is Announced

In 1953 J. S. Weiner, Kenneth F. Oakley, and W. E. Le Gros Clark, distinguished British anthropologists, published the news that Piltdown was a fake.[36] Despite a great deal of sarcastic quips and laughter, the British House of Parliament was not amused at how easily the great experts of the day at the British Museum had been taken in for more than forty years. They came within inches of passing the motion that the House has no confidence in the Trustees of the British Museum. Cooler heads succeeded in having the devastating motion tabled.[37]

After all this, *Time* could look back and sum up the whole mess as a triumph of science due to the new techniques being applied to the skull.[38] The guilty party was merely a prankster.

WHODUNIT?

How Was the Hoax Carried Out?

No one can guess how many Piltdown relics were planted, retrieved, repaired, and replaced in the gravel before the great discoveries were made. Someone undoubtedly had a hilarious time trying to second guess the day-to-day progress of the workmen with their picks and shovels. We can assume that many a handsomely filed and dyed specimen was smashed or buried because the workmen could scarcely be expected to use any particular scientific care in shoveling sand and gravel. The same patient hoaxer had to keep putting the specimens not too obviously where more careful searchers like Dawson, Woodward, and others would be most apt to find them.

Saving Face with Stiff Upper Lips

The entire Piltdown saga is an interesting insight into the history of science. In the aftermath, since the topic of the hoax could not be avoided, the British tried somehow to make the best of a dreadful situation for the British Museum. Face-saving explanations generally were of four kinds:

First, the exposure of the fraud was a great triumph for new techniques developed by scientists. The actual sequence of events described above leading to finding out the hoax gives faint support to that notion. Second, the uncovering of the hoax greatly clarified the picture of man's evolution. This outrageous twist is beneath comment. Third, the culprit must be identified, and he must be someone outside the scientific community. This solution was attractive since it would absolve the scientific community from blame in the matter.

The fourth kind of explanation followed a different tack. According to M. H. van der Veer it is not true to say that our account of man's descent was seriously misled by this so-called important find.[39] Piltdown was in fact rejected as dubious evidence, and to speak of a deception that lasted for forty years is to put things too strongly. This is another outrageous statement. Van der Veer ought to tell us all about the many doctoral dissertations and other formal studies written about Piltdown that analyzed everything except the fresh modern stained bone, the chewing gum, file marks, and barn paint. All the hundreds of studies, we may assume, concluded that here was the great discovery of the great missing link, and of course it was found on British soil.

Who Was the Hoaxer?

A new breed of experts arose seeking fame by telling who the guilty party was. Yet, a half century after the hoax was exposed, there is still no consensus on who was guilty, and whether one or more than one person was involved.[40] Here are the nominees:

It was Dawson. After all, he was the discoverer. J. S. Weiner believed the circumstantial evidence pointed to Charles Dawson acting alone. Dawson certainly had a motive. As the official finder of Piltdown remains and a mere amateur fossil collector, Dawson became the favorite scapegoat. The local antiquarian society ridiculed Dawson's demonstration of how so-called Dawn Man's artifacts could have been formed by natural events. Piltdown Man brought Dawson the fame he longed for and at the same time he revenged himself on those who had rejected his ideas. So went the thinking.

Yet few found this acceptable, even when no better candidates were suggested. The general consensus was that an over-eager Dawson had been duped by someone else. In turn, Dawson's good friend, Arthur Smith Woodward, Keeper at the British Museum, was deceived and lent his immense prestige to the authenticity of the finds. Most of those acquainted with the Piltdown hoax believed that neither Dawson nor Woodward were the actual hoaxers. For example, we have such cryptic statements as that of Alfred Kennard, died 1948, who said he knew who the hoaxer was, but it was not Dawson.[41] Dawson, however, long remained the favorite suspect because he was not a scientist, and the scientists at the Museum could not bring themselves to believe that one of their own had done the dastardly deed.

It was Sollas. Out of the blue in 1978 another candidate was nominated by 93-year-old James A. Douglas of Oxford University. On the most slender evidence, Douglas apparently convinced himself that the real culprit was his predecessor at Oxford University, William J. Sollas, Professor of Geology. As motive, Douglas, who had worked with Sollas for 30 years at the Museum, suggested that Sollas sought to destroy the reputation of Arthur Smith Woodward, who had the most prestigious position in England as Keeper of Geology, British Museum. Smith Woodward had publicly ridiculed Sollas, and Douglas believed Sollas created the Piltdown hoax to destroy Woodward. Despite the dramatic deathbed taped "confession" of Douglas, few were able to take his account seriously.[42]

It was Smith. Ronald Millar has just as little tangible to go on, but his nomination as the hoaxer is Grafton Elliot Smith, an accomplished human anatomist and an expert on prehistoric human skulls. Piltdown Man would have fit into his theory of ancient waves of immigrations of man. His professional work in northern Africa would have given him easy access to some of the planted fossils and bones. In addition, he could have easily settled important reconstruction questions for the Piltdown remains that baffled Woodward, yet he remained silent. Smith is the same person who in 1922 reconstructed Nebraska Man, his family, and his culture on the basis of one tooth. Unfortunately for Smith, the tooth turned out to be that of an extinct peccary (pig), and the incident underscores the absolute obsession of men of science to "prove" the descent of man from ape-like ancestors.[43] Some months before Piltdown Man was found, Smith published a scientific paper in which he predicted what the missing link would look like. It would have a large human-sized brain, but its features would be ape-like. It seems most unlikely that Smith would then follow up his paper by constructing and planting the Piltdown forgery, which was exactly as described in his paper. It is obvious that the actual hoaxer did use Smith's description to create Piltdown Man.

Still another suspect—the creator of Sherlock Holmes. Other investigators offered a formidable array of evidence linking Sir Arthur Conan Doyle with the Piltdown hoax.[44] Doyle was an avid collector of fossils, he loved hoaxes, adventure, and danger, and was a practical joker. He bore a grudge against the British science establishment. He lived only a few miles away from the Piltdown site. He had free access to it, visited it frequently, and was a friend of Charles Dawson. He was a physician and chemist and in his travels had visited all the Mediterranean areas where the planted fossils had originated. Authors John Winslow and Alfred Meyer point out interesting parallels between Doyle's book, *The Lost World*, and the Piltdown hoax. For example, one of the characters says that if you are clever and you know your business you can fake a bone as easily as you can a photograph.

It was Keith. In 1990 another candidate emerged. Sir Arthur Keith was ambitious and knowledgeable. As a noted anatomist he had the background to create the hoax in an effort to humiliate a rival at the museum.[45] Perhaps he was in league with Dawson. Curiously, Keith wrote an article about Piltdown before the discovery was first announced in which he revealed details never discussed at the meeting. Further, when Keith learned that the hoax had been discovered, he immediately burned all his correspondence with Dawson.

It was Hinton. In 1996 a new candidate emerged from the shadows. Martin Hinton, a brilliant curator at the British Museum, who became disgruntled in a pay dispute around 1910. He set out to trick his boss by chemically treating relatively young fossils to make them look old. After his death, Hinton left behind a trunk containing old bones, other fossils and teeth, all stained with the same chemicals used exactly as with the Piltdown bones. He had worked with Dawson before the hoax, and he was an expert in the geology of the Sussex area where Piltdown Man was found.[46]

But it was really Teilhard. Plausible as some of the above suspicions seem, there was another candidate who fit the great hoax much more precisely. For many years after 1953 there were somewhat mysterious undercurrents about the whole affair with hints that the real culprit was known by a good many people, but because of his position or for other personal reasons, his identity would not be revealed. More recently, persons in high places, such as Dr. Louis S. B. Leakey and Stephen J. Gould of Harvard, have been willing to bring the man out of the shadows into the light, and the candidate turns out to be Father Teilhard de Chardin. It could have been a youthful joke that quickly got out of hand and made confession or undoing impossible, because in so doing, Teilhard's own promising career would have been destroyed.

Looking at the matter in terms of the age old rivalry between France and England, Piltdown was the ultimate French prank against the stereotype of the stodgy inept British. Need we remind ourselves that Teilhard was French? There is a second possible explanation that is even more probable. Teilhard was totally obsessed by the necessity of bringing religion and doctrine into harmony with "science," with bringing evolution as a welcome partner into the church. What better way than to invent a perfect missing link to break down the last lines of resistance. Piltdown could greatly accelerate what otherwise would continue to be a long and perhaps debilitating struggle between science and the church. Neanderthal remains were not quite convincing enough. Piltdown was perfect. To Teilhard this would not really be a hoax because it would be only a matter of time in his thinking until a genuine missing link would be discovered that would show the same transition between ape and man. After all, the end justifies the means.

We gain insight into Teilhard's mind from the way he described his strategy in bringing evolution into the church.[47] He advised the following: Keep your language orthodox, but change the meaning of the orthodox words to fit the "new" teaching during a period of transition from old beliefs to the new and "scientifically" enlightened beliefs. In this way the people could be lulled and reassured until the transition had been completed.

Those who have taken a closer look at the so-called "minor" role of Teilhard in the Piltdown discovery have noted such items as the following: According to Bowden, Louis S. B. Leakey described the hoaxer in terms that could only fit Teilhard.[48] Leakey's wife, however, prevented the publication of her husband's book after his death that connected Teilhard with the hoax. Stephen Gould of Harvard charged that Teilhard was the hoaxer. A key point made by Gould regarding the two Piltdown sites seems to incriminate Teilhard. Mary Lukas, a defender of Teilhard, did not rebut many evidences that came to light and which point to Teilhard as the hoaxer.[49]

Note all these amazing coincidences that could tie in with the Piltdown hoax, which began in the summer of 1912. In 1908 Teilhard began the study of theology at Hastings in Sussex, but his primary interest was rocks, fossils, and geological curios. He loved the Sussex area and much of its appeal rested in his freedom to wander about the English countryside. As a result of significant fossil finds he soon came to the notice of Charles Dawson, and their lasting friendship was well established by May 1909. For years Teilhard was not considered a possible candidate as the hoaxer, because it was thought that he had been out of the country at key times. A more careful study of the matter shows that this was not the case. Hastings was about 20 miles away from the Piltdown finds. Dawson could have been the perfect foil for an unscrupulous person interested in speeding along the acceptance of human evolution.[50]

Before Teilhard arrived in Hastings, he had been in Egypt. The *Elephas planifrons* molar plays an important role among the planted fossils. Tests of radioactivity place its origin at Ishkul, Tunisia, as the only possible site in the world it could have come from. Again before coming to Hastings, Teilhard actually stayed near Ishkul, Tunisia. His stay in Cairo, Egypt, gave him additional opportunities for fossil and artifact collecting of the kinds associated with the Piltdown remains. The elephant bone artifact, shaped so much like an English cricket bat, aroused an unusual amount of interest. Dawson learned that only one other fossil like it had ever been found. This was in Dordogne, France, which happened to be Teilhard country. One must grant that Teilhard had an uncanny knack of having been at all the sites that had significance for Piltdown.[51]

Teilhard's Qualifications as a Hoaxer

The hoaxer had to be an expert paleontologist who would need to know quite precisely the toleration limits of his fellow scientists in accepting a missing link far different from what most experts had forecast. This fits Teilhard. The hoaxer had to have enough prestige to keep the hunt alive for another seven years after the initial find by quietly suggesting just the right place at the right time without arousing suspicion. Furthermore, as the planter of the items on numerous occasions, we can assume, the hoaxer had to be one who could roam the hillsides and meadows without arousing suspicion. It is difficult to picture dignified English scientists coming out from London or Oxford to plant fakes, especially without drawing any attention to themselves. The whole business seems contrary and out of character for an Englishman.

Much has been made of bone staining. Dawson, in supposed ignorance, openly stained the Piltdown bones thinking the chemicals would harden them. In another curious coincidence we find that Teilhard was a lecturer in chemistry in Cairo. In a letter he speaks of seeing a butcher bone in a stream stained like the finds at Piltdown. As a competent person in chemistry, Teilhard would know how to "age" the faked fossils and bones.

By hindsight, we note that unusual things happened at the Piltdown sites when Teilhard was around. When Dr. Woodward from the British Museum appeared to lend prestige to the search, the fake flint tool, the elephant tooth (from Tunisia), and the altered ape jaw are quickly "discovered," the first two by Teilhard himself. The most startling discovery of the three was part of a lower jaw with two molars. Astonishingly, the lower jaw precisely fit the skull fragment found some years earlier. Although the pieces were separated in the gravel, no one thought it strange that the supposed 500,000 years of erosion, weathering, and movement of the gravel by water action had not affected in the slightest the fit that the fragments made.[52]

Teilhard's incredible luck continued. In August 1913 he found a human canine tooth in the gravel that not only fit the mandible previously discovered, but amazingly confirmed Smith Woodward's predictions of what Piltdown Man's canine tooth would be like in shape, size, and amount of wear. Teilhard's keen eye spotted the fossil tooth where Dawson and Woodward had already carefully searched. In a letter of June 3, 1912, which Malcolm Bowden calls highly incriminating, Teilhard speaks of the famous human skull *before* the faked jaw had been found. By itself, the brain case would have meant nothing. How could Teilhard know the skull would become famous?

When Teilhard was informed in 1953 that the Piltdown finds were discovered to be a well-planned hoax, he offered a possible explanation that "someone innocently threw the bone fragments from a neighboring cottage into the ditch."[53] A world famous paleontologist is speaking here who had studied the finds in great detail and who knew firsthand exactly where they had been found. Teilhard would know better than most scientists that his own statement was an absolute impossibility. Granted that the English are famous for a wonderful array of eccentrics, what British housewife discards human skullcaps and ape jaws in a ditch outside her cottage? On this strange and irrational statement alone, suspicion must rest on Teilhard's role in the hoax!

Uneasy Teilhard

In 1953 Teilhard wrote Oakley to congratulate him on solving the puzzle of the Piltdown Man, and now for the first time we learn that Teilhard "had always felt uneasy" about the find and had once termed it an "anatomical monstrosity." One must say that his uneasiness was well concealed for more than forty years, until after the hoax was exposed. It seems strange that Teilhard would send a letter of congratulations to the man who had helped expose a fraud in which Teilhard was deeply involved. Again we have an irrational action that raises a serious question.[54]

In 1954 Teilhard wrote that he was not often allowed to leave his cell and that he knew nothing about anthropology. But C. Cuenot said he was allowed to go frequently on scientific excursions. While official reports list him four times as accompanying Dawson and Woodward to the site, Sir Arthur Keith records that Teilhard shared in all the toils at Piltdown. Leakey reports that many persons were aware that Teilhard was a frequent visitor to the Piltdown area.[55]

OUR CONCLUSION

Led by an Obsession

The above remarkable series of coincidences is impressive. The whole matter is moot, however, and Teilhard cannot defend himself unless he has left written statements not yet published. It does seem reasonable, however, that Teilhard must be given a special place of honor among all the nominees who could have perpetrated the hoax. The whole episode may well be a case study of how far an obsession will drive a person. As Rupert Furneaux points out, the whole comical series of impossible finds may have been the earnest attempts of the hoaxer to expose his own humbug—but without avail.[56]

The most eminent authorities, however, vied with each other in proclaiming the absolute certainty of the whole thing. Apparently no one ever asked why an ornithologist (one who studies birds), Pycraft, headed the British Museum's anthropology division (those who study man's supposed evolution from the ape world). But no one could use venom and sarcasm better than he in putting down any and all opposition to Piltdown Man.

Over the years a few had guessed the truth, but scathing attacks and ridicule reduced the protesters to silence or ineffective statements, and the mainstream ignored them. Those who take the Bible seriously should know that a Christian scholar, Theodore Graebner, cited much obvious evidence that the skull and jaw never belonged to the same creature.[57] But how could a theologian know anything about paleontology? It took distinguished scientists another twenty years to make the same discovery.

An Evaluation

Loren Eiseley, an evolutionist, wrote an epitaph for Piltdown Man.[58] Eiseley was roundly criticized by fellow-evolutionists for being so candid. He had this to say: A skull, a supposedly very ancient skull, long used as one of the most powerful pieces of evidence documenting the Darwinian position upon human evolution, has been proven to be a forgery, a hoax perpetrated by an unscrupulous but learned amateur. (*Comment*: The culprit was a learned but misguided scientist— not an amateur.)

The famous Piltdown cranium, known in scientific circles all over the world since its discovery in a gravel pit on the Sussex Downs in 1911, was jocularly dismissed by the world's press as the skull that had "made monkeys out of the anthropologists."

Eiseley further noted that for seventy years after the publication of *The Origin of Species* in 1859, there were only two fossil finds that seemed to throw any light upon much controversy about the evolution of man. One was the discovery of the small-brained Java Ape Man (who later turned out to be 100 percent ape); the other was the famous Piltdown or Dawn Man (a 100 percent hoax). Both were originally dated as lying at the very beginning of the ice age, and, though these dates were later to be changed, the skulls, for a very long time, were regarded as roughly contemporaneous and very old. Eiseley commented that two more unlike missing links could hardly be imagined. Though they were supposed to share a million-year antiquity, the one indeed was quite primitive and small-brained, the other, Piltdown, despite what seemed a primitive lower face, was surprisingly modern in brain.

Eiseley evaluated the Piltdown Man hoax. Many of the world's leading authorities on early man were taken in and contributed to the vast amount of literature about Piltdown. Eiseley states that Piltdown can now be viewed historically as a remarkable case of history in self-deception. It should serve as an everlasting warning to scientists (that is, evolutionists) who may exhibit irrational bias or give allegiance to theories with only the most shaky basis in fact. It is very embarrassing to see the rapidity with which scientists embraced the specimen solely because it fell in with preconceived wishes and they saw that it could be used to support all manner of convenient hypotheses.

Eiseley concludes that the enormous bibliography in many languages that grew up around the skull is an ample indication, also, of how much breath can be expended fruitlessly upon ambiguous or dubious materials, not to mention outright fraud.

A Final Thought

Oh yes, the doctoral dissertations and countless hundreds of learned papers about Piltdown turned out to be more acts of worship of the false god of evolution than anything scholarly.

Time writes off this monumental scandal as Teilhard being accused of "a little playful tampering with evolution."[59] Would *Time* be equally forgiving if a comparable hoax had been done by a "well-meaning" Christian?

For Discussion and Reflection

1. Piltdown has gone into the language as the name of a very ancient pre-human fossil. For one thing, there was intense rivalry between England and France to discover man's most ancient ancestor. England came to be a very poor second to France where wonderful fossil skeletons of man and amazing cave paintings were discovered. How did Piltdown Man become the great showpiece for England "proving" human evolution?

2. The Piltdown fossil remains were guarded for over fifty years as more precious than the crown jewels of England. What would give this fossil such an incredible value to the British?

3. The world's most illustrious paleontologists were not permitted to handle the Piltdown fossils and could study only plaster casts of the original. Can you think of any scientific reason for protecting the fossils from any close examination?

4. More than a century ago a psychologist spoke of a "need to believe" among people. Since we now know that the Piltdown remains were an outlandish hoax, how does this hoax illustrate the need to believe? What did the British need to believe about this fossil?

5. As explained in the chapter, why do you think it took a dentist rather than an expert paleontologist to uncover this dreadful fake?

6. Is Piltdown Man another example of itching ears that want to hear what they desire to be true (2 Timothy 4:2)? Piltdown Man is only one of a long series of mistakes and frauds over the past 150 years that attempted to bolster the belief of evolutionists that man evolved out of the animal world. Why is this effort so important to evolutionists?

CHAPTER NINE

SPEAKING TO THE ANIMALS

INTRODUCTION

What We Are Looking For

National Geographic gives us an example of "simple" life that vividly contrasts the beliefs of creationists and evolutionists.[1] Scholars from two research institutions combined resources to analyze the genetic structure of a worm barely visible to the eye. Each one of its thousand cells consists of 17,000 genes and 100 million base pairs of DNA. These cells control how the worm reproduces, grows, eats, moves, and performs the myriad functions of a living creature. Stated baldly, the evolutionist believes that this life gradually evolved by chance first out of nothing, then from non-life, then from a one-celled creature to more advanced forms of life over billions of years. The creationist believes this worm came from God's original creation some thousands of years ago.

As this is a book about the search for the Genesis world, full of great mysteries, we shall be looking for evidence that expresses the biblical framework of a young, created earth. On the other hand, the lifeblood of the evolutionist is time, chance, and other natural processes in vast amounts, so we need to consider what this belief has going for it. These two models of what the early earth was like are gradualism for the evolutionist, and catastrophism, such as Noah's flood, for the creationist.

If creation is true, the location of the earliest animal domestication could be significant. Early sophistication in the oldest cultures, including early extensive sea travel, would also support the model of a young, created earth. Can we discover evolutionary steps in the use of domestic animals? Are there serious flaws in the evolutionary sequence of Paleolithic to Mesolithic to Neolithic to Chalcolithic to Early Bronze ages? These will be some of our explorations as we examine interesting puzzles and mysteries in the animal world, both domestic and wild.

The Biblical Message

This is what we learn from the first book of the Bible.

Genesis 1 explains creation briefly but clearly. Day 5: All sea life and all bird life were created. Day 6: All animal life, wild and domestic, and all creeping things. Adam and Eve were given dominion over the fish of the sea and over the birds of the air and over every living thing that moves upon the earth. Every green plant was to be food for every creature on land and in the air. Everything God had made was very good.

In **Genesis 2** Adam gave names to domestic and wild animals and to the birds.

In **Genesis 3** Satan, acting through the serpent, led Adam and Eve into sin. As one consequence, the serpent was cursed, and to a lesser degree, all animal life, wild and domestic.

In **Genesis 4** Abel became a keeper of sheep. The first burnt offerings of his flock are recorded, and the Lord was pleased with his offering. A descendant of Cain, Jabal was the father of those who dwell in tents and have cattle.

In **Genesis 6** God determines to destroy man and beast and creeping things and birds of the air. The earth in this time was filled with violence and *all* flesh had corrupted their way upon the earth. Two of every sort that breathe the breath of life will come into the ark with Noah.

In **Genesis 7** seven pairs of all clean animals were to enter the ark. When the flood came, all birds and flesh that moved on the earth died.

In **Genesis 8** the creatures are brought out of the ark that they may breed abundantly on the earth, to be fruitful, multiply, and replenish the earth. God established this triple blessing for the rapid repopulation of the earth.

In **Genesis 9** the fear and dread of man is now upon every beast of the earth, the birds, the creeping things, and all fish, and they are delivered into the hand of man. All moving creatures (but not their blood) are now food for man.

ANIMALS OF THE WILD

Spin—A Glimpse of the Fossil Record

Whatever geologists may be faulted for, they cannot be accused of a lack of courage. Great courage is required to spin a coherent yarn from a smattering of conflicting evidence, mysteries, and perplexing puzzles. A few illustrations will show something of the dimensions of the problems faced by evolutionists.

How Many Species Were There/Are There?

We need to take a careful look at the estimates splashed in the media today. The situation is very confusing because *species* is not defined (no easy task), and more importantly, which forms of life are and which are not included in the counts.

Around 1975 biologists estimated a total of one million species in the world. More recent studies of the rain forest and ocean bottom have now resulted in a total estimate of 130 million.[2]

Another estimate is that 982 million species have existed during earth's history. Of these, so far scientists have recognized about 130,000 in the fossil record. There seems to be something grossly off with these numbers.

Still another estimate of 13.6 million total species by type includes the following: 8 million insects; 50 thousand vertebrates, and most of the remainder are tiny, often invisible, things.

There is little point in continuing this absurdity. No one has much of an idea of how many species there are. The popular thing is to keep making all estimates much larger than previous ones.

Species Lumpers and Splitters

There is a never-ending dispute among paleontologists who want to name a new species for every slight difference in fossil bone material found. In contrast, other scientists recognize that differences in age, gender, environment, diet, and various factors will result in significant physical differences in fossils discovered, yet they may all belong to the same species. Hilbert Siegler, for example, observed that taxonomists like to create an unnecessarily large number of genera.[3] Richard Finennes reduced fourteen genera of canids (wolves, e.g.) listed by Ernest Walker to only five. The vast range of interbreeding among the canids raises many questions about the validity of many supposed species designations. This suggests that large numbers of so-called fossil species may be fictions in that differences may only represent varieties of the same species or stem from the ego of the finder who gets the privilege of naming the new species, often inserting his own name into the official designation.

Living Fossils

The supposed rate of evolution is in an embarrassing muddle. All that evolutionists are left with is to say that some organisms evolved very little over long periods of time; some apparently have not changed at all over vast stretches of time; others evolved at tremendously fast rates. Using evolution time, one writer observes that land animals evolved faster than marine animals. Examples given were the lamp-shell (unchanged for 500 million years) and the horseshoe crab (unchanged for 180 million years). In the same article, the observation is made that the possum has not changed in 60 million years).[4] Whether species are supposed to be unchanged or to have changed very rapidly, it is embarrassing to say that they all appear in the fossil record fully developed.

A great deal can be written on peculiar gaps in the fossil record. Extinction is assumed when no given fossil has been found in a given period or epoch. The coelacanth is a dramatic example, for it was assumed to be extinct for 400 million years in evolution time until living examples were found off the coast of Madagascar some years ago. Other writers say the fish had been extinct for 70 million years, which is illuminating in itself on the degree to which vast ages are arbitrarily tossed about in the literature. Assuming the coelacanth had been known only in fossil form in rock 400 million years ago, but that the species lives today, perhaps 10 million generations of this fish supposedly lived and died without a trace in the fossil record. The coelacanth, and other species that are unchanged from ancient fossil specimens, are called "living fossils." This is hardly a satisfying explanation.

Another living fossil is the Tuatara reptile, known as the beakhead, which today is found only in New Zealand. The only known fossils of this reptile are dated to the early Cretaceous, dated to 135 million years ago. Again, millions of generations supposedly lived and died without leaving a trace in the fossil record, but they are somehow here today. There are many other examples of gaps in the fossil record. A deep-sea mollusk was recently found in an ocean trench off Central America at a depth of more than two miles. It was believed to have been extinct for 280 million years. Crustaceans found in waters of the beach sands of northern England closely resemble fossils thought to be extinct 300 million years ago.

There are impressive lists of "living fossils," each with a living look-alike. The list includes starfish, bacteria, coelacanth fish, crustaceans, sharks, mollusk, squid, numerous insects, and centipedes. These fossils, as much as 600 million years old in the evolutionist's timetable, show no change whatever. Why did they not evolve into more advanced forms, given all the time they had to do so? The list would be much longer except for a common slight of hand practice among paleontologists. George Price noted the obvious when he pointed out that when fossils are found that are strikingly like living species, a new name is invented for them, so as to be able to call them extinct.

Look at a very strange creature, the aardvark. It has a powerful kangaroo-like tail, the digging claws of the ant bear, bristles of the pig, incredibly long snout, and enormously oversized ears. If any animal might be tagged as in the middle of evolving into something else, this is it. Yet aardvark fossils, estimated by evolutionists to be 10–15 million years old in France and Greece, show that this animal has remained unchanged through the ages.

Extinction Tallies vs. Theory

There is another problem. On the basis of elementary logic, new species must emerge at a faster rate than extinctions, or there would be no life left on planet earth after the passing of so many hundreds of millions of years.

Evolutionists speak of five or seven great dyings in the history of the world, but most of the discussion centers on the end of the Permian period, supposedly about 250 million years ago, on the end of the Cretaceous period, thought to be about 65 million years ago, and the end of the ice age, Pleistocene epoch, thought to be around 12 thousand years ago. There is even more emphasis on supposed extinction rates today in the world. All dates such as the above are the product of evolutionary thinking.

The large number of extinctions would follow logically for a young, cursed earth. It does not fit into an evolving concept of the world where creatures keep on evolving into better and better organisms.

At the end of the Permian we read estimates of extinctions such as these: 85 percent or 90 percent of ocean species; 70 percent or 90 percent of land vertebrates; 80 percent reptiles; 75 percent amphibians; 75 percent or 95 percent of all species.

At the end of the Cretaceous we note estimates of extinctions like these: 96 percent of all ocean species; 75 percent of all animal species; 50 percent of all species. Included are all dinosaurs, sea reptiles, flying pterosaurs, and ammonites.

The end of the ice age resulted in the extinction of an estimated 96 percent of all large animals, such as the mammoth and the mastodon.

Estimated Total Extinctions in the World

The above estimates treat extinctions that supposedly occurred at three points in time. There is no lack of "experts" who estimate the total of all extinctions thus far in the world.

Here we read such assessments as the following: There are many wildly different estimates, such as up to 99.9 percent of all species that ever existed are now extinct.[5] If, as we noted above, there are about 130 million living species, we now must believe that there have been 130 billion species in the history of the earth. Another estimate is that 98 percent of all animal species died out, often but not always with man to blame. If we understand this correctly in evolution talk, man was already causing dreadful numbers of extinctions long before he had evolved out of the ape world.

Extinctions Today?

Some comment will be called for about the following estimates. We read ominous statements like these: Mammal extinctions are 40 times as great as fossil records show they were millions of years ago according to a new U.N. report.[6] A total of 30,000 plant and animal species now face possible extinction. Others give us statements like these: 100 species per day are now becoming extinct (36,500 per year). We will lose 20 percent of all species in the next 30 years. A fifth of all species were slated for extinction by AD 2000, a date now already past. A fourth of all species will be extinct by AD 2015.

We are to believe that one species a day became extinct by 1981, but this became one per hour as of 1990. The U.N. report, using scientific data, concluded that 484 animal species had become extinct over the past 400 years. This is a very shocking contrast to the wild figures quoted above.

What Caused all These Extinctions?

The favorite target is man as the cause. No responsible person will advocate destruction, waste, and pollution that may cause extinctions and other grave environmental problems, but there are many differences of opinion about what is appropriate and what the priorities should be.

These catastrophic extinctions in the past have caught evolutionists between a rock and a hard place. They have persistently denied catastrophes in the past, because that would support Noah's flood. Their central doctrine of gradualism is that nothing ever happened in the past that is not going on today, and this simply does not look at the evidence. Finally, man, as the current favorite politically correct whipping boy, hadn't evolved yet when almost all the extinctions took place. As evolution places man very late in the history of the earth, however, there is more and more realization among evolutionists that one or more catastrophes caused the great dyings. There are hundreds of gigantic impact sites on the earth and in the oceans that indeed reveal these as the source of great destructions. Creationists will differ on whether these are all to be associated with the flood, or that there may have been catastrophic events before, during, and after the flood, a view that this writer holds is fully in harmony with the Bible. (See my books *Noah to Abram* and *Genesis and the Dinosaur*.)

A Needed Comment

It is no surprise to note that decades ago an analysis revealed that about 50 percent of all scientific data published in journals is unusable and full of defects.[7] Certainly research on the environment is high on the list. A hero to environmentalists, Farley Mowat, was found to misrepresent and fabricate some of his research about the environment. Noted political leaders and scientists are unembarrassed to say that it does not matter if the data are wrong, because the problem is so important that we must alter our way of life in case it might be true. One analysis describes a backlash against warped school propaganda that consists of false information blended with a cloudy mixture of New Age mysticism, Indian folk-lore, and primitive earth worship, all presented as good science.

The Special Problem of Extinctions

It is very difficult to distinguish between fact, falsehood, and hysteria on this very important problem. The evidence for extinctions, of course, is found in the fossil record. Explanations of fossil formation by evolutionists emphasize the word *gradually* in contrast to crediting Noah's flood with most fossils.

Thus it was unexpected to read in *Discover* about fossils forming unusually quickly long ago in northeastern Brazil: pterosaurs, fish, insects, amphibians, and plants. "For reasons geochemists still don't understand" this material began to fossilize within hours.[8] Another miraculous discovery reported in the *New York Times* consisted of one-celled creatures, perfectly preserved in stone, caught in the act of cell division. This also indicates very rapid fossilization. Termite wings decay within two days, yet a wing in stone was reported in Ural Mountain deposits.

Consider this account of a jumble of fossils found in a Red Cloud, Nebraska, quarry, which again indicates instant death and rapid fossilization: mastodons, rhinos, camels, horses, tapirs, flying squirrels, more than a dozen snake species, frogs, toads, lizards, giant tortoise, seeds, and grass stems—more than one hundred vertebrate species.[9] There are many hundreds of similar locations around the world.

In northern Siberia and on the off-shore islands there are millions of frozen bones and bodies of animals, such as the woolly mammoth, modern bison, horse, and many other species. Scientists of the National Fish and Wildlife Forensic Laboratory estimate that more than 600,000 tons of ivory are still awaiting mining there. This estimate comes after many centuries of ivory collecting there.[10] That adds up to more than 10 million mammoths plus countless other animals. Evolutionists typically focus on a handful of mammoths that they say became encased in ice after falling into a crevice.

Other Evidences of Catastrophe

Here are some examples of past catastrophic events: Seals from the Arctic Ocean living in Lake Baikal, Siberia, more than 2,000 miles away; many thousands of sea monster fossils (ichthyosaurs) near Holzmaden, Germany in a confined area; ferocious sharks in Lake Nicaragua, 110 feet above sea level, a large lake suddenly formed by massive earth movements; a whale buried with elephants in northern Siberia; nearby are remains of cattle, bison, horses, sheep, great fishes, and numerous shells; painted scenes of elephants, sheep, giraffes, hippo in the Egyptian desert and central Sahara.[11] It is no exaggeration to say that there are many thousands of additional examples.

None of this is surprising to the creationist who accepts the awesome reality of Noah's flood and other ancient catastrophes. The evolutionist cannot stomach the possibility of catastrophes. Isaac Asimov, for example, despite mountainous evidence to the contrary, tries to maintain that there never has been a catastrophe. He will, however, grant the use of the concept of catastrophe as long as each lasted for many millions of years. To him the flood is only a myth. Here is his mother of all statements to explain away catastrophes: "What I have eliminated as inadmissible catastrophes are therefore changes that happen rapidly."[12]

Time Anomalies

Don Eicher states that any single line of fossil evidence is subject to greater error than several lines interpreted collectively.[13] This is an interesting assumption. The same kind of assumption has been used in personality testing, where the validity of one test could not be demonstrated, and therefore the researcher used several more invalid tests, arguing that therefore the amount of error was reduced. This is complete nonsense, but when one's options are reduced, one tries to make do.

This state of mind leads to an endless parade of statements such as this: traces of creatures were discovered that lived "about 1.9 billion years ago, 1.4 billion years earlier than the earliest known multicellular animals existed."[14] And these pellets found in shale are no different from those of today's microscopic animals. Or we read that one-celled fossils from South Africa are 100 million years older than previously thought. Fossil shrimp uncovered in Arizona were dated 1.2 billion years—several hundred million years too early). Or *Science News* reported that traces of worm-like animals in Precambrian rock in southern Wyoming have left scientists in a quandary. They are a billion years older than any such fossil and much too old to be possible! Note what Don Eicher says: "At present there is no scientific support for the geologic dating system, but some day there might be."[15]

New Mexico paleontologist Jerry MacDonald discovered fossil tracks of creatures who had lived in strata identified as Permian, that is, around 280 million years ago. Among the tracks were unexplained bear-like tracks, not to mention simian- or ape-like tracks: "I comment that they look just like bear-tracks. MacDonald says reluctantly, 'They sure do.' "[16]

The problem for the evolutionist is that bears will not evolve for another 200 million years in their time table! In the meantime, the dating system is still universally used, most often presented as fact. Which university student or what professor would dare to question geologic dates based on nothing at all? Academic freedom does not include questioning the evolution belief system.

The Index Fossil

The so-called Index Fossil idea has long served as the almost infallible magic marker to determine the age of a rock sample. Index Fossils are distinctive and abundant species supposed to have lived within a narrow time span, and which were widely distributed around the world. As time has gone on, however, this sacred measuring stick has crumbled. Some of the Index Fossils turned out to be unreliable and currently serve only a subordinate role in determining age. Whatever it is that has replaced the Index Fossil for age determination is certainly not radiometric methods. Actually it is quite vague just what has replaced the Index Fossil.

Note that hope springs eternal. With little if anything to clutch at in support, Eicher bravely states:

> Eventually we may develop composite standards for all systems in important areas, and ultimately we might hope to attain a composite standard covering the whole of the Phanerozoic Eon (that is, all the fossil zones from Cambrian down to the present).[17]

Strange Land Pockets

Intriguing clues to dramatic events in the past were illustrated when I visited a region of southern Alberta, Canada, called the Cypress Hills. In this peculiar and sheltered pocket of land, we find scorpions, horned toads, and vipers. These species are separated from the rest of their kinds by many hundreds of miles to the south in the United States.[18]

An intriguing problem is the discontinuity of certain species. We find the tapir in Central America and in the East Indies, but nowhere else. Fossil tapirs have been found in North America (Nebraska, for example) in Asia, and in Europe. The spread of the tropic-loving tapir must have preceded our present climatic zones. It could hardly have joined the so-called big animal march on the Bering Land Bridge.[19]

Effects of Environment

The antlers of the extinct Irish elk were monstrously large. There are many ludicrous attempts to explain why certain species became extinct. The tired clichè "survival of the fittest" is an embarrassment more often than not. Presumably the elk became more fit for survival by developing large antlers. It is seriously proposed that this elk then became extinct because its movements were impeded in dense forests.[20] It is well known, however, that dense forests do not provide food for grazers. Rather than feeding in open meadows and other grasslands, this elk apparently insisted on going into dense forests where there was no food and where its antlers kept getting caught among the trees. Many other major species of mammals also suddenly became extinct, but they did not have this handicap. Nothing plausible has been offered to explain these extinctions.

Another oft told tale is that the giraffe escaped extinction by adapting a long neck to feed on tree tops while those less fortunate starved below. This raises all kinds of interesting questions. Why did the equally tall, horned giraffe become extinct, as well as much bigger and taller giraffe species? What did the shorter female feed on in a time of severe drought? This question is even more appropriate for the young of the species. What did they eat until they could reach the tree tops? The home of the giraffe is also the habitat of a great many smaller species of grazers. How did they survive without growing tall?

One fact generally overlooked in the literature is mentioned by Eicher. Strata of the same age may differ in fossil content simply because of differences in the environment. Yet fossils are held to be the most accurate means of establishing the geologic age of a given stratum It is very obvious that gross errors can occur in dating a stratum (layer), because it may represent a different environment from another rock layer of the same age. The epochs of the Tertiary are particularly vulnerable to dating errors. If 49 percent of the fossils in a stratum represent living marine species, the rock is Miocene. If 51 percent represent living species, the rock is Pliocene, and the two samples may be dated as much as 22 million years apart in age. Such a courageous scheme recognizes no environmental differences in the world. Further, the assumption is made that all samples are representative of arbitrarily set boundaries. Imagine the strain on the species of those epochs. Every day they had to count noses to make sure that exactly the right proportion of species was intermingled.

The effect of geographical factors has been too little noted. Shull experimented with butterflies raised at abnormal temperatures. Those raised in a cold environment were formed like a northern variety. Climate undoubtedly has a marked effect on the form of species. Sumner raised white mice at two different temperatures and found marked physical differences between the two samples. Glandular flow, which in turn is affected by diet and the environment, alters stature, limb length, jaw size, nose shape, hair growth, skin texture, and other aspects of form. Dietary deficiencies account for a variety of physical abnormalities, including a brutalization of form. Similarly, isolation and resultant inbreeding can cause strange effects on the form of a species. In the turbulent history of our planet we cannot begin to estimate the far-reaching effects of these and other factors on the earth's flora and fauna. We are not speaking of species changing into other species, but profound changes within species.

The extent of the potential variability with species is little known. No doubt a great many so-called species simply represent differences in food, habit, and environment. Fish from Denmark were moved a few degrees of latitude. It was found that one result was an increase in the number of vertebrae, though these fish bred only among themselves. The dramatic role of trace metals, rare earths, and other chemicals on plants and animals is only dimly understood today. Much study is needed to examine effects on life of these variables.

Fish are believed by evolutionists to have had paired eyes in their heads since the Silurian period, 400 million years ago. If a little magnesium chloride is added to the water in which fish are developing, an abnormal fish with a median cyclopean eye results. Such environmental effects may dramatically affect size and form of species.

Evolutionists have not done well in their attempts to explain any of the above.

Domestic Animals

Genesis vs. the Spin

We can rely on Genesis 4:2 as the earliest domestication: "And Abel was a keeper of sheep." We can hardly expect to find any trace of pre-flood animal domestication today among the estimated 135 million cubic miles of sediments left by the flood. The confusion in the literature stems from attempts to piece together domestication in the early post-flood world.

The evolutionist has a difficult time with domestication. There are large areas of the world at this time where hunter-gatherers thrive and would never change their way of life unless forced to do so. Charles Reed cannot explain why ancient man in various parts of the world all of a sudden began to grow plants and domesticate animals at about the same time, after "millions of years" of hunting and gathering.[21] How could ancient man suddenly visualize the delayed advantages of growing crops and raising animals as the foundation of settled village life? The biblical framework described above gives an entirely different picture of the early centuries after the flood.

Dating the Earliest Domesticated Animals

In 1973, regarding domestication, another shoe dropped. With a new method of dating old bones by means of extracting collagen and analyzing the remaining carbon, we learn that the cow and the pig were domesticated in Europe about 9,000 years ago. The old view that agriculture and animal domestication began in the Zagros mountains of Iran and Iraq is now supposed to be outmoded. In a curious display of logic, C14 is said to be discredited because specimens from the same stratum differed by as much as 3,000 years. Yet the new dating method is naively accepted as valid. The conclusion then is drawn that first domestications took place at about the same time in both the Near East and in Europe—another triumph for independent invention.[22] The possibility that both dating methods may be strongly suspect is not considered.

Although C14 indicates that sheep also appear first in Europe, it is conceded that despite the evidence one might continue to believe that they were first domesticated in the Near East or in Turkey. Dogs are believed to have been domesticated in both Europe and the Near East at about the same time. Asia in west central Iran is currently believed to be the first center of goat domestication about 8000 BC.

More Problems of Time

Problems of time are inseparable from any study of ancient plants and animals. If the geological column is followed, it is accepted at the expense of the historical details of the Old Testament. Geologists love to make sport of the James Ussher chronology with the very clear message that therefore the Bible is largely myth and legend. There really ought to be greater honesty in admitting the frail structure of assumptions on which the geological column is based, and which is taught as fact.

It seems appropriate to examine a few problems of dating to illustrate the fact that neat and tidy textbook charts are far from pure.

On the basis of studies in Wales, Roderick Murchison and Adam Sedgwick in 1835 postulated the Cambrian and Silurian periods. For forty years thereafter, geologists quarreled about where the Cambrian left off and where the Silurian began. With all the wisdom of Solomon, Lapworth solved the dispute in 1879 by taking all the disputed territory, which he called the Ordovician, and which lasted, depending upon which textbook one happens to read today, 50 million or 70 million or 75 million years or in just one moment of terror. To the novice, this might seem to be a strange way of establishing truth, but one does not become popular by asking questions in geology. One geologist, whom we shall leave anonymous, parries such questions with the clever but wholly meaningless counterthrust, "Your view or your question betrays a misunderstanding of the problem." It is much easier and more enjoyable to attack James Ussher than it is to defend the virtues of the geological column erected in the early nineteenth century.

Don Eicher acknowledges what is seldom confessed in texts about the natural boundaries of geological periods and epochs.[23] Commonly, the boundary of a period like the Devonian for a specialist in one area differs from the boundary for the same period held by another specialist in another area. Controversies among geologists exist for nearly all the periods and epochs from the time they were first derived down to the present day. It seems shocking to state that there is a lack of agreement for any standard system of boundaries.

Long ago Thomas Huxley groaned over the incessant disputes as to whether a given stratum was Devonian or Carboniferous (up to 125 million years' difference in age), Silurian or Devonian (up to 80 million years' difference in age), or Cambrian or Silurian (up to 195 million years' difference in age). No definite line could be found separating the fauna of the Pliocene from the Pleistocene (a span of 13 million years). The St. Cassian beds were found to be a mixture of Mesozoic and Paleozoic types (span of up to 537 million years).

A most interesting device was derived in order to distinguish the various epochs of the Tertiary Epoch. The epoch is determined by the percent of living and extinct marine fossils found in a given stratum. For example, Pliocene formations are established if the rock contains from 50–67 percent of living species. A generation ago, the range was 50–90 percent. The Miocene is identified if the rock contains about 17 percent of living species. In the 1930s the range was 20–50 percent. It must be confessed that the system lacks a convincing ring to it.[24]

P. V. Tobias concedes that dating the past used to be sheer guesswork. Yet when he reviews the dating of the old Taung Skull, he resorts to guesswork after all. The skull was believed for many years to be about 2.5 million years old. It is now believed to be about 750,000 to 800,000 years old. Tobias extols the virtues of new scientific dating techniques, yet he derives the younger date for the skull on the basis of when the cave was thought to have been formed, gradually, of course, and the time believed to have elapsed for a sequence of geological events there.

If one wishes to accept C14 evidence, Reginald Daly observes that many reported dates are shocking to accepted beliefs in geology.[25] He also suggests that there have been some rather desperate efforts in recent geological literature to blur the significance of C14 dates, though the method is universally praised for its accuracy. Daly cites such data as the following: C14 shows that the ice cover in the Antarctic must be less than 6,000 years old because algal remains on the latest terminal remains give this date. The oldest ice on Greenland is estimated at only 10,000 years, and Greenland was ice free just before then. Cretaceous layers containing dinosaur bones have been C14 dated at 34,000 years old, a totally unacceptable age for the evolutionist. Natural gas deposits in Eocene layers have been dated at about 30,000 years, but the Eocene Epoch is believed to have been from 16 to 58 million years ago.

Around 1915 estimates among experts varied from 2 million to about 80 million years for the duration of the Quaternary and Tertiary periods. Today, depending on which source is taken, the combined age of the two periods is given as 63 million or 70 million years. One can hardly say that there is any scientific basis for such ages. Certainly the various epochs making up these periods were not derived on the basis of any convincing radiometric tests. Rather, it merely results from a fervently felt need of sufficient time for mammals to evolve into some half-human form. If creation is rejected, there is nothing left but this kind of reasoning. Similar examples could be cited almost indefinitely.

How Can Paleolithic and Neolithic Mix?

Walter Fairservis speaks of the puzzle of Siberia and the northern China plain. Paleolithic and Neolithic evidence lies comfortably together at the same period of time. The remains are dated about 6000 BC. Were it not for the evidence of domesticated goat, sheep, and dog, the other evidence might be dated as 100,000 years old or even older. James Mellaart is puzzled at Catal Huyuk that Paleolithic and Neolithic are linked, that is, happening at the same time. New research in the late 1990s there expresses the same puzzlement. Robert Braidwood refuses to use the terms Paleolithic, Mesolithic, and Neolithic because they have been so misused. He states that they are used to hide fuzzy thinking, and that there are violent disagreements about the interpretation of materials found at the earliest villages and ancient campsites.[26] He pictures men going out on the hunt after crops had been harvested. Could these be the seasonal ice age hunters not far to the north of Catal Huyuk? The astounding thing is that artistic styles, themes, and traditions of Catal Huyuk are the same as in much ice age cave art. These and many similar evidences are ignored because they contradict the supposed stages of evolutionary development.

The Geographic Spread—Was it Here? Was it There?

Writers on ancient history are none too helpful. Jacquetta Hawkes waxes lyrical by saying that in southwest Asia men were inspired to counterattack against nature and to make the momentous revolution in human history that accompanied the domestication of cattle, pigs, sheep, and goats. She states that by 5000 BC there were peasant communities in Palestine, Iraq, and Iran. Shortly thereafter such settlements spread to India, Egypt, and the eastern Mediterranean.[27] Analysis of this statement leaves one completely in the dark. According to the prevailing point of view about the ascent of man, how does a half-human creature announce at breakfast that he plans to revolutionize history that morning by going out to domesticate the first animal?

C. W. Ceram gives the nod to Mesopotamia as the first center of animal domestication, but he marvels at two things. Apparently all the important domestications were concluded between 3000–2000 BC with no more progress until the nineteenth century AD.[28] This is an unsolved riddle. It is sad that modern man merely added the ostrich to the list of domestications for the sake of its tail feathers. According to Carleton Coon, linguistic evidence shows that Indo-Europeans did not domesticate one useful animal or plant. All they had and used was borrowed from Hamitic peoples. The second incredible fact was the speed with which domesticated plants and animals spread around the world. It is important to say that if one would take the Old Testament seriously, much of the mystery disappears.

And the Winner Is . . .

When all is said and done, the lower hills of the Zagros Mountains of Iran and Iraq, just to the south of Mount Ararat, are still overwhelmingly accepted as the site of the first animal domestications. Europe will have to try harder before it is seriously considered as a contender for the honor. Highlands extend to the northeast from Iran all the way to northwest China, where early domestication of sheep, goats, pigs, and cattle is believed to have occurred in that huge country.

Diffusion from One Source

Will Durant satisfies no one by saying that the Neolithic age began about 10,000 BC in Asia and about 5000 BC in Europe. He further believes that the dog was domesticated about 8000 BC. After 2,000 years of experimenting with the dog, man then domesticated the goat, sheep, pig, and ox about 6000 BC. Durant expresses astonishment about the very evident degree of cultural contacts among diverse groups of widely separated early men. Technology was amazingly uniform over vast geographical areas.[29] Once more a common source of this technology is strongly indicated.

Animals Moved by Sea

E. K. Victor Pearce points out the astonishing fact that by 5000 BC Neolithic farmers sailed the length of the Mediterranean Sea bringing with them their cattle, sheep, and pigs. But the Mediterranean is only the beginning of the spread of sophisticated early man. Pearce believes it feasible that the first culture to reach America had a common cultural ancestry with Catal Huyuk, and that the similarities found much later in the Pueblo culture are not accidental.[30] The Bering Land Bridge may be a relatively minor part of the story of the spread of early man. For the real story we must look to the sea. We are only beginning to grasp the fact that early man navigated all the oceans.

Juergen Spanuth provides some astonishing evidence for early extensive ocean travel. Ancient rock art of southwest Sweden shows unmistakable images of the giraffe, ostrich, camel, and elephant, many aboard ship. As we might expect, they have aroused "storms of indignation," but there they are![31]

The oldest boat images in Europe are found on the Norwegian Island of Soroya, 250 miles above the Arctic Circle, and other islands even farther from the coast. The evolutionist pictures primitive man of that time afraid of water. Yet he crossed "the great oceans, the fearsome wide and lashing sea, from one hemisphere to another."[32] Ancient man came to Australia by boat even earlier. In 1999 museum and university anthropologists at last pointed to new research that challenges traditional theories that the first visitors came to the Americas by way of a land bridge to Alaska. The first settlers could have been Polynesians or southern

Asians who arrived by boat. Other very early remains have features typical of Europeans. This makes every textbook on the subject for more than a century obsolete, and it will still take many more years before rigid scholars will be dragged, kicking and screaming, into accepting evidence that has been published but ignored for many decades.

Signs of Catastrophe

There are interesting finds in India and Burma of uncertain date in cairns, circles of stone, and barrows. Evidence of domestication includes the buffalo, horse, sheep, camel, elephant, pig, and bullock. William Perry notes that some terrible event robbed the land and converted it to a wilderness of swamps and forest.[33]

The great Sahara Desert poses complex problems of dating. There are strong pressures by evolutionists toward dating the formation of this desert far into the past, and that it all happened very gradually, inch by inch. This rationale, however, cannot be defended. Far in the interior of the Sahara there are rock drawings of cattle that wore discs between their horns just as pictured in Egypt. There are paintings of war chariots drawn by horses in remote desert areas. A large human population raised cattle in fat pastures here. The sand is of recent origin. A stupendous catastrophe created the desert. The rock art of the cattle breeders in the Sahara shows an undeniable relationship with the Nile region, with the Puel people, and with the present Bantu tribes of Africa. Many authorities give a very late date to the domestication of the horse and the camel. This view runs headlong into depictions of the domesticated camel and horse found in the interior of the Sahara desert.

Too Much Sophistication?

In 1904 about 800 remarkably well-preserved rock drawings were discovered in four adjacent little valleys close to the summit of the Maritime Alps near Tenda, Italy. They are of uncertain date, as are most rock drawings. Most figures represent a rural, peaceful, agricultural life. Men plowed with two or more yoked oxen. The oxen were accompanied by a boy leading them. It is remarkable that the same kind of yokes and the same kind of plows are used today in more primitive parts of Italy. Other domesticated animals are also shown in the carvings. An unusual feature is that most of the scenes are shown as they would be seen from above.[34] The assumption here, of course, is that the drawings are old because there is no tradition of such art work in historical times.

Sophisticated stock breeders lived in eastern France about 3000 BC. On the bank of Lake Clairvaux near Besancon, eight well-preserved wooden houses were excavated under 15 feet of mud. They were made of precisely-tooled oak with hallways, rooms, and conduits in the roof for chimneys. Evidence of tamed dogs was also found.

Cattle in Ancient America?

It would seem very far-fetched to speak of domesticated cattle in the Americas before Columbus, yet Edward Vining reports some curious information. A Chinese historical account of a voyage shows that cattle with very long horns had been domesticated. The description would not fit the bison. According to Friedrich von Humboldt, cattle horns of monstrous size were found in ruined monuments near Cuernavaca in southwestern Mexico. Michel De Castelnau reported that cattle with very long horns as well as another species with small horns lived near the Amazon and in Paraguay that could not be Spanish related. Von Humboldt also stated that two species of oxen with large horns ranged in herds on the plains of Rio del Norte before the Spaniards came to America.[35] It is known that Montezuma showed the Spaniards enormous horns as curiosities. Conceivably the long-horned extinct bison still lived some centuries before this time and had been domesticated in some areas. Coronado reported seeing sheep with long, fifty pound horns near Cibola. This is all rather nebulous, but perhaps further study will be undertaken some day to pursue the possibilities noted in these early accounts.

Mysterious Variations

The strange story of Jacob who altered the color patterns of the flocks by placing objects at watering places must seem to most as far-fetched. Yet Arthur Custance reports studies that show even greater peculiarities in animal behavior.[36] When horned herds of cattle are moved into areas where hornlessness already occurs, a similar hornlessness occurs without interbreeding. This phenomenon has been carefully noted in Europe, Africa, and South America. Swiss cattle moved into Hungary and developed longer horns and legs like the Hungarian cattle, again without interbreeding.

Other cattle moved from the Bavarian Alps to Hungary and developed the longer horns common there, including the lyre-like form, while the skull became narrower. The dramatic changes were probably due to changes of climate and the change in food drawn from the soil, not from interbreeding.

The Sacred Pig

The earliest mention of the pig or swine in the Old Testament is in Deuteronomy 11:7 where it was pronounced unclean to the children of Israel. Although many other creatures were similarly declared by God to be unclean, we may conjecture that one reason for this designation was that the pig had been made into a sacred animal long before this time by heathen nations around the world. This animal was tied in with the worship of the great mother goddess. It is more than a little startling to find life-like clay models of the pig in early Neolithic remains. Curious relics of the memory of the sacred pig have come down to our day.

Four thousand years ago and more the sacred pig was eaten only after a castration ceremony. In jest we sometimes think that the ultimate in feasting is bringing in the roast pig with an apple in its mouth, without realizing the very ancient roots of this practice. The bringing of the boar's head with pomp and dignity at Christmas is a bizarre carryover from the dim past, and it is curious that a well-known English carol is known by that name. The pig festival is still a major event on many Pacific islands today, including Hawaii and Tahiti.

John Cohane stresses the enormous role in prehistory of both the pig and barley.[37] Our language carries the memory of this role, for in older English both pig and barley were called by the same name, that is, "bar." Our word *boar* comes from this origin. At a very ancient time the bar (or the pig) was spread to many parts of the world by navigators. The pig came to the British Isles about 2000 BC. It is very significant that wherever European explorers traveled, the pig was already there, even on islands where no one lived anymore. The pig was found on the Canary Islands, Bermuda, Barbados, Puerto Rico, the Americas, and on the islands of the Pacific.

The bier on which the dead were laid and the board on which religious ceremonies and sacrifices were performed were both also called bar at an early time in England.

Ancestry of the Pig

All domestic pigs are derived from the ancient cross of the Chinese pig with the European boar. Charles Reed identifies the southeastern Asiatic subspecies that is the ancestor of all domestic pigs.[38] Because the oldest remains of domestic pigs are found in the Near East and much of Europe, he tries to imagine herds of these pigs being driven many thousands of miles to the Near East. However, he is unable to picture pigs being driven across vast stretches of desert. We discuss the evidence for widespread transoceanic travel and related trade networks in very ancient times. This is the apparent means of the rapid and extensive spread of domestic animals. And there may well be a simpler explanation—a rapid spread into the world from the plateaus in the Ararat mountain region.

Homer speaks of the companions of Ulysses as trapped in the snare of Circe (our word *cereal*). It is said that Circe mixed barley meal, and that the men were penned up in pig sties—an allusion, dimly understood, to the ancient and sacred roles of barley and the pig.

It is clear that the Island of Borneo was heavily populated in ancient times, particularly in the uplands, which until recently were almost inaccessible. It is not surprising to learn that Borneo was noted for its hordes of wild pigs when European explorers arrived at the island.

Sebastian Englert comments on early travel between islands in the Pacific involving boats up to 80 feet long and carrying more than 60 people.[39] Among the amazing provisions in these double canoes connected by decks was the pig. As a footnote to the subject of the pig, we note that the guinea pig (no relation) was domesticated, along with the llama and alpaca, in pre-pottery Peru as early as about 2500 BC.

OUR CONCLUSIONS

What We Have Learned

No one described the blind leading the blind more vividly than evolutionist Charles Reed:

> We can easily get stuck in a rut, going round and round, reinforcing our preconceived notions by the happy process of talking only to those who agree with us and avoiding any dangerous new thoughts that might expose us to critical comment. Surely this is one of the greatest blocks to the creative process that should infuse the scholarly world. . . . Many conferences are made up only of "accepted" scholars who blandly sweep over the most fundamental questions and plunge on with their "accepted" lines of inquiry.[40]

While we who hold the Bible to be the true framework for ancient history find it easy to point fingers at evolutionists for faulty thinking, we must be careful not to substitute wishful thinking of our own for actual evidence. All sides to difficult problems and issues would be well served if fact is labeled fact and speculation is labeled speculation.

We must not convey the impression that creationists are agreed on every detail about the early world, even though the biblical framework is fully accepted by them. On many questions there is simply inadequate evidence, or we do not know how to read the evidence that lies before our eyes. It is freely granted that many questions may never be satisfactorily answered, nor is there any real need to do so, other than to satisfy a curiosity about the past. Many valuable and interesting insights have been gained from a study of scientific facts, as opposed to conjecture. This is a highly valuable undertaking that shows that the Christian who takes Genesis seriously certainly has nothing to fear from scientific data. We must never confuse explanation imprisoned by evolutionary theory with the observations we can make in the world around us.

As this chapter illustrates, there is no lack of fascinating problems for careful study. When all factors are considered, a biblical framework makes a good fit

Glacial groves on Kellys Island east of Toledo, Ohio, in Lake Erie show the action of large stones trapped in moving ice during the ice age after the flood.

for the data as well as the mysteries that we face. The Bible shows the sequence of a perfect created world, followed by a cursed earth ending in the flood of Noah. After this event new beginnings were made on a partly "healed" earth that remained turbulent for some centuries after the flood, including the rapid rise of the ice age to the north, followed by a tumultuous meltdown. The world we know today was preceded by many dramatic events, mighty acts of God.

What is the Message of this Chapter?

Prominent archaeologists have devoted their professional lives to searching for the origin of agriculture. The above shows how little success has resulted, and we believe the problem has always been the faulty assumptions that governed their search. If we permit the Bible to furnish the framework for investigating the history of domestication, we are rewarded with insights such as those identified throughout this chapter.

For Reflection and Discussion

1. Scientists have spent their professional lives to discover where agriculture began and then gradually developed over long periods of time. Already in

Genesis 1:24–25 livestock is specifically mentioned. What does Genesis 4:1–5, 20 have to say about this matter? May we say that agriculture was one of the many gifts God gave to man from the earliest time? May we say that Noah and his family carried agricultural wisdom with them for the new beginning after the flood? See Genesis 8:1 (livestock) and 9:20 (vineyard).

2. One expert (Charles Reed) concluded after a lifetime of study that agriculture began in the Zagros foothills and mountains of present-day northwest Iran. See a map of the Middle East and locate the mountains surrounding Mount Ararat and the Zagros Mountains. How well does this fit what we know of early life after the flood?

3. May we relate the trillions of fossils world-wide to Noah's flood? If not, what would be an alternative?

4. The lifeblood of the evolutionist is gradualism over millions and billions of years. Creationists look at the plain evidence of catastrophes. One evolutionist stated that he believed in catastrophes, as long as they happen slowly over vast stretches of time. Agree or disagree? Why is the word *gradually* so important for the evolutionist?

5. Discuss this statement by an evolutionist: Dating the past used to be sheer guesswork. Evaluate the fact that the strata of the earth was essentially locked into place 150 years ago, long before radiometric dating—the method that has its own host of problems.

6. What is the value of the so-called Index Fossil? Why do you think these are seldom mentioned any more today?

7. It has been often demonstrated that the appearance of a species and even the structure of a species may change dramatically from a change in the environment. Are such changes variation or evolution (microevolution or macroevolution)?

8. Can you think of problems that would come with believing in very gradual change from one species to another? For example, Charles Darwin was troubled about how an eye could gradually develop and what would be the value of half an eye?

9. Christians are called to be good stewards of the earth and should never contribute to practices that would lead to extinctions of species. What do you think is the motive for wild figures given as science and as fact, for the number of extinctions each day or year? Some extinctions are blamed on man before he was thought to have evolved!

10. What do you think is the logic in calling some fossils "living fossils"? Related to this question is the well-known practice of giving a fossil a new name even

when it cannot be distinguished from living species? Why is this practice followed today?

11. What do you think is the payoff for discovering a "new" fossil that then will include your name in its official designation?

12. What can animals, birds, and fish teach us? See Job 12:7–8: But ask the animals, and they will teach you, or the birds of the air, and they will tell you; . . . or let the fish of the sea inform you.

CHAPTER TEN

THE BIZARRE COURSE OF THE HORSE

THE CONVENTIONAL STORY OF THE HORSE

A Familiar Story

In the fall of 1985 the Government of Canada opened the new world-class Tyrrell Museum of Palaeontology, said to be the largest fossil museum in the world, in the heart of one of the great dinosaur fossil beds of the world at Drumheller, Alberta. There in the great museum, surrounded by a fabulous dinosaur collection, is the centerpiece of the "proof" for evolution—the horse series from little *Eohippus* to modern *Equus* in five easy stages—the century-old illustration endlessly copied and recopied billions of times in four generations of textbooks, and just as endlessly in other media for the education and edification of the masses.

The Horse Series

This so-called series is the claim to fame of Yale paleontologist Othniel Marsh, who discovered fossil bones of the horse and of the *Hyracotherium* (incorrectly and commonly called *Eohippus*) in Wyoming and Nebraska. His reconstruction and arrangement of these fossils are still said to be on display at Yale University, and have been copied with some variation in the sequence in museums everywhere as a prize exhibit of evolutionists "proving" their theory.[1] It has become axiomatic that the horse (*Equus*) evolved in the midwest of the United States, and if fossil *Equus* is found in Africa, or anywhere else, it migrated there from Nebraska or thereabouts. As George Simpson, considered by many to be the dean of paleontologists for an entire generation, expressed it: "The beautiful series of ancient and modern horses displayed in many museums are still the simplest way to convince any open-minded person that evolution is a fact. You can see it with your own eyes."[2]

Wyoming geologist Michael Hager, who works in the heartland of fossil horses, is one of many evolutionists who believes that the fossil record of horses is compelling proof.

> Through study of hundreds of specimens, a very complete picture of the sequence of gradual change in the horse lineage has emerged. The history of the horse family is one of the clearest and most convincing demon- strations of the fact of evolution. A number of fossils, from the oldest known horse to the modern horse can be laid in a series in which adjacent fossils of slightly different age are very difficult to tell apart, but fossils fur- ther apart in the series and in age can readily be separated. From such a series, the fact of gradual change with time becomes evident. Horses lived in North America from the Eocene to the Pleistocene Epochs. Fossil horses from rocks of each succeeding epoch show a gradual increase in size, a reduction of toes from five to one, and a change in tooth pattern.[3]

Evolutionists believe that creationists have no defense against the scientific evidence for the evolution of the horse. For example, Michael Ruse throws down the gauntlet: "It would be nice to see the Creationists take on the question of the horse, which is one of the best documented cases of evolutionary change."[4]

Herbert Wendt is typical of the enthusiasm of evolutionists when he stated that it was in connection with this animal that the Darwinists enjoyed their great- est triumph.[5]

No one has written a more devastating critique on the merits of the parade of so-called human fossils over the past 125 years than William Fix. He believes that though the fossil evidence for the emergence of man from apelike ancestors is a total nothing, "We still have the evidence pertaining to the horse."[6]

This chapter will attempt to separate fact from wishful thinking that sur- rounds the fossil horse. The results are surprising. No attempt will be made here to repeat much of what is said about the supposed evolution of the horse in text- books, but we shall examine and evaluate this very familiar story.

1. WAS *EOHIPPUS* THE ANCESTOR OF THE HORSE?

How it Began

It all began, as many texts recite, with *Eohippus* or *Hyracotherium* in the Eocene Epoch, which begat the *Orohippus* late in the Eocene, which begat the

Mesohippus of the Oligocene, which begat the *Hipparion* of the early Pliocene, which was the father of *Equus*, the modern true horse.[7]

We are indebted to Frank Cousins for providing an English translation of Heribert Nilsson's historical overview of the fossil horse and his evaluation of the so-called family tree of the horse. Nilsson, not a creationist, was a leading scholar of the biological world in the twentieth century. Little but ambitious *Eohippus* was the size of a large rabbit and had the peculiar characteristic of three hind toes and four front toes. Strangely, the modern hyrax or coney, with the same toe pattern today, still hides in the rocks of the Middle East. Its skeleton is just like that of the fossil *Eohippus*. One might ask whether it would not be reasonable to suggest that *Eohippus* is merely a slight variant of the modern hyrax.

One of the arguments for pushing *Eohippus* as the ancestor of the horses was its "primitive" brain as inferred from a fossil brain case. It was recently discovered that this fossil brain case was from an entirely different animal. Study of a genuine *Eohippus* skull revealed a more "advanced" brain than expected, thus adding further problems to constructing a supposed chart of horse evolution.[8]

George Simpson described the first *Eohippus* finds. The great British paleontologist, Richard Owen, first identified the teeth as belonging to a monkey. With more finds of fossils, he gave it the name of *Hyracotherium* and observed that the large size of the eye must have given it a resemblance to the hare and other timid rodents.

Simpson acknowledges that the name of the *Eohippus* (horse-like) should be restored to *Hyracotherium* (rabbit-like), but can't bring himself to do so. Later, Simpson explained the problem this way. The genus *Equus* (domestic and wild horses, asses, zebras) has numerous characters in common that distinguish it from all other living animals. But not a single one of these characteristics, so far as they can be seen in fossils, occurred in *Eohippus*, which nevertheless was the common ancestor of the whole horse family. Something may escape the inattentive reader in this flow of logic.

For a time the five-toed *Phenacodus* from the order of *Condylarthra* was thought to be the ancestor of the horse. The bases on which it was disqualified are interesting and vulnerable: It was too large for the supposed sequence including *Eohippus*; it was too well developed, that is, it was not primitive enough; its existence in time did not allow enough time for it to evolve into *Eohippus*. *Phenacodus* was replaced by *Hyracotherium* (*Eohippus*) as the ancestral horse. This creature was removed from still another order, the *Hyracoidea*, or the hyraxes and coneys, rabbit-like rodents. For more than a century, this animal has been placed at the beginning of horse evolution.[9]

2. Geographic Factors in Horse Evolution

Where the Horses Are

Most of the story of the fossil horse is based on the study of fossil horse collections excavated in western North America, especially Wyoming and neighboring states.

Fossil remains of horses are found on all continents except Antarctica and Australia, but strangely these finds as yet do not play an important role in what is believed to be chiefly a North American story. Thus the supposed evolution of the horse ignores most of this fossil record, which is clearly poor scholarship and a very parochial view.

3. When Horse Toes Went from Five to One

When Less is Better

Similarly, evolutionists believe that the number of toes decreased systematically from five down to one. This notion is so deeply ingrained that some evolutionists lose their ability to count. For example, N. Berrill can hardly contain himself with his feeling of wonder about the evolution of the horse: Little five-toed (actually four in front and three in back) *Eohippus* went step by step with clockwork precision over 60 million years to become the large one-toed horse.[10]

There is a supposed five-toed, but as yet unknown, ancestor. *Eohippus* is caught in a rapid transition between four and three toes. Two of the remaining toes later reduce themselves to barely noticeable splints, after which the modern one-toed horse emerges as the great showpiece of evolutionary development. Because of the imaginary but supposed sequence in horses of five toes to one, there is a pervasive view in paleontology that *five* is primitive and *one* is specialized, that is, more modern. If only we could ask the animals why they went to all the trouble of first evolving five toes when their goal throughout was to become one-toed.

Paleontologists struggle with that assumption. By way of illustration, E. B. Branson gamely states, "Man is primitive in respect to number of toes."[11] We are apparently to look forward to the great unshackling when we will have one highly evolved toe, and presumably also one finger on each hand for pressing important buttons. Imagine the beauty of the piano concert millions of years from now, after man has evolved as far as the horse already has! Man is not all primitive, however, according to Branson. He is somewhat specialized, that is, modern, with respect to teeth, because the supposed ancestral supply of forty-four teeth commonly found in "early" mammals has been reduced to only thirty-two.

We can be certain that the one-toed and three-toed species of the horse are related in the same way the house cat and the lion are related. This has nothing to do with evolution. Perhaps some elements of alleged toe reduction are in part a matter of selective breeding, as well as variety found among related species. Much stranger things are true within the dog species. It is not uncommon today for horse handlers to comment on clear indications of three toes on some of the horses they work with. Instances occur of domestic horses being produced with a small additional toe with complete hoof, and sometimes three or more toes may be present.[12] It is interesting that a 1922 British Museum publication discusses three Shire horses that were three-toed. Other museums display evidence of modern three-toed horses. Perhaps the distinction there is largely one of selective breeding. Julius Caesar rode an extraordinary charger. Each of its hoofs was cloven in five parts. Only he could ride this animal. No one suggests that this horse was changing into a new species.

4. Fossil Horse Associations

Unexpected Companions

Alan Brodrick observed that fossils of the modern true horse are found associated with long extinct forms in East Africa. He can only suggest that animals became extinct at very different times in different places. Sonia Cole expressed amazement that the remains of long-extinct ancestors are found together with completely modern forms in East Africa. Similarly, D. S. Allan and J. B. Delair reported that modern horse remains in Algeria were in deposits of supposed long extinct ancestors from Miocene and Pliocene times, that is, up to 25 million years out of place. The three-toed horse grazed side by side with the one-toed horse, which is embarrassing, to say the least, for the evolutionary sequence of horses illustrated in millions of textbooks.[13]

Or so I thought—until I visited the remarkable fossils gathered around a long extinct water hole in northeastern Nebraska in the summer of 1993. In a video at the visitor's center I heard an evolutionist speak with all the fervor of a snake oil salesman that if anyone did not believe in evolution up to now, this was the place to be. The actual split off from three-toed to one-toed horse did not occur millions of years apart after all. It must have happened at the very moment in time when a volcano exploded in that region, burying all life there in seconds. Three-toed and one-toed horses grazed together and died together here.

5. THE AGE OF HORSE FOSSILS

Dating the Horse

The conventional view of the horse is that older forms first evolved in North America, then roamed there for millions of years. That horse died out suddenly around 10,000 years ago. The modern horse evolved in Asia and was introduced into the Americas around AD 1500 by the Spanish explorers. Little attention has been given to horse chronology elsewhere in the world, except that it must all fit into the story of the horse in North America.

But chronology as usual is a problem. There was an embarrassing find of the horse genus among Miocene fossils in Siwalik beds in Nepal. This discovery showed that this horse was older than its ancestors. Therefore, the age of the rocks was declared to be about 12 million years younger (Pliocene) and the dilemma was solved.[14]

An even greater problem of time was noted in Texas. Remains of the giant horse (*Asinus giganteus*) were excavated in Texas in close association with camel, rodent, elephant, and other mammal remains. These fossils lay directly on top of the red-beds containing dinosaurs. We can hardly agree with evolutionists who would place millions of years between the two layers.

In 1984 Soviet paleontologists discovered eighty-six fossilized tracks with equine (horse) imprints of hooves. The location was in southern Uzbekistan north of the Afghan border. Specialists all confirmed the fact that the sandstone was formed in the Cretaceous period, and dated the prints at 90 million years. This is 25 million years before dinosaurs became extinct, according to evolutionists, and nothing horselike had yet evolved. The choices are to ignore the evidence or to look for a reptile with hooves.

It is not worth the bother of documenting, but it is nevertheless interesting to note that one authority gives the horse 60 million years to evolve, another times it at 50 million years, and still another states 40 million years. Yet most major mammal orders are supposed to have arisen in a short period of time between the Cretaceous period (65 million years ago) and the Eocene epoch, which followed shortly after. Tomorrow we may see another time frame for the supposed evolution of the horse. At any rate, about every 7,500,000 years a new horse genus supposedly emerged.[15] How fast did these changes happen? Steven Stanley informs us that the average change in size and other characteristics was estimated at 0.0001 inch per thousand years![16] This is classic gradualism and must be an embarrassment for even the most rabid defender of the theory. There are many examples of just the opposite "gradually" happening. I have not yet seen a discussion of why species over millions of years gradually grow larger, while other species gradually grow smaller, yet both "prove" evolution.

C. Gilmore evaluated reports for the Smithsonian Institution of fossil horse-like tracks found in the Grand Canyon. In his view, because they were found in Permian strata, that is, about 250 million years old, they could not be regarded as animal-made, but are interesting because they had a "superficial" resemblance to tracks made by horse hoofs. All tracks here pointed in a common direction. The Supai Indians there believed the imprints were tracks made by a band of horses.[17]

Other tracks had been found in the Triassic of Connecticut, supposedly about 200 million years old, which bore a striking resemblance to a horse's hoof, and thus were named *Hoplichnus equus*. Other hoof-like tracks were reported from the New Red Sandstone of Scotland with the same hoof-like shape and showing a distinct pace and uniform alternate progression. The evolutionist, of course, is compelled to ignore or discard such reports, even though the discoveries were made by scientists. The reason is obvious: In evolutionary theory horses had not yet evolved. The prints were up to hundreds of millions of years too old for them. Creationists are not hemmed in by a theory that does not describe the real world.

6. The Tapir/Hyrax/*Chalicotherium* Problems

The Tapir and Hyrax Problem

No one has much at all to say about the very horse-like tapir. Throughout all the supposed millions of years that the horse was developing, its tapir cousin was timidly going along unchanged. For the horse, the fourth front toe supposedly was a very temporary stage in toe reduction. Somehow the tapir and the modern hyrax did not get the message. The tapir is one of a great number of living organisms called "living fossils" as though this is an explanation. Instead, it points to inadequate theoretical constructs. William Scott, one of the great paleontologists of the twentieth century, has little to say about the tapir, except that despite very little fossil evidence, they must have been plentiful throughout the Tertiary period (supposedly the last 60 million years), and that the known history of the tapir family continues to be in a very unsatisfactory state.[18]

The tapir in fact has one of the oddest geographic distributions in the world. It is found in Middle America and in the Malay Peninsula, Borneo, and Sumatra. It belongs to the same order as the horse and the rhinoceros. No one seems to make a point of the fact that the tapir (as well as the hyrax) has four toes on each front foot and three on each hind foot. How do these animals fit into the supposed reduction of toes from five to one?

The *Chalicotherium*

A very strange large creature, the *Chalicotherium*, once lived in America and elsewhere in the world. This animal was supposed to be extinct for many thousands of years. It is assigned to the same order as the horse, but is clawed and bizarre in appearance.[19] Strange reports and sightings of a fierce unknown animal in East Africa have been persistent for many years. One authority from the British National Museum has suggested that this animal may be a late-surviving *Chalicotherium*. Many more horse-like animals were removed from this order because of a slight difference in the hoof. This is a remarkable example of arbitrary misclassification of species.

7. Was Horse Domestication Recent?

Horse Domestication Time Line in Historic Periods

The horse is believed to have been first domesticated in the steppes of the Ukraine. Certainly the reputation of that area for horse breeding came down into historical times. At best, however, we can only guess about when and where the first domestication took place. A traditional date of 4350 BC is often given for the first horse to be domesticated. Setting a date in the past for this event, however, is an invitation to disaster.[20]

Various writers note the following about the introduction of the horse in the Middle East. In 1928 two chariot wheels were found at Kish, a pre-Sumerian city, dated about 3200 BC. In other tombs there, complete four-and two-wheeled chariots were found.

Considerable energy has been spent in attempting to determine the date the horse was introduced into Egypt. Various authorities place this event somewhere between 1800 and 1500 BC. The earliest suggestion of the horse noted in the Old Testament is Genesis 41:43 where the Pharaoh gave Joseph the second chariot and made him ruler over Egypt. In Genesis 46:29 Joseph rode his chariot to meet his father Israel. During the seven years of famine, Joseph exchanged bread for horses among the Egyptians (Genesis 47:17). This would indicate that the horse was already common among the Egyptians at that time. In his blessing to his sons, Israel spoke clearly of the horse and rider in Genesis 49:17. After the death of Israel, Joseph led a great company of chariots and horsemen for the burial of Israel in the cave of Machpelah at Hebron in Canaan. Much later the new king arose who did not know Joseph (Exodus 1:8). This is often taken to be the mysterious Hyksos invaders.

Egyptian chronology is far from secure, and we may expect other estimates about the arrival of the horse in the future.

The Domestic Horse at an Earlier Time

In Europe early cave drawings and rock pictures show both the heavy-boned northern type of horse and the lighter southern type. The pure Arabian horse appears in the rock paintings of Arabia. It is often shown galloping, mounted by a rider holding a spear. A granite boulder found in Arabia illustrates three horsemen with long spears. In east Jordan another rock drawing shows a horseman with a long spear in his right hand. Such finds are difficult to date. Inscriptions in Egypt show the Arabian horse as both ridden and driven.[21] An immense number of rock paintings and engravings have been found in the heart of the Sahara desert at Tassili n'Ajjer. Among both archaic and modern animals depicted are the horse and the camel. The pictures date to a time before the region became a desert.

Sensational finds were reported in Victoria Cavern (England) about 1879. Among the bones of long-extinct animals were fine bone tools, a perfect bone needle, awls, and other tools. One bone, which appeared to be sawed, had the outline of a horse's head engraved on it. The horse is described as hog-maned, which means that humans had clipped the mane and therefore presumably had domesticated the horse far earlier than is conventionally recognized. In the report of the find it was also suggested that only a metal shears of some kind could perform this clipping function. In 1889 scientists discovered horse carvings in the cave at Mas d'Azil, France, that showed horses with nosebands and others with carved head collars and bridles. The discoveries were ignored because these ice age dated finds contradicted everything believed about the domestication of the. Alexander Marshack also reported evidence of domesticated horses in ice age times. A cave engraving in France appears to show a horse with halter and a leather muzzle.[22]

Jacquetta Hawkes discusses rock engravings and occasional rock paintings found in Scandinavia.[23] These include engravings of the horse that are all but identical to Upper Palaeolithic drawings of the horse found in southern Europe and in Africa. The strange conclusion is drawn that the Scandinavian drawings must be tens of thousands of years later than the same drawings farther to the south. Hawkes is forced into this conclusion because Scandinavia was thought to be ice-covered during the time when man was making cave drawings in the south of Europe over a period of many thousands of years. The long chronology Hawkes works with, though commonly accepted, is not justified by the evidence.

More cave paintings are being discovered. A cave at Escoural, Portugal, located in 1963, had paintings covered with a thick layer of flow stone. Among the covered scenes was one featuring the horse.[24] Such finds have helped revise wildly improbable guesses about the amount of time required for stalactites and stalagmites to form.

The fascination that man has always had about the horse is illustrated by the Great Stone now kept in the village of Plessis-le-Fenouiller, France. A large block of quartzite was covered with carvings of cup marks, human footprints, a man's head, and the marks of horse hooves. Across the channel in England is the famous huge White Horse cut in the chalk scarp at Uffington, Berkshire, near an old Iron Age hill fort. This site had sacred cultic significance associated with some kind of deification of the horse. The search for rock engravings is just beginning in many parts of the world. There is a recent report from the valley of the Helmand River in southern Afghanistan of rock engravings. At one site ancient painted pottery was found along with baked animal figurines including bovines and a dubious horse.

Paleontologist H. Martin found several pairs of worn horse teeth at a prehistoric site in southern France.[25] They showed signs of crib biting, exhibited only by captive animals, never by those in the wild. The meaning is clear: ice age hunters had domesticated horses long before it was known in more civilized areas. Horses were corralled, ridden, and kept for meat during ice age times.[26]

Finds Elsewhere

The horse-drawn covered wagon seems as American as apple pie, but even this practice dates far into the past. The Karasuk stone stele engraving in northwestern China shows a horse-drawn covered wagon, which was apparently the home of each small family of a pastoral people in the plains of that region.[27]

Dartmoor, described as a large patch of misplaced tundra just to the northeast of Plymouth, England, is remarkable in being the site of prehistoric remains of all kinds that stretch back an estimated 5,000 years. The special allure to man of this locality was the presence of metal resources, particularly tin. Apparently early miners brought in horses from the continent as beasts of burden for the mining operation. Wild ponies, whose ancestry is far from clear, still live in the Dartmoor area, and they are mentioned in British records as early as AD 1012. In the Somme Valley, France, at an incredible depth of about 200 feet, horse bones were found associated with both trimmed flints and fire. Again, the find does not fit conventional ideas about dating.

As the textbooks have it, the very last of the Neanderthals regretfully gave up the ghost roughly 45,000 years ago, when the world was still filled with many now extinct mammals. The find at Teshik-Tash or Pitted Rock, which lies about 5,000 feet above a valley in Uzbekistan, is enough to make an anthropologist give up his profession. In this cave, about 64 feet deep and 60 feet wide, many typical Neanderthal artifacts were found. Here also a Neanderthal child had been buried, circled by pairs of Siberian mountain goat horns in a ritual pattern. The problem is that the associated fauna bones are all wrong. They are modern forms, e.g., the boar, the mountain goat, horse, leopard, marmot, and others. Not a mastodon or saber-toothed tiger was to be found. One might be understandably reluctant to believe that at this place, not knowing that they had become extinct, Neanderthals continued to live on quietly until modern times. Again conventional chronology is found wanting.

8. How the Horse Came to America (Again)

Finds in North America

Horses in North America that were contemporary with the mastodon elephant were called *Equus fraternus*, but it is important to note that these are not distinguishable in size or in details of form from the present domestic horse. It is universally assumed that the horse was long extinct in the Americas until reintroduced by the Spanish in the sixteenth century. Yet several tantalizing reports raise some questions about the matter.[28]

In the 1930s a curious report emerged from Wisconsin. A horse skull, dated about AD 490 by C14, was found in the Spencer Lake mounds in northwestern Wisconsin. A planted skull had been found in 1928, which, of course, supports the thought that a hoax was involved. Yet another skull was found *in situ* (in its original place) by four reputable archaeologists, and this favors the view that at least some horses survived in America well into the Christian era. Dr. James B. Griffin of the University of Michigan knew the above archaeologists personally. Because of the prevailing view that the horse had been long extinct, however, no follow up of this mystery seems to have been done.

Robert Marx refers to frescoes found in 1957 on two walls of ruined stone buildings in Yucatan, which showed horses grazing, frolicking, and running. Some carried riders. The frescoes are dated many centuries before the Spanish arrived.

Some rock carvings and gravel outlines of horses, sometimes with riders, are unmistakably to be dated after the Spanish arrived in the Americas. In the desert near Blythe, California, is a horse effigy 53 feet long. The Indians living in the area have no tradition about its origin. It is very difficult to assign a date to this figure, and estimates vary widely.

One of the most curious mixes of old and new was reported from the Thunder Bay area in Ontario by T. L. Tanton in 1931. He found a copper spear point with horse and bison bones on a blue clay deposit covered by 40 feet of cross-bedded sand. The mammal bones were considered to be modern species. No satisfactory explanation has been offered, but it is obvious that the horse survived into more recent times than is conventionally stated.[29]

Another "horse problem" occurred in connection with Midland Man (actually a female skeleton) excavated in Texas. The human remains were found with flint tools and the fragment of a horse leg bone. This bone had a series of sharp grooves incised on it, and the cuts were made when the bone was fresh. Also associated with this find were the bones of many long extinct species. The thought was expressed that the geological history here is far more complex than had been realized, but this is far from offering any sort of explanation or rationale of the find.

Another upsetting find was the Tranquility Site in California. Here were found mineralized skeletons of man along with those of extinct horses and camels. There can be no doubt that all remains are of the same age, and the evidence points to a great age for the remains. The problem occurred with the associated artifacts. At most these are a few thousand years old, and there is no way to fit the find into conventional thinking about ancient man in North America. Again we see something radically wrong with ancient chronology.

Gordon Smith reported the find of a fossilized skull of a modern horse (*E. equus*) *in situ*, partly mineralized, in the Borrego Badlands of California. Geologists conventionally date other fossils in the same location at hundreds of thousands of years old. Mammoth bones were in the strata just above and below the horse fossil. The conventional view is that there were no modern horses in America until the Spanish brought them around AD 1500. If the fossil is not ignored, we may assume it will be redated or classified as something other than a modern horse.

Some of the mysterious discoveries above may have an interesting solution. A new dimension to the mystery of horse remains in the Americas before Columbus is offered by Gavin Menzies.[30] He documents early Chinese explorations in the Americas and far beyond. These large fleets included so-called horse ships. Horses brought ashore for exploring regions may well have escaped from their riders. This could explain some of the mysterious modern horse remains discovered that are dated when there were supposedly no horses in the Americas.

Finds in South America

Ferdinand Berthoud in 1881 studied reports of horses found in South America soon after the first Spanish explorers entered the continent. His conclusion was that the horses were found at a time and in places that make it very difficult to believe they could be Spanish related.[31]

Harold Wilkins reported that the figure of a horse appears in a pre-Inca inscription at Tiahuanacu off the shore of Lake Titicaca, Bolivia. If this is verified, the whole assumed sequence of the horse in the Americas, its extinction, and its reintroduction by the Spanish cannot be sustained.

The bones of primitive horses and men were found together in caves on the Straits of Magellan in 1938, and presumably the horse was used for food at the southern tip of South America, though it had been extinct for many millions of years.

In 1933 a cave site in Brazil yielded Confins Man, fully modern but not Indian, that was dated at only a few thousand years in age. As we shall see, this dating defies every notion of dating ancient man. In the layers above the skeleton were more than six feet of alluvial soil and a layer of stalagmitic deposit material. Above *Homo sapiens* (modern man) were the fossils of the horse, giant sloth, and the mastodon.

EVOLUTIONISTS FLAY THE HORSE STORY

The Rest of the Horse Story

But there is more to come. Does the rest of the beautifully drawn and illustrated horse series survive the demise of *Eohippus*? George Simpson states that, contrary to his statement early in this chapter, the horse series we see in textbooks is a serious mistake, caused by both inadequate evidence and by superficial and erroneous methods of study. False conclusions were accepted and endlessly repeated to others down to the present day. There was no steady increase in size as illustrated, there was no systematic reduction of toes to one as illustrated, and the same is true of all other such illustrated changes. The diagram is not only oversimplified, but also essentially false. It never happened, Simpson concluded: "The continuous transformation of *Hyracotherium* (that is, *Eohippus*) into *Equus* (the modern horse), so dear to the hearts of generations of textbook writers, never happened in nature."[32]

The Swedish scholar Heribert Nilsson,[33] though a prominent defender of evolution, had no patience with those who promoted *Eohippus* as the ancestor of the horse. He concluded that *Eohippus* was a rabbit-like hyrax, and had nothing to do with the horse: "Horses they are not!"

Generations of students were and are taught that the horse series is factual. Although some noted evolutionists deny any validity to the false sequence, the "impressive" diagram continues to be used. All that is needed is a pointer, and then one can say that sometimes evolution went this way, and sometimes it went that way. Straight-line evolution is out, and zigzag is in. More elegantly, the process is described as mosaic evolution. Curator D. Raup, Field Museum, Chicago, an evolutionist, of course, sums it up very gently: "some of the classic cases of Darwinian change in the fossil record, such as the evolution of the horse in North America, have had to be discarded or modified as a result of more detailed information."[34]

OUR CONCLUSIONS

What We Have Learned

With the quickness of the cat, the evolutionist always lands on his feet with a new *ad hoc* explanation, regardless of what the evidence actually indicates. But what have we discovered in examining the literature about the fossil horse? We have examined above eight fictions surrounding the supposed evolution of the horse in order to beef up a faltering case for evolution.

- *Eohippus* was <u>not</u> the horse ancestor.

    The misnamed *Eohippus* is *Hyracotherium*, a form of rabbit/hyrax still thriving today unchanged. It is unrelated to the horse. We do not know if anyone ever told some evolutionists that *Eohippus* never had five toes as claimed.

- Geographic distribution of fossil horses is ignored.

    The fossil record of the horse from other continents is largely ignored. The supposed evolution of the horse is based on faulty assumptions regarding Wyoming fossils. Paralyzing assumptions about the fossil horse record have caused many distortions of the actual record.

- Fantasies about toe reduction in the horses

    The three-toed horse should have been long extinct with the emergence of the one-toed horse. The whole presumed sequence of five toes down to one toe has fallen on hard times. No five-toed species is known; the three- and four-toed ancestor belongs to the rabbit family and to the "primitive" tapirs (who still thrive today); and no two-toed variety has ever been reported. Thus we have had only three-toed and one-toed forms, and they lived at the same time.

- Can remote ancestors live with distant descendants?

    When the belief in gradual toe reduction was invented, these evolutionists did not know that three-toed and one-toed species lived together in various parts of the world. Despite the shattering discovery that remote ancestors lived alongside distant descendants, evolutionists gamely stick to their story and pretend that the evidence does not matter. Evolutionists hold that the horse had to be an evolved sequence over many millions of years, but they ignore other similar differences in the dog, cat, deer, monkey/ape, and other families, where species of all sizes and other differing characteristics live at the same time. To our knowledge no one claims that the pussy cat is the remote ancestor of the lion.

- Fossil horse dates are full of contradictions.

    We have documented a few of the many absurdities about the supposed age of different discovered horse fossil remains. These discoveries speak dramatically for themselves. The chronology of the evolutionist is pure fantasy.

- The order that includes the horse has serious problems.

    The order that includes the horse includes strange bedfellows. Species that ought to be included, such as the so-called South American "false" horse are excluded, and other species, such as the *Eohippus* are falsely included. Some believe the hippopotamus would fit better into the order than the rhino, which is included. No one quite knows what to do about the tapir in this order, whose toes are hopelessly "primitive."

- Evidence of early horse domestication is ignored.

Evidence for horse domestication during the ice age is ignored. The assumption is that early man, emerging out of the ape world, was not intelligent enough for this achievement. The evidence, however, speaks convincingly for itself.

- Accounts of how the modern horse came to America are in error.

There is much evidence of true horses in the Americas before the arrival of the Spanish, but this evidence is ignored.

A Final Thought

Grossly uninformed evolutionists still point to the horse sequence as a compelling proof for their theory. Over the past 150 years evolutionists have announced many great triumphs that prove their theory. Among those of highest rank were Neanderthal Man, Piltdown Man, the speckled moth, and the horse series—a pitiful array. Neanderthal is simply a variety of modern man, *Homo sapiens*; Piltdown was an outrageous fraud; the moth story was faked at no less a place than Oxford University; and even evolutionists are embarrassed by a horse sequence that never happened.

The above eight fictions on the supposed evolution of the horse show vividly how wishful thinking replaced the scientific method in attempts to support the theory of evolution. In addition, we have learned that though some prominent evolutionists have openly stated that the textbook evolution of the horse never happened, this false story long ago developed a life of its own, and continues to be displayed in textbooks and museums as proof of the "fact" of evolution. We need to understand that in our age those in power hold that certain beliefs, such as evolution, are so important that telling lies in support of the theory may be strongly defended! Those who point out these lies may expect to be violently attacked, and truth is swept aside. The story of the horse illustrates how many evolutionists are incapable of embarrassment.

For Discussion and Reflection

1. One author has stated that one might demolish all the arguments supporting evolution, but we still have the horse! What is your experience in texts and other media about so-called horse evolution?

2. It is no exaggeration to say that illustrations of the horse series from tiny to large have been published many billions of times. As this chapter points out, the horse series has not fared well in recent years. Even distinguished evolutionists have stated that the series has never existed anywhere on the earth.

Why was this so-called proof of evolution prominently featured by evolutionists for such a long period of time? Is this another example of the need to believe?

3. *Eohippus*, the Dawn Horse, was presented as the ancestor of all the horses, except that evolutionists still required an unknown five-toed ancestor of *Eohippus*. *Eohippus* already possessed four-toed front legs and three-toed rear legs. Its name has become *Hyracotherium* because it is now recognized as identical to the modern rabbit-like hyrax. It has no relationship to horse species at all. Do you think the hyrax will some day evolve down or up to a one-toed animal?

4. Why do you think museums and texts still picture the fossil horse series when leading evolutionists admit that the series is not true?

5. Fossil discoveries show that one- and three-toed horses, as well as other horse species, lived at the same time in the past. Evaluate the argument that such discoveries fix the exact time that horses evolved into one-toed animals.

6. Could one line up the skeletons of other related creatures from small to large, such as cat, dog, deer, as "proof" of evolution? Why or why not?

7. The small-to-large belief is a companion to the large-to-small belief. Do you think evolution theory can predict which direction the fossil species would evolve? Do you know of *any* predictive power that the theory of evolution is able to demonstrate?

CHAPTER ELEVEN

THE STRANGE STORY OF THE "FALSE" SOUTH AMERICAN HORSES

WHEN SCIENTISTS STUDY FOSSILS

The literature of paleontology, the inexact science of fossils, is full of strange stories, but we need to remember that we are really speaking of two fields of knowledge, not one. A good parallel is descriptive and neutral *physical* geology versus *historical* geology, which is an interpretation of the evidence within the boundaries of a theoretical framework, almost exclusively that of evolution.

One may have the greatest admiration for those who extract and reconstruct fragments and other pieces of fossil bone from the dirt and rock. We can be very understanding when we read about the wrong head placed on the fossil skeleton of a dinosaur by realizing the condition of the material boxed up in the field for restoration and study.

It is only when the interpretation begins with mountains of speculation based on molehills of evidence that one has every right to be skeptical. Paleontologists are not famous for distinguishing between the two, and many scholarly battles are fought between the various "armed camps" of experts.

Although the creationist may sharply and categorically disagree with the evolutionary assumptions of a paleontologist, this discussion in no way reflects on the fascinating but often excruciatingly difficult work of that field of study. Collecting and restoring fossils is not evolution. The interpretation of the evidence, as we see especially in the case of horse and human fossils, most often is.

The Horse-like Creatures

The above introduction is background for perhaps the strangest story of all in interpreting fossil bones. Meet the very horse-like Litopterna, found only in South America, and the Perissodactyla, which include the horses, found on all continents except Australia and Antarctica. Originally the Litopterna fossils were classified among the horses.[1] As time went on, however, not all experts agreed.

Though we need to use some technical names in this chapter, you may freely pass over them by thinking "horse" when you see the word *Perissodactyla* and "false South American horse" whenever you see the word *Litopterna*.

The Basics of Classification

In the process of examining the decision to remove the Litopterna from the order containing the horses, we gain a rather illuminating lesson into the basics of taxonomy—the elements that go into the sometimes agonizing decisions about placing a woefully incomplete or puzzling fossil into this or that category.

The Litopterna Revealed Serious Problems

Here are several horrors the scientists had to contend with. They noted that problems of dating abound everywhere. This always means just one thing. All the evidence clearly showed a sudden catastrophic destruction, yet evolution demands a slow, gradual time frame over many millions of years. It is easy to understand the stress this caused a scientist who attempted to describe honestly what he observed within that framework.

Large numbers of fossil horse-like bones were found in South America that were one-toed. The problem for the evolutionist was that the bones were found in strata clearly much too early for the modern horse to be present. What does one do when irresistible bones are found in immovable but wrong strata? The solution was amazingly simple. The fossils were called "false horses," placed into a separate order, and the problem was solved.[2]

What is Taxonomy?

For the benefit of the reader with little or no background on taxonomy in zoology, the following somewhat oversimplified points should be noted as we examine some aspects of the false-horse story. Every textbook in biology and or zoology tells us the following. The animal kingdom is made up of twenty-two phyla, of which the last one, *Chordata* (spinal cord), includes the horses. Of the four great classes of *Chordata* (amphibians, reptiles, birds, and mammals), the latter includes the horses. Mammals then are divided up into about thirty-four orders that classify about 4,400 living species and numerous fossil forms. Orders are quite broad. For example, the primate order includes every kind in all the world of lemur, monkey, ape, gorilla, and even man, following evolutionary thinking.

In this chapter we focus on two of the thrity-four orders: the extinct Litopterna (meaning "smooth heal") and the Perissodactyla (meaning "odd fingered"). The Litopterna consist of three very different families. The Perissodactyla, more familiar, are made up of ten families and includes a remarkable span of creatures, such as horses, asses, zebras, tapirs, and rhinoceroses. Bizarre extinct creatures are also included. By definition, the horse is more closely related to any other creature within this order, such as the rhinoceros, than to any other animal in any other order.[3]

How Classification is Determined

The key to classification is phylogeny or the supposed evolutionary history or ancestry of a group, such as small to large, more primitive to more modern structure, and the like. Some of the beliefs about evolutionary development are examined below.

Why the Litopterna were placed into a separate order

That the litopterns presented a problem in classification is unquestioned. Made up of very different kinds of creatures, one group seemed very camel-like while another is sometimes described as out-horsing the horse. Two paleontologists in Argentina, Dr. F. Ameghino, with his brother, Carlos, are described by William Scott as providing services to paleontology that are quite inestimable. They discovered the formations of unique South American fossils and arranged them in chronological order. F. Ameghino, who first distinguished and named the order of Litopterna, concluded that the Litopterna belonged with the Perissodactyla, that is, with the horses. Even Charles Darwin, when he discovered the camel-like, least horse-like Macrauchenia in the Pampas of Argentina, declared unequivocally that it belonged to the order now called the Perissodactyla. What more would he have said if he had found the horse-like kind?[4]

The Horse and Litopterna Part Company

In 1910, however, Scott removed those fossils from any connection with the horses.[5] He placed the Litopterna into a far-removed and separate order. This decision on classification, supposedly highly controversial and very much open to debate, has remained a closed question up to the present.

How the Classification Game is Played

As we observed earlier, the Litopterna exhibited unwelcome traits. It seemed to appear too early in the fossil record, and the one-toed species also seemed to appear in the fossil record too soon.

Scott observed that the order of the Litopterna has been the subject of much debate because of supposed convergence. And what is convergence? It is a device in taxonomy that declares that though two animals may closely resemble one another (for example, the horse and the Litopterna), their ancestors did not. We must add another term—homology. When it is convenient, the slightest resemblance between two fossils proves common ancestry. Thus the evolutionist cannot be cornered.

In convergence, likenesses are merely coincidental and may safely be ignored. This device, innocent of any supporting evidence, has gotten many a taxonomist out of many a sticky wicket. As we have observed, little interest is shown today in taking a second look at how the litopterns are classified. George Simpson notes the striking similarity between horse and Litopterna:

> In spite of the various collocations of the typologists, horses are not so convergent toward any other living group that a modern taxonomist would be likely to mistake their homologies. There is, however, an extinct South American ungulate, Thoatherium, between which and horses there is strong convergence, notably in the fully one-toed feet.[6]

If we get past the jargon of this paragraph, Simpson is saying that one might argue that Litopterns and horses ought to be in the same order, not in two. Let us examine the situation more closely.

How are Litopterns Described?

They are very *like* a horse! It is interesting and instructive to see how paleontologists describe the litopterns. The statements below are edited so as to focus only on the expressions of various authors on how the horse and the litopterns were alike. An important scholar and authority, Scott makes these statements:

> The teeth and skeleton of the modern horses are extremely characteristic and unlike those of any other family, except for one group of the South American Litopterna.

> . . . the incisors are arranged as in the horses.

> The shape of the hoofs and the whole appearance of the feet are most surprisingly like that of the three-toed horses.

> The extraordinarily interesting (Litopterna) with completely monodactyl (one-toed) feet.

> Limb-bones, approximating much more closely to the proportions seen in the horses.

> The head would seem to have had some resemblance to that of a small horse.

Thoatherium from Argentina is the most completely monodactyl mammal known, surpassing even the modern horse in the complete reduction of the splint-bones.[7]

Earlier in this century, the litopterns are described thus:[8]

The Litopterna, exclusively in South America, are perhaps more nearly akin to the Perissodactyla (that is, to the horse).

The Litopterna show a curious parallelism to the equine line; the feet are very like those of Hipparian (a horse).

The Litopterna have notably horse-like fore and hind limbs.

The Macrauchenids of the Litopterna, whose record with regard to high crowned teeth, almost exactly parallel that of the horses.

A three-toed form of the litopterns, Diadiaphorus, lived alongside Thoatherium, which seems more horselike than any true horse, for it was single-toed with splints more reduced than those of modern equids.

The Proterotheres were the "horses" among the South American ungulates. They never became very large, but some of them evolved in ways that were remarkably similar to horses, especially in the adaptations of the feet for running.

In a way the litopterns are easy for us to comprehend, for they are more directly comparable to the hoofed mammals with which we are familiar. To put it another way, there were close parallelisms that make the litopterns seem to us like reasonably orthodox hoofed mammals.

In these Litopterna the skull is just as in the horses; the incisor teeth, the cheek teeth paralleled to some degree the horses of the same age, and a molarization of the premolars as in the horses.

The hind feet were especially horse-like; the ankle bone was similar to that of horses.

It seems reasonable to suppose, therefore, that the habits and the mode of life of these Litopterna were similar to those of middle Tertiary horses in North America.

How are Litopterns Described?

They are very *unlike* a horse! After reading all of the above, we are quite nonplussed to find that things are not at all as they seem. The same paleontologists show another side of the coin:[9]

It is totally different from them (the horses) in dentition. The dentition and the rest (except limbs) of the skeleton show, however, that instead of kinship, parallel evolution took place.

The feet of three-toed forms were strikingly (if only superficially) similar to those of horses.

Thoatherium is said to be more horse-like than true horses. This pseudo-horse was, however, comparatively unprogressive in other respects, for the cheek teeth were low crowned, and the carpus was poorly adapted for monodactyl running.

We leave the final comments to Scott, the authority who "removed" the Litopterna from the Perissodactyls and placed them into a separate order:

> The horse-like Litopterna are a remarkable instance of convergence, for with all their resemblances, they were not remotely related to the true horses ... [10] [Comment: This is an amazing statement because Scott is the one who removed the Litopterna from the Perissodactyls.]

The members of this family (the Litopterna) are remarkable for the many and *deceptive* resemblances to the horses which they display.

The Literature on the Litopterns

If indeed the litopterns are a most remarkable case of convergence, that is, the animals are very similar, but their ancestors were not, one would suppose that such a case would attract widespread attention in the scholarly world and a large number of studies. However, such is far from the case. An extensive computer search of a number of scientific data bases brought forth very little—three journal titles in English and four in Spanish—but none of these were of any substance in dealing with the theoretical issues supposedly involved. Surprisingly, there does not appear to be any book in English or Spanish devoted entirely to the litopterns, nor does there seem to be a book about the horse and its apparent twin, the litopterns. Another extensive computer search more recently showed no improvement.

The Non-treatment of an Unwanted Order

The litopterns are touched on in some expected places. P. S. Martin treats Pleistocene extinctions in *Pleistocene Extinctions: The Search for a Cause*.[11] This Yale University publication promised slight to moderate coverage of South America. Litopterna is misspelled as "Liptoterna" several times in the text, and only the more camel-like varieties are briefly noted. The author failed to list this order in the index. Similarly, Bernhard Peyer's authoritative reference on animal dentition, published by the University of Chicago, carries the same misspelling of Litopterna, and at least one of the brief references to this order is omitted from the index.

It seems more than passing strange that there is a great curtain of silence around this problem described to be "of great significance for the philosophy of evolution, full of far-reaching consequences, and supposedly the subject of much debate."[12] Is the subject too embarrassing to treat?

A Lesson in Taxonomy

On the surface it would appear that a good deal of soul-searching by lead-ing experts went into the decision to remove the litopterns from among the horses. It is important to let paleontologists describe how decisions are made about diffi-cult classification problems. Certainly the average reader at this point would not have a clear picture on why the separation took place.

Just how does the taxonomist do the extraordinarily difficult task of classi-fying animals when part of the animal resembles one kind of creature while other features are much like a completely different kind? Our first insight comes from George Simpson in the introduction to his pioneering book on animal taxonomy, where he quotes A. J. Cain: "Is it not extraordinary that young taxonomists are trained like performing monkeys, almost wholly by imitation, and that in only the rarest cases are they given any instruction in taxonomic theory?"[13]

Ad hoc Fragile Beliefs

The following material illustrates the state of such taxonomic theory as it applies to the horse. It will quickly become apparent that we are dealing with quite fragile beliefs used in an arbitrary manner with respect to the horse at least, rather than demonstrable principles, and that one belief may contradict another, thus forcing an arbitrary choice in the direction chosen by the taxonomist. Here are some of the beliefs of the taxonomist:

The splitters versus lumpers beliefs. The first problem that should be men-tioned is the almost legendary one of the lumpers versus the splitters. Lumpers focus on the remarkable degree of variety found within species and are sparing about identifying "new" species. Splitters tend to magnify minor differences among specimens and clutter the field with an inordinate number of new species. Most of the outraged comments on the subject appear to come from the lumpers, and no satisfactory solution is in sight, especially with the fossil record that often must deal with fragmentary remains. One of the best examples of the work of the splitters is the human fossil record where every new fragment is hailed before the media as a sensational new breakthrough, and of course a new species. William Fix has observed that such news conferences tend to be held just before new funding is requested from foundations and other donors.[14]

The embryo belief. It is believed that embryos represent or repeat ancestral adult structure, thus showing the evolutionary stages that a species went through.[15] It is well known that this strange belief, known as recapitulation theory, still per-sists in high places among evolutionists, even though leading evolutionists dis-carded the idea long ago.

The subjectivity belief. Simpson states that it is virtually impossible to be completely nonarbitrary for any taxonomy other than species, and frequently also for species, but he declares that there is nothing wrong with being arbitrary in the practice of an art, including the art of classification.

The evolutionary taxonomy belief. How does one decide which resemblances among animals count and which ones do not? According to Simpson, the only way is by the use of evolutionary taxonomy. The taxonomist creates a canvas of who the ancestor was, and the changes that must have occurred through the ages for the modern form to evolve. Whatever fits this totally speculative blueprint is taken as a homology—a resemblance that counts. For reasons not made clear, some creatures apparently suffer from arrested development. Thus the rhinoceros, and still more the tapir, are more "primitive" with respect to their toe development, or closer to the supposed ancestor than the horse. If little or no change is evident between modern forms and ancient fossils, such occurrences are tagged as living fossils—an undefined evasion of the issue.

The anatomy belief. Scott stated that in proportions and general appearance, the three families have little in common (tapirs, rhinoceroses, horses); but so alike in anatomy are they that placing all in the same order is obviously the only course to take.[16]

The teeth versus skeleton belief. The classification adopted will depend upon the relative importance given to the teeth, on the one hand, and the skeleton, on the other. For example, we are told that the dentition of the Litopterna shows that these extinct mammals of the South American Tertiary cannot be related to the horses, despite notably horse-like fore and hind limbs.

The tooth belief. Taxonomists believe mammals originally had forty-four teeth but through specialization over a long period of time, the number was reduced.

The toe beliefs. The number of digits on each foot of those belonging to the order of perissodactyls is usually odd, one or three, but may be four, as in the front foot of the tapir. No five-toed perissodactyl has yet been found. There is the belief in a principle that many animals originally possessed five digits—a primitive stage—and that through evolutionary processes the number was reduced, for example, to a perfect one in the case of the horse.

Time belief. There is a belief that Hipparion (three-toed horse) became extinct three million years ago and was replaced about two million years ago by *Equus*, a true horse. Donald Johanson relates with some relish that his rivals, the Leakeys, were in trouble trying to establish a date of 3 million years for a "human" fossil find when it was found with 2-million year old *Equus* teeth. "You can't turn a 3 million year old Hipparion into a 2 million year old *Equus* just because you want to."[17] Curiously, this belief and the famous pig studies of Basil Cooke set aside any K-Ar dates (a popular radiometric test for age) in Africa in establishing dates for the so-called human fossils found there.

Uniformitarian belief. Wright comments on a gradualistic view of evolutionary change, for example, that in the lineage of horses from Eohippus to a modern form, the average change in size was .0001 inch per thousand years.[18]

Environmental beliefs. South American ungulates (the litopterns) illustrate very nicely the close correlation between animals and their environments, and indicates how similar environmental conditions will lead, by genetic processes, to the evolution of remarkably similar animals, quite unrelated except through their very remote ancestors.[19]

Geography beliefs. Environmental beliefs are offset by beliefs related to geography. With or without evidence, taxonomists invoke land bridges, mountains, or other barriers, drowned lands forming an island, such as South America at times, in order to attempt to explain the fossil record. Paleontologists believe they have located the point of origin where a type of animal evolved, and that similar creatures found in other parts of the world migrated from the place of origin. There is considerable aversion to suggesting two or more evolutions of the same species, despite the environment belief noted above. We are familiar with such declarations, for example, that man evolved in East Africa, the horse originated in Nebraska or Wyoming, and the litopterns originated in South America. Such statements are of course based on where the fossils first were found along with estimates of their age. Scott observed that geography is difficult to apply because of the uncertainty concerning the manner in which the evolutionary process operates.[20] While it is easy to announce that many genera and families of mammals traveled in both directions across the Bering Strait, Scott speaks of unexplained facts and the difficult to understand absence of crucial fossils to support such beliefs.

Where Is the Scientific Method?

The above by no means exhausts the list of beliefs that taxonomists use for explanation and classification. What is missing from these principles is any logical, scientific method of choosing among them. Taxonomists sometimes acknowledge the hazards of their profession, though such uncertainty is seldom alluded to in texts and popular literature on the subject. Scott rightly says that ancestor charting is the most inexact of the sciences because it has such a large subjective element and depends so much upon the judgment of the individual naturalist. It is for this reason that paleontologists differ so often and so radically in the answers they give to very basic questions. Their fundamental preconceptions are so irreconcilable that one regards as quite impossible what another believes to be usual and normal.

There are Many Difficulties

It seems appropriate to give several examples of the kinds of difficulties that taxonomists have identified in their work.

Horse and rhinoceros. They are evidently related to some degree. On the other hand horses differ from rhinoceros quite strikingly in having only one toe and in numerous other respects, such as thinner and more hairy skins. In those respects and other characteristics, rhinoceros resemble hippopotami more than they do the horse.[21]

Missing fossils. There is the problem of expected but missing fossils. Tapirs lived in the Pleistocene along the Pacific and in the eastern forests, but apparently not on the Great Plains. There is not a half-way complete skeleton between the Eocene and Pleistocene, yet they must have been plentiful to persist to the present.

Strata that do not match. Scott points to the difficulty of relating fossils from South America to other regions. For example, he notes that no Paleocene deposits, marine or continental, have been found in any part of South America, and that most of the Tertiary formations in South America are little more than names. Only one formation, the Patagonian, is thought to relate directly with a formation in the northern hemisphere. The neat drawings of the geologic column in the textbooks do not describe the real world.

Dentition dilemmas. Tooth structure has received much comment. Bernhard Peyer observed that the significance of tooth characteristics in determining relations among animals has been differently interpreted in the course of time.[22] Many assignments or groupings based on dentitional features have proved to be untenable. Peyer then gives examples where animals with similar teeth are placed in different orders, while some with different dentition go into the same order. Limbs, too, may follow the same pattern in classification. He cautions that the obviously great significance of tooth features should not lead to their one-sided over-evaluation. It should always be remembered that dentitional characteristics represent only one aspect of an animal's structure. Scott chimes in that too exclusive a dependence upon the dentition has more than once led to unfortunate error. This is all well and good, but there is no scientific basis for using or ignoring tooth structure to classify a mammal. We discover then that the choice made is arbitrary and subjective.

Scott and the Unwanted Litopterna

As we observed earlier, William Scott described the classification of the Litopterna as a problem of great significance for the philosophy of evolution, and his solution had far-reaching consequences.[23] Thus it came as a surprise to find his pivotal monograph in storage in one of the great research libraries of the world, the University of Michigan, and to note that no one had ever opened it or checked it out during the seventy-seven years it lay in this library after it was received as a gift from Princeton University. We know this is so, because we had to cut many pages apart to learn what Scott actually had to say. Further, the writer originally assigned to write the monograph had died, and the actual author, Scott, devoted almost all his time in the La Plata Museum in Argentina studying other groups of Santa Cruz fossils. Finally, Scott stated that his analysis was done without any detailed stratigraphical knowledge of the formations or any record of where the specimens were located in the strata. Such comments are hardly the stuff for such a "pivotal" work.

A Hasty Action

Scott devoted only several pages to explain why he concluded that the Litopterna should be placed into a separate order. He listed some differences of dentition and skeletal remains. One could ask why he did not do the same with the tapir, rhinoceros, and other forms in the Perissodactyla order that differ far more radically. He acknowledged that other taxonomists differ:

It is not surprising that students of the Litopterna should have reached opposite conclusions regarding the systematic position of the group, for this is merely another case of the oft recurring problem, as to how far certain resemblances are offset by differences of structure.

It is no secret why Scott's very arbitrary conclusion was quickly and generally accepted without objection or even discussion, at least in print.

Other Problems and Questions

From the foregoing discussion, it seems reasonable to offer some comment about the horse and the litoptern—some of which has already been said by others in other contexts:

The horse story is still based on old uniformitarian principles now repudiated by leading evolutionists, but if this outworn concept is replaced by punctuated equilibrium, which operates in such a way as to leave no fossils, then in terms of paleontology the new thinking rests on negative or missing evidence, which Darwin and many other evolutionists declared worthless.[24] The litopterns were removed from the order because they upset a now repudiated view of horse evolution.

There is a troubling dishonesty in the way the showpiece of evolution is displayed in museums and in texts. Many prominent authorities in evolution have declared the horse series to be fictional. No attempt is made to sort fact from speculation, and contradictory evidence is ignored.

Charles Hapgood is one of numerous authors who has documented the obvious fact that the great fossil beds of North and South America, Siberia, Africa,

India, and elsewhere, were the result of catastrophic events, much more recent than conventional thinking permits, yet interpretation invariably follows the old uniformitarian path of slow gradual change over many millions of years.[25]

Darwin observed in South America that extinct animal bones, when heated in the flame of a spirit-lamp, exhaled a very strong animal odor and even burned. Yet he could not draw the obvious conclusion, because in his thinking there would be no time for present-day species to evolve. Note how catastrophism is treated in Alaska by David Hopkins. The author stated that unfortunately most of the Alaskan Pleistocene vertebrate material now in museums have no known stratigraphic context.[26] Speculations about age are therefore based entirely upon the morphology of the skeletal material and upon assumptions of probable trends with time of changes in such critical dimensions as horn width and tooth size. We may add that such guesses are built on now discarded uniformitarian notions.

While granting the great difficulties inherent in developing a taxonomy for animals, the arbitrary and subjective path taken in orders such as those including the horse and the litopterns should be fully confessed. It is easy to document such excesses where the grossest differences are shrugged off as irrelevant, while the most minute differences are taken as crucial. A perissodactyl may have claws (the Chalicothere) but the litoptern described as more horse-like than the horse is not even a perissodactyl when among other things it has a smooth spot on the heel.

One kiss of death for the litopterns in South America was that the three-toed form seemingly lived after the one-toed form became extinct, according to interpretations of the different strata in which the fossils were found. This would have by itself made the assumed horse sequence untenable.

Contradictory evidence is routinely discarded to make life possible for speculative animal family trees. As William Scheele observed, it takes an expert to decide if horse bones are of recent vintage or ten thousand years old.[27] In Africa, bones are collected that can scarcely be distinguished from fossil forms that are known from late Pliocene times, that is, up to at least several million years old according to the conventional time scale. Skeletal material has always been endangered, that is, destroyed, if it looks too modern. In many ways we can see that evolution is a theory that prohibits thinking rather than stimulates it.

An isolated island is invoked for South America at times in an attempt to explain horse distribution or lack of it. Yet during the same periods other animals such as the giant sloth and the armored glyptodont—both extremely slow-moving creatures—apparently traveled freely between the two continents. The horse people do not talk with the sloth people. The possibility of land bridges between South America and Africa, much more recent than the conventional geological time table would allow, ought to be re-examined. This concept was argued by F.

Ameghino, who was the discoverer and namer of the litopterns among his many other achievements in paleontology.[28]

It seems reasonable to conclude that if the horse story were done over from scratch by evolutionists today with no preconceptions other than their commitment to evolution, it would be very different than the fraudulent illustrations we are confronted with *ad nauseam*. Despite the tapir, which no one wants to deal with seriously, little *Hyracotherium* would not be misnamed *Eohippus*, and it would be placed with the hyraxes and conies where it has always belonged. The litopterns would be placed with the horses where they belong, but of course another equally speculative horse ancestry would have been created by imaginative evolutionists, innocent of any substance. The creationist can see relationship among creatures, but the evolutionist can see relationship only in terms of supposed ancestry.

The gross subjectivity of taxonomy is well illustrated by the horse/litoptern story and the many contradictory statements of taxonomists about the skeletal parts, ancestral forms, and the like. The statements given on why the litopterns are not horses are vague and unconvincing. It borders on the ludicrous that an authority would accuse the litopterns of practicing deception for looking so horselike. If the reasons are so obvious that litopterns are not horses, it ought to be possible to communicate this clearly in the professional literature. A careful examination of reconstructions and illustrations of the fossils is interesting. *Eohippus*, closely related to the hyrax or coney, and which bears no resemblance at all to the horse, is made to appear precisely like a miniature horse, while the litoptern branches, so very much horse-like, are made to appear more like antelopes. Tails are made to look quite hairless, when structurally they are no different from that of the horse. This follows the same biased path as that taken by those studying supposed human evolution. Here ape fossils are made to look as human as possible. Fully human fossils are drawn to appear as ape-like as possible. This exercise in futility and self-deception makes the transition from ape to man more believable to the evolutionist.

OUR CONCLUSION

In this chapter we have dealt with only a tiny aspect of the horse/false horse story. Much more remains to be sifted. But we must point out again that evolutionists do not read their own literature. Michael Ruse believed that he had made a crushing statement against those who believe in our young, created world when he said, "It would be nice to see the Creationists take on the question of the horse, which is one of the best documented cases of evolutionary change."[29] We suggest that Ruse ought to begin reading the evaluations of the horse written by his own evolutionist colleagues. In this brief examination of the Litopterna and the horse, we must say again that we have found nothing to contradict the statement early in this chapter that young taxonomists are trained like performing monkeys.

In her book *Out of Africa* Isak Dinesen gives a remarkable description of the behavior of natives in her area:

> I learned that the effect of a piece of news was many times magnified when it was imparted in writing . . . but if a mistake was made in writing, which was often the case, as the Scribes were ignorant people, they would insist on construing it into some sense, they might wonder over it and discuss it, but they would believe the most absurd things rather than find fault with the written word.[30]

Déjà vu! Despite the fictional nature of horse and false horse evolution, believers cannot let go of the idea because every text and countless other materials endlessly and reassuringly repeat the fictions. Students copy down and regurgitate this fantasy endlessly as though it has some basis in the real world.

For Discussion and Reflection

1. When Charles Darwin saw certain fossil remains in Argentina, he said they were horses, that is, members of the order Perissodactyla. Long afterward a paleontologist, who stated he was not qualified, placed these fossils into a separate order. Why was there no debate or outcry from evolutionists?

2. A major study of the "false horses", sponsored by a distinguished university, was described as controversial and open to debate, Why do you suppose its report was left untouched for more than seventy-five years by paleontologists at another major university?

3. Evolutionist paleontologists believe that the fossil record shows five toes are primitive and one toe is a highly evolved creature. One geologist stated that humans are primitive in this respect since they still possess five digits. Agree or disagree? How would your life change if you had evolved into a one digit species?

4. What do you suppose is the scientific basis for stating that the increase in the size of the horse was .0001 inch per thousand years, that is, one inch every ten million years.

5. Although the South American horse was described as more horselike than the horse, it was disqualified and moved to a separate order because it had a smooth area on the heel of the fossil. What was the real reason for this decision?

6. Normally when one digs down into strata (layers of the earth or rock), each layer is older than the one above it. There are many exceptions, however. Can you think of a situation where older strata lie on top of younger strata?

7. Classification (taxonomy) of fossil animals and other species can be extremely difficult. One basic rule (phylogeny) is to imagine how a specimen fits into the supposed evolutionary history or ancestry of seemingly related species, such as small to large, or more primitive to more modern in form. A second rule (homology) is that the slightest resemblance proves a common ancestry. Another basic guide (convergence) is that while two specimens may look alike, their ancestors did not. A fourth principle (convergence) rules that likenesses are merely a coincidence and thus may be safely ignored. Would you agree that decisions may be very arbitrary? A leading authority stated frankly that they must train paleontologists like monkeys.

CHAPTER TWELVE

ANCIENT PLANT ODDITIES AND MYSTERIES

INTRODUCTION: PLANTS AND PREHISTORY

Mysteries Among the Plants

Oaks and baobabs, cotton and reeds, pineapples and bananas, barley and sweet potatoes, buttercups and peanuts seem like a rather ordinary list of plant life. However, behind each of them, and others as well, lie strange facts, mysteries, and oddities. Some plants play highly significant roles in prehistory and help to shed light on the ancient past.

We have a special interest in how all this ties into the early chapters of Genesis. Noah and his family lived for centuries after the flood. I believe these were the Old Ones spoken of in ancient civilizations. They carried with them the culture developed during the pre-flood centuries. Thus they were the teachers of the early generations after the flood. Some of the implications of this belief will be discussed below.

Two Unlikely Announcements

Among the unshakable beliefs of archaeologists and anthropologists for at least a century are: (1) Early man first came to the Americas only via the Bering Strait in Alaska; and (2) Except for the grudging admission of several possible temporary Viking landings, no one from the Old World of Europe and the Mediterranean civilizations came to the Americas before Columbus. It is no exaggeration to say that professional university careers could be and were destroyed by anyone who dared question these "truths", or worse yet, submitted evidence against them. However, after decades of underground activity, scholars by the end of the twentieth century were finally brave enough to come out into the open to disagree publicly with these long held beliefs. One occasion was the annual meeting of the American Association for the Advancement of Science.[1]

The reason for the resistance is obvious. Early man, supposedly evolving out of the ape world, was simply not intelligent or advanced enough to master the seamanship required in those early times. Compelling evidence that this belief is false has been ignored, attacked, or ridiculed. This chapter illustrates that the ancient world was very different from textbook accounts, and furthermore the real truth fits the framework of early Genesis. Several examples at this point will suffice to illustrate the need for a radical new view of the past.

A toxicologist is an expert on the detection of poisons in the body. One of these scientists discovered fragments of New World tobacco leaves in the stomach of the pharaoh Ramses II, conventionally dated 1290–1224 BC, roughly 2,800 years before Columbus.[2] The obvious implications are staggering but still almost completely ignored today.

The curator of the Smithsonian Institution, among other scientists, announced that several waves of prehistoric immigrants came to the Americas across the Arctic, the Pacific, and the Atlantic. Other scientists are beginning to challenge the Alaskan land bridge route. They point to new evidence of ancient arrivals from Polynesia or South Asia. They also stress ancient remains in the Americas typical of Europeans and the Ainu of Japan, among others.

Plant Names Conceal History

When the descendants of the Old Ones spread over the earth, especially after the Tower of Babel event (Genesis 11:1–9), they carried with them the names of animals, birds, fish, flowers, trees, and plants, just as the Phoenicians and others did many centuries later. Such names are found, sometimes half concealed, in spoken and written languages of widely separated peoples in the world. Clues like these are of help in piecing together some aspects of ancient history. As one example of such research, Ignatius Donnelly massed over 650 evidences for the common origin of Old and New World cultures, including large numbers of fruits, vegetables, and animals that could have been transported from a common source in prehistoric times.[3]

When We Munch a Carrot

In Neolithic times (that is, in the earliest villages) in Germany and Switzerland, the people munched on carrots. We have evidence of this despite the fact that traces of root crops are very difficult to find due to their rapid decay. The carrot is a technological triumph, a hybrid of two wild species. The Neolithic Swiss lake-village people also cultivated a variety of apples. The early Danubians probably brought the cherry-plum into Central Europe. In Europe this tree was crossed with the aloe plant to produce the cultivated plum. Walnuts, which were not native to Switzerland and Germany, were brought in by Neolithic people and cultivated there. In Neolithic times olives were grown in southwestern Spain. Much technical knowledge was required to make them edible. Garden peas were introduced into Europe by the Danubians at an early date. In northern Mesopotamia, in addition to cultivating peas and lentils, we note with some degree of envy that the people of Jarmo, and we might add, Sodom, cultivated the pistachio nut. It is a most surprising fact that no Neolithic culture is known that did not grow both wheat and barley.[4]

Ancient Technical Wisdom

We have good reason to believe that sophisticated people in very ancient times spread a high level of technical wisdom about domesticated plants and animals. Sebastian Englert speaks of primitive sea travel in the Pacific in which boats up to 80 feet long carried more than sixty persons.[5] Double canoes connected by decks were used for astonishingly long voyages. These boats carried an amazing assortment of provisions both for use on the voyage and for transplanting on other islands and mainlands. They carried many seedlings, pigs, dogs, and chickens. Sometimes transplanting was successful, sometimes not.

Where Did Invention Come From?

Independent discovery and invention are almost always given as the reasons to explain the mystery of early plant cultivation. Yet the mind is insulted by this kind of guesswork. In conventional thinking we have preceding this remarkable period of history only a picture of a human hunter who knows how to chip rocks in order to make tools and weapons, but little else. It is very difficult to picture him at breakfast one day, gnawing on a bone, saying that he is tired of hunting and has decided to develop some hybrid plants for food instead.

Mystery of the Sewn Plank

There is scarcely a more clever invention in the ancient world than the sewn plank tradition in boat building. Half-round pieces of wood were carefully fitted on both sides where two planks came together, edge to edge. Drilled holes were made just above and below where two planks came together and through the half round pieces, which made "sewing" possible with fibers or sinew. The holes then were caulked with asphalt or another waterproof substance. The joints were extremely strong and made large seagoing vessels possible. This ancient tradition was known in Europe and in China, and extended all the way across the Old

World. Independent invention seems most unlikely, and hence it is of particular interest that the sewn plank was also found in Chile, on the southern California coast, in Hawaii, and in the South Seas.[6]

Carrying Plants to New Lands

Most important in navigation were the living plants, carried to be planted in the new lands, for example, the sweet potato or yam. The first European visitors to Easter Island were given gifts of sugar cane, chickens, yams, and bananas, but none of the gifts was native to that island. The Polynesians were carrying on an old, old tradition. They worshiped Taama, the god who first brought the coconut, fruit trees, vegetables, and other foods to the islands.[7]

Where Did Agriculture Begin?

Most authorities hold that the cradle of agriculture, that is, the oldest area where agriculture was practiced, was in the uplands to the south and east of the mountains of Ararat. This region covers an arc marked by the western Zagros mountains (Iraq and Iran), the Taurus (southern Turkey), and the Galilean uplands (northern Israel). As one piece of evidence for this belief, scholars point to the wild ancestors of our cereal grains in this region and the remains of farming communities that predate any known civilization. All cultural roads seem to lead out of the Iranian Highlands in early history.[8]

Dating the Earliest Agriculture

Of course, new candidates for the earliest agriculture are proposed from time to time. Recently Thailand was suggested because a Thai site, called Spirit Cave, was found filled with relics of ancient farmers, including crude stone tools and seeds of peas, beans, cucumbers, and Chinese water chestnuts. All had been domesticated, and the finds were dated earlier than anything in Mesopotamia. As has often been noted, however, C14 dating is a highly subjective business, and one set of C14 dates is not likely to sway many opinions, especially in the scholarly world.[9]

C. W. Ceram is typical of writers when he marvels that all the principal domestications of animals and plants took place by 3000 BC, or at least by 2000 BC.[10] No further significant steps took place until the nineteenth century of our era, an enormous and mysterious time gap. Ceram marvels even more that the principal domestications took place everywhere in the world at about the same time, or that they spread around the world with incredible speed. We see what a remarkable fit these scientific discoveries make with the early chapters of Genesis.

Early Agriculture in Peru

Civilization about as old as anywhere else in the world reached Peru, and this included domesticated plants. What is surprising is that the earliest intensive agriculture in South America is believed to be east of the Andes in the Amazon lowlands before it was carried over the Andes, perhaps even before the Andes mountains existed. Much of the Andes chain is very recent—shockingly recent!

At the same time that predynastic villages were thriving in the Nile delta, agricultural villages were built along the well-watered coast of Peru. The Peru coast is now a barren desert. It is interesting to observe at this early date that the staple crop was lima beans. Fishhooks were cleverly formed by shaping thorns on living trees into a hook shape. The people knew how to weave, and they made fish nets out of coarse fibers. They used mortars to grind seeds, and made ornaments of stone, shell, and bone. Well-preserved flutes were found, indicating a love for music. This surprisingly high degree of culture is thought to have disappeared by 2700 BC, many centuries before the birth of Abraham.[11]

The Lake of Great Mysteries

Lake Titicaca, bordering Peru and Bolivia, and the ancient city of Tiahuanaco on its shore, raise many mystifying questions. The site of the city ruins is too isolated, too cold for most crops, and too empty of game to support people, yet it once was a thriving cultural center. The lake, which lies on a plateau with an elevation of 13,000 feet, is used for navigation today. Jacques Cousteau photographed remarkable massive stone blocks fitted into a wall under the lake waters. The wall points directly to the city of Tiahuanaco. There are large expanses of terraced hillsides in the area, but today nothing can be grown on them due to the high elevation.

The age of the city is a controversial question. We do know, however, that drawings of an extinct animal, the toxodon, have been found on pottery fragments there. The Andes rose abruptly in historical times when man was already sailing ships. There was a sea harbor in Lake Titicaca. Rings for ship cables on piers are so large that they would be appropriate only for ocean vessels. Yet today the site is 200 miles from the Pacific Ocean and lies at an altitude of 12,500 feet. Traces of seaweed are found at the lake, and Pacific Ocean sea shells are still seen in this part of the Andes. Numerous raised beaches may be seen, and the water in the southern portion of the lake is still salty. The lake has a chalky deposit of ancient seaweeds with lime, about six feet thick, which indicates that the ridge where it is found was once an ancient seashore.

Modern geologists say that the present shore of the lake was once immersed in the ocean. Ocean fauna are found in the lake. A salt line appears on the mountains surrounding the lake. Lake Titicaca was formerly an inlet of the ocean. A rise of about 2.5 miles in elevation sounds wildly implausible, yet technicians of a Western Telegraph ship searching for a lost cable in the Atlantic in 1923 found that the cable had been lifted 2.25 miles during a 25-year period. The ocean floor was and is far more plastic than anyone had realized.[12]

The Earliest Farmer's Almanac!

A carved bone found in Spain was dated as Upper Paleolithic. Evolutionists believe this was a long period of time before primitive man built the earliest villages. The bone not only showed artistic representations of plants and animals, but some kind of notation was also engraved on the bone. The conclusion was that the plants and animals indicated various seasons, and the notation system revealed a precise record of lunar phasing, spanning a period of almost nine months. This incredible find, supported by many other similar discoveries, has changed current thinking about the level of sophistication of men in this ancient time.[13]

Neanderthal and the Plant World

Evidence has been brought forward of ancient man's relationship with plant life. Great surprise was expressed when a Neanderthal burial was found in a cave on the western slopes of the Zagros mountains in Iraq. The man was buried on a bed of wild flowers. Pollen analysis indicated blossoms from hyacinths, hollyhocks, and bachelor's buttons.[14]

Early Medicinal Use of Plants

Ancient man knew a great deal about plant life. In an excavation in Soviet Armenia dated approximately 1500 BC, it was found that delicate head surgery equal to modern techniques was being performed. Remnants of more than fifty kinds of flowers and herbs were found at the site, and the conclusion was reached that the plants were used for anesthetic and other medicinal purposes.[15]

Clever Fishing Technique

One of the native tribes in Brazil uses a very unusual method of fishing. Branches from a certain kind of bush are cut and the river water is thrashed with them. The fish are stupefied and easily caught. The strange thing about this is that African tribes are known to follow the same method of fishing. This is one of a great many hints of very early contacts between the old and new world.[16]

Sudden or Gradual Climate Change?

The idea of slow, gradual change is so deeply embedded in modern thinking that clear evidence is often overlooked that points sharply to sudden change. This is part of the legacy of evolution that imprisons the mind. Unquestionably dramatic and sudden changes of climate took place in historic times, accompanied with radical political consequences, including the massive migrations of ancient populations. Mesopotamia was once the most fertile country in the world. Herodotus speaks of its crops yielding 400-fold. Israel was truly a land of milk and honey. Isaac reaped 100-fold, as we read in Genesis 26:12, in what is now a very arid region. Except for the river valleys, no one viewing such regions today would find it easy to accept the productivity of the past. In fact, Francis Voltaire and others several centuries ago denied that Judea could ever have supported the great and numerous cities that history tells us once flourished there. The vast desert areas of Mesopotamia are now dotted with the ruins of once flourishing cities and kingdoms.[17]

A Sudden Change in California

Tuolumne Table Mountain in California was formed by a massive lava flow. Plant remains from the Tertiary period were found under this mountain, along with the fossilized bones of extinct animals. When gold miners tunneled into the base of this mountain, they found that others had been there earlier, but there was no evidence of a cave in the area. Tools, mortars and pestles, and spearheads lay among the plant and animal remains.[18] If this discovery is considered at all, the usual explanation is that the remains are from ancient and possibly pre-Indian miners, which is remarkable enough in its own right. Yet the bones of men and the artifacts are reported as being embedded in gold-bearing gravel under lava at the foot of the mountain. Russell Brubaker, who has studied the problem, concluded in private correspondence that the finds are exclusively of ancient miners. This is prime evidence for catastrophic change. More recently Edward Lain and Robert Gentet concluded that the bones and artifacts testified to pre-Indian and early post-flood gold mining before the ice age, a truly astounding discovery!

More Recent Climatic Changes

Marked changes in climate are not limited to the distant past. James Enterline has assembled convincing evidence on the Vikings to support two theses that run contrary to conventional thinking about these early explorers of America.[19] The Vinland of AD 1000 had nothing to do with grapes, but the word meant a land of pastures in Arctic Canada. The Vikings eventually vacated Greenland but explored throughout much of North America and as far to the north and west as Alaska. Traces of the Vikings are being found in very unexpected places on the North American continent, which lend credence to Enterline's conclusions. Those who hold to an unchanging world violently attack the evidence offered for the above interpretation.

The Finds are Controversial

Conventional texts and literature ignore discoveries such as these. The reason such discoveries are so controversial is because they do not fit the narrow mold of what anthropologists think today. Evolutionists have long held that nothing happened in the past that is not happening today. Thus, loyal to the doctrine of evolution, the standard response for all such discoveries is that they are frauds or misinterpretations of the evidence. There is a fear about any evidence that seems to conflict with such conventional thinking.

Trees Full of Surprises

Unexpected Discoveries

Who would have predicted that scientists would pick up fossil beech wood and layers of southern beech leaves 300 miles from the South Pole? There are 3-foot diameter fossil tree trunks on Seymour Island in the Antarctic, now barren of trees. Who would have thought that one could mine from deep under the earth sound tree trunks for lumber in the Dismal Swamp, Virginia, the Hackensack area in New Jersey, northern California, the marsh area of the isthmus connecting Nova Scotia and New Brunswick, vast areas in northern Siberia, and numerous other areas in the United States and around the world?[20]

A Strange Time Gap

A wood spear dating from the so-called Great Interglacial period was found in Europe. Another wood spear was excavated in deposits of the so-called Last Interglacial. By conventional dating the two finds are separated by several million years. The finds are called "curious." They are similar enough that they could have been made by the same man. Both are made from the wood of the yew tree. It might be more appropriate to call the dating system "curious".[21]

Sudden Changes

Who would think of forests thriving on the barren Sinai peninsula? At one stroke great lava flows burned down the forests leaving desert behind. In Palestine a volcanic eruption filled the valley of Jezreel. This event was dated far into the past until a vase, dated about 1500 BC was found embedded in the lava.[22] Visitors to the Dead Sea region in Israel today would find it very difficult to imagine that in early biblical times this was an area of rich farmlands, that present desert areas in India, Arabia, Mongolia, Palestine, and North Africa were water wonderlands of thick forests and supported an astonishing variety of animal life.

Problems of Classification

Trees have special classification problems because leaves, roots, and bark often are found separately and no relationship can be established. One classic example is the *Lycopsids*, parts of which were assigned to four different species until the day came when an entire fossil plant was found together. The trunk had been named *Lepidodendron*, the roots were called *Stigmaria*, the leaves were named *Lepidophyllum*, and the cones were called *Lepidostrobus*.[23]

The Age of Rock Strata

Great age is assumed from successive strata in rocks that are believed to have been laid down very slowly and gradually. This is the heart of uniformitarianism in establishing the dating for the geological column from Precambrian to modern times. This principle, however, is not faring well in geological research. As just one example, the trunk of a tree at Craigleth Quarry in England was found to intersect from ten to twelve successive strata of limestone. It became fossilized without even losing its bark. Presumably the tree and its bark survived patiently for many millions of years as the strata covered it millimeter by millimeter.[24]

Carbon 14 Dating Problems

Plant remains often figure in C14 dating for geological and archaeological dating. The results are often less than welcome, as we shall see. For example, C14 dating of living trees near an airport came out to more than 10,000 years old, because the trees contained so much inactive "fossil" carbon from airplane exhaust.

The most renowned spruce trees in all the world were buried underneath the Mankato glacial drift at Two Creeks near Manitowac, Wsconsin. First, the C14 dates from this sample forced geologists to reduce the date of the final glacial advance from 25,000 to 11,400 years ago. Actually, the literature of that time more often dated the last glaciation at 35,000 or more years ago. Having found that result, the findings were then used to help calibrate the C14 dating method, a procedure that sounds suspiciously incestuous. Ernst Antevs held that the date ought to be 18,500 years on the basis of varve (supposed annual layer of sediment) studies, but this method, since discredited, could not stand up against the prestige immediately given to C14 dating. Then H. E. Suess of the United States Geologic Survey reported that other wood found at the base of interbedded blue till, peat, and drift outwash from the last glaciation yielded a C14 date of 3,300 years, but that was rejected out of hand. Is it possible that such a date fit the biblical record too closely for comfort? More and more puzzles are predicted if tests continue.[25]

Pollen Studies

The Precambrian Hakatai shale in the Grand Canyon has produced pollen from conifers. If these findings are verified, the news is very upsetting for any kind of evolutionary development of plant life. The pollen of flowering plants were also found, which are even more difficult to explain. All formations beginning with Permian on down to Precambrian yielded essentially the same types of spores. Study is continuing and some kind of contamination is not yet ruled out, but "impossible" pollen grains are being found in Early Paleozoic sediments.[26]

Laclerc and Axelrod reported the discovery of spores and fragments of woody plants belonging to dozens of genera in Cambrian rocks, supposedly 200,000,000 years too soon. A geological report from Venezuela in 1963 reported similar findings. Well-preserved pollen and spores were found in Precambrian deposits. The study was repeated, eliminating any possibility of contamination of the samples. Similar findings were reported by the British Guiana Geological Survey. At the Iron Ore Company, Ontario, Canada, fossil wood identified as white pine and redwood was reported in Precambrian deposits. C14 tests on the wood, however, indicated an age of a mere 4,000 years. To relieve a little of the embarrassment, the wood was later called Late Cretaceous, which fits neither the deposit it was found in nor the C14 dates.[27]

The Curious Redwood Tree

It was long believed that the dawn redwood, *Metasequoia glyptostroboldes*, had been extinct for around twenty million years. In 1944 a Chinese forester reported that the species still lived in Szechwan in central China. Later, a grove of these trees was found in the Shuihsa valley.[28]

A curious report appeared in *Science* in 1934 about the sequoias of California and Oregon. Somewhere between 1300 and 2000 BC all giant sequoias then living were wiped out by some catastrophe.[29] There is a remarkable absence of evidence of any such trees of a generation previous to those now growing, except for buried trees described below.

The fossil specimens of the redwood *Sequoia sempervirens* found at Atanekerdluk in northern Greenland indicate an entirely different environment in the past. One estimate is that the average temperature had to be 30 degrees higher than at present to support this kind of plant life. This raises interesting questions about the permanence of the poles and of the inclination of the earth's axis in the past.

In 1963 a well digger found a redwood tree trunk in a state of good preservation at a depth of 80 feet in the low rolling hills near Napa, California. At the surface above this find was the stump of a redwood thousands of years old. Similar finds to a depth below the surface of 120 feet have been reported by the California Highway Department in connection with road excavations in the Sacramento valley. Nearby, on the surface, is the long unbroken trunk of a petrified redwood. In this immediate area are huge living redwood trees. This may be the only place on earth where all of the above are found in close proximity.

The tree burials occurred in deposits of clay, mud, and sand. Some of the buried trunks have been cut and used as lumber, and they have been found in areas of California where they will not grow today. Such finds are explained away on the basis of the amount of tannin contained in redwoods that permits them to remain unchanged and undecayed for millions of years. Some redwoods are buried in deposits that appear to be Mesozoic along with materials of Tertiary and Pleistocene times, undeniable evidence for catastrophic events in earth's history.

The Bristlecone Pine

The bristlecone pine tree has achieved fame in recent decades as the oldest living species on earth. Several other plants have been suggested for this honor in recent years, however. The pines are found at elevations of about 10,000 feet in California interior mountain regions. Dating of the tree rings has run into unexpected difficulties for two reasons. More than one ring may grow in one year, and annual growth rings cannot be accurately distinguished in periods of extreme drought. Botanists report that three to four rings may form in a year where a tree grows on a slope with the ground alternately wet and dry from rapid water flow several times during a given year. Estimates vary on the age of the oldest tree examined; for example, one was reported as 4,600 years old.[30]

One discovery of interest is that a fossil timberline was found above the present timberline, and that all bristlecones above the present line died about 700 BC. This calls for study of the event that caused the dying at that time. In 1954 Herbert Sorensen announced that bristlecone pine chronology had been extended back to at least 7485 BC with the hope that it would soon extend back to about 10,000 BC. However, no information was published on how well specimens cross-matched, nor are data provided allowing such a check or independent assessment of the dating. Most people respect factual data, but so far this chronology is suspect. The problem of so-called missing growth rings during dry periods and the possibility of multiple rings forming in the same year raise serious questions about this chronology. In addition, climate and rainfall can vary significantly from hill to valley, creating major problems for this method.

Can Trees Date Carbon 14?

Another report in 1972 indicated that every year back to 6200 BC had been specifically identified from bristlecone pine rings. Curiously, C14 tests with samples from this wood showed apparent errors in the C14 calibration ranging from a few centuries to a thousand years. Not recognizing the problems noted above about tree rings, the C14 scale was then hastily recalibrated, and now C14 dates around the world are being reissued up to a thousand years earlier than previous announcement, with some results bordering on the ludicrous.[31]

Suddenly it is fashionable to say that Europe's earliest tombs, monuments, temples, and tools predate their counterparts in the Middle East by as much as 1500 years. Although this blueprint of the past is being run up the flagpole, hardly anyone is saluting. The revised C14 is due for another revision.

The ginkgo tree. This strange tree has a curious history. Fossil remains were found in Miocene deposits in Greenland, and the tree was declared to have been extinct for millions of years. Then, to the amazement of botanists around the world, the ginkgo was found thriving as a garden ornament in China, though none are known to exist in a wild state anywhere in the world. In a similar fashion, the extinct water-cypress, found only in fossil form at locations around the world, was discovered recently living in Szechwan Province, China. Another mysterious survivor is the Asiatic magnolia. Other plants have been found that were previously thought to be long extinct.[32]

The mighty oak. Almost everyone knows the oak tree, but few know something of this tree's past. The Sahara Desert seems to be the last place on earth one would look for an oak. At the Oasis Kharga, however, fossils of the moisture loving oak have been found, pointing to an entirely different climate in the past.[33]

Six-ton petrified oak trunks were found standing upright in a deep coal mine near Oak Ridge, Tennessee. The trunks, up to 45 inches in diameter, are thought to be 300,000,000 years old (in evolution time, of course). The bark is remarkably well preserved. It seems clear that something happened very quickly rather than that long periods of time were involved.[34]

Submarine forests are known off the coast of England, for example, off the Isle of Man. The Strandhall submerged forest yielded a surprise. The trunk of an oak was removed from the sea. On its surface were the marks of a hatchet. This poses something of a quandary. Scientists are reluctant to give a recent date to these submerged forests, but they are equally hesitant to date hatchet marks far back into the past.[35]

Oak was used in highly sophisticated carpentry dating back to about 3000 BC, before the beginning of the Old Kingdom in Egypt. On the bank of Lake Clairvaux near Besancon in eastern France, eight well-preserved wooden houses were uncovered under 15 feet of mud. They were not on stilts like the famous Swiss lake dwellings. The houses were made of precisely-tooled oak, with hallways, rooms, and conduits in the roof, possibly for chimneys. All the homes were oriented in the same way, with the least exposed walls to the north and the doorways facing south. From preliminary studies we know that these unknown people knew stock breeding and had domesticated dogs.[36]

The oak played a highly significant role in the religious life of the ancients. The oak was sacred and a symbol of fertility. John Cohane presents evidence that the word *oak* itself (like the word *egg*) is derived from prehistoric man's vocabulary of the sacred and is found in a great many old place names.

Ancient Village Architecture

Visitors in Europe are struck by the architectural use of timber and bricks together. Between heavy timber frames, including diagonals, brick work was laid in these very old and sturdy buildings. It is striking that in one of the oldest cities known in the world, Catal Huyuk, in southern Turkey, the same style of architecture was used in the very earliest levels, except that mud bricks were used. Somehow the tradition was carried unbroken into modern times.[37]

Palms and Other Trees

Fossil palm wood can be clearly identified both as petrified wood and in coal formations. The palm tree is believed to have evolved 100,000,000 years later than other trees and vegetation found in Carboniferous coal beds. Then large amounts of palm wood were found in Carboniferous beds. This, of course, created a problem, but the problem was solved by changing the label. Any strata containing fossil palms are now placed in late Cretaceous or early Tertiary times instead of in the Carboniferous period. If the wrong fossil is found in a formation, the formation is relabeled, and the problem neatly disappears.[38]

Kenneth Macgowan notes that the coconut palm was carried as a food plant by the ancients to Middle America and also to the islands of the South Pacific. Carroll Riley makes a similar point, observing that the coconut, along with other Old World plants, is found in Peru. He, too, accepts the idea that the Old Ones carried seedlings with them on long sea voyages to plant in the new lands.[39]

In the spring of 1971 the British Museum processed palm kernels and mat reeds from the tomb of Tutankhamen. Dr. Edwards, Curator of the Egyptian Department, reported to the museum at the University of Pennsylvania that the results gave a C14 date of 899 to 846 BC. Without explanation, however, the results have never been formally published, because the dates support the "wrong" Egyptian chronology. In a follow up of this matter, Mr. Burleigh, director of the laboratory of the British Museum, stated that he expected the results to be published shortly. Then he admitted that results that deviate substantially from what is expected are often discarded and never published. In 1973 a Mr. Barker of the British Museum stated that the laboratory had made no measurements on material from the tomb of Tutankhamen. Deviating dates are not only not published, it is even denied that they had been found.

There are a good many puzzles in the distribution of some species of plant life around the world. Each of the two known species of the travelers tree is found only in one location: one in Madagascar, and the other in tropical South America. There are four living species of the strange-looking baobab tree. Two are found in Africa, one grows in Madagascar, and the other is found in northern Australia. Where and how any tree originated is mystery enough. Assuming the baobab is native to Africa, no one knows how the seeds could have been carried naturally to Australia.

For many years children have built model airplanes and other objects from pieces of balsa wood. The first Europeans to reach the Pacific coast of South America made sketches of another application of balsa. The natives built large balsa rafts complete with a square sail, a centerboard, and a long steering oar astern. Thor Heyerdahl successfully tested such a raft for its ability to sail on long ocean voyages.

How Does Coal Develop?

One of the sacred tenets of geology is that vegetable matter covered in swampy areas converts into peat. So far, so good. As time passes, and under the right conditions, peat changes into lignite, and later into bituminous coal. Under great pressure the bituminous coal is metamorphosed into anthracite coal. A curious fact, however, is that nowhere in the world is there any evidence of peat now being transformed into lignite or coal. No location is known where the lowest level of peat has begun to grade into the typical coal or lignite bed. It is fundamental to the doctrine of uniformitarianism in geology that no process occurred in the past that is not occurring now. Something is clearly wrong in the textbook sequence of peat to coal.[40]

Some have suggested that forty million years of the Carboniferous age can be reduced to a matter of a few hours on the basis of the fresh, unwithered leaf marks on the limestone roofs of coal mines in England and Germany. While great age is attributed to lignite deposits, there is hard evidence that lignite deposits are young. Lignite from Senckenberg, Germany, was dated to the Miocene age. Yet the lignite contained both lignin and cellulose, which was not at all expected, and hence was evidence of a young age.

Otto Stutzer, the great authority on coal formation, stated categorically that coal forms in a short time under the right conditions. In support of this statement is the remarkable observation by Petzoldt in 1882. A railway bridge had been constructed near Freiburg, Germany. Rammed wood piles had been made to support the bridge. Later these piles were enormously compressed by overriding blocks of stone. In the center of the compressed piles a black coal-like substance was found. The color changed from the center to the surface from black, to dark brown, to light brown, to yellow, showing all the stages from brown coal to anthracite. Pressure, rather than time, was the factor.

In England up to thirty layers of clay alternating with coal seams have been described. In America up to eighty such alternating beds have been noted. According to geologists, 300,000 to 800,000 years are required for such layers to develop one by one. Yet in such beds the vegetation of the highest layer is similar to the

A fossil elm leaf that shows no sign of wilting or decay. It contradicts the belief that life forms gradually became fossils over vast periods of time.

lowest layer. Trees have been found embedded in the coal beds with fruits still hanging to the topmost branches and the tree limbs are expanded as perfectly as in a herbarium. These are erect trees, not rotted away, covered over by different sediments. Leaves are impressed in the matrix with veinules perfectly preserved. Burial was very sudden.

Stutzer demonstrated that much coal was born in a time of violence, including violent water action. Coal balls up to two tons in weight and four feet in diameter are found with plants perfectly preserved inside. This again points to quick burial and that the coal balls floated for a time. In the Shane Mine, England, coal balls were uncovered four times as thick as the coal seam in which they were found. Boulders up to 400 pounds in coal beds are erratics from far away sources. In the Pittsburgh area there are 2,000 square miles of fossil charcoal beds, and there is no explanation how they were formed because forest fires have been ruled out. In coal beds in New York and in the Saar Basin in Germany, stumps were all broken off at a uniform height. The crumble coal in Saxony consists of crushed, frayed, splintered wood, and small nests and lenses of sand are found everywhere. Trees that extend up through overlying strata are so numerous that a listing of the many known examples would be tedious. According to conventional thinking, the trees extend through millions of years of slow formation of deposits without any trace of decay.

Dating Coal Deposits

Coal should be too old to be dated by the C14 method because the age assigned to it is far beyond the range ascribed to this method. Yet fossil coal and wood from Spain was carbon-dated at 5,025; 3,930, and 4,250 years. Coal from Kirgizia yielded a C14 date of 1,680 years. Samples of coal, petroleum, natural gas, and lignite, which should be at least 100,000,000 years old, were all dated at no more than 50,000 years.[41]

Coal and a Catastrophe

The coal beds at Joggins, Nova Scotia, show strong evidence that the petrified trees and coal there were not formed where these plants originally grew. Some terrible event tore trees and plants loose. Clusters floated, became saturated, and sank. During their drift, tube worms and mussels fastened themselves to the mass and fish swam among the debris. Sedimentation was very rapid. No soil zones can be detected. Unusual plant fossils are found in stump hollows, and there is remarkable preservation of delicate fossils. A C14 test on coal from this formation gave an age of about 3,700 years old.[42]

A Glimpse of an Earlier World

Bituminous beds in the south of Australia, in Tasmania, and in New Zealand contain fern trees similar to those now flourishing on the banks of the Yarra Yarra River in Australia and in Tasmania. Similarly in South America, fossil trunks of conifers and fern trees are found similar to those which now grow in that zone. On the eastern flanks of the Andes, leaf impressions with the green and yellow colors still partially preserved are found in white clay seams. Tropical vegetation, including fern trees, is entombed in Nova Scotia.[43]

Trees can be a great source of wonder. We have been educated to think that thousands, if not millions, of years went by before the gradual process of fossilization was completed. Yet we see fossil wood that reveals the most delicate structures and fibers preserved without any sign of decay or injury. We read of logs that petrified in less than one year, contrary to all conventional thinking.

Fresh Chlorophyll in Lignite

If fossilization occurs at all, it happens very rapidly, as we have illustrated. The lignite beds at Geiseltal, Germany, support this view. Scientists were astonished that leaves were preserved in the lignite in a fully fresh condition. The chlorophyll is so well preserved that the two basic types can easily be recognized. The finest details of cellular structure were preserved as well. It is evident that some event occurred that covered the material rapidly. The Geiseltal beds pose another mystery that points to a catastrophic event. At this location are found plants from all climatic zones and from all recognized geographic regions.[44]

Man and Fossil Formation

Scheuermann discovered a number of sticks, apparently artificially pointed, that had been found in a lignite bed.[45] Another recent discovery in India yielded evidence that something happened faster than is usually thought. Small pieces of wood had been worked by man before they became fossilized. This find was similar to the sharpened sticks turned into lignite described above. The choice for the evolutionist is grim. Either the man who sharpened the sticks must be given a preposterous age, or the lignite must be unacceptably young.

The Mystery of the Tar Pits

The millions of broken, mashed, contorted, and mixed bones in the La Brea Tar Pits, Los Angeles, are well known. The conventional explanation is that the animals were trapped in the asphalt as they came to drink or to feast on other trapped animals and birds. Less publicity is given to the fact that many trees also were caught in the tar pits and were apparently swept in by some overwhelming force. Some catastrophic event trapped both trees and animals. After the event, of course, other animal life came and was caught in the tar deposits.[46]

Mysterious Buried Forests

The Norfolk forest bed in Cromer, England, near the North Sea coast, is another exceptional puzzle. Some fantastic moving force brought together the remains of species of mammals and other animal life from all climatic zones and mixed them with flora of a temperate zone. In this forest bed are many upright stumps with the roots interlocked. Drift is evident. The roots were broken off from one to three feet from the trunk. Bones and stumps were thrown together into a fantastic jumble. Along the coast of Nova Scotia and New England stumps of trees stand in water, telling of once forested land that suddenly became submerged.[47]

Arctic Plant Mysteries

The New Siberian Islands off the coast of Siberia in the Arctic show that luxuriant forests ended there in a disastrous moment of time. Here are remains of enormous forests. Trunks are partly standing and partly buried in the frozen soil. The islands contain remarkable wood hills up to 180 feet high of sandstone alternating with bituminous trunks of trees. Fossil charcoal is found everywhere along with ashes, while some of the wood has petrified. Trunks of trees have broken and splintered ends. In this wild disorder are found leaf impressions, fruits, and cones.[48] Where inch-tall willows now grow on the New Siberian Islands, Toll found on this frozen wasteland a fallen 90-foot tree still bearing ripe fruits and green leaves. Once this was the land of the tropical breadfruit tree. The South Pole also once had a temperate or subtropical climate. Admiral Byrd examined rock fragments within 200 miles of the South Pole. Invariably the fragments included plant fossils, leaf and stem impressions, coal and fossilized wood.

The Plastic Ocean Bed

In 1949 beach sand was brought up from a depth of 3.5 miles in the Atlantic. The location was 1,200 miles from any continental shore. In the sample brought to the surface were twigs, nuts, and bark fragments. This area was once above sea level, and some event caused it to drop far into the deep.[49]

The Yellowstone Petrified Forest

The famous petrified forest layers in Yellowstone National Park are cited in most texts to support the notion of long periods of time. Layer upon layer of forest is found, each in turn destroyed by lava. Yet, as Andrews has noted, no cones have ever been found in the layers. This raises the question of whether the forest remains are in place, or whether these, too, were rafted in or drifted into place.[50] New light on how this formation came to be has now been discovered in connection with the eruption of Mount St. Helens in 1980. Multiple layers of vertical tree trunks and fragments have been discovered in the lake bed just below the mountain.

A Violent Happening in Canada

Catastrophic events also happened in recent historic times. For a number of weeks in the St. Lawrence River region there were unusual aerial phenomena, such as fiery phantoms, lances of fire in the air, shapes like fiery serpents, and flaming wings. Then, on February 5, 1663, a tremendous earthquake struck the region. A forest was submerged, deep fissures spread out for hundreds of miles, and hills of mud heaved up from the river bed to heights of a hundred feet or more.[51]

Iceland Then and Now

Iceland has had a violent past. Lignite strata have been found on this island that show its tropical climate in the past. Extensive forests in the form of fossilized tree trunks are found in many places on the island. In Tertiary times the forests were overwhelmed by pumice, ashes, and flowing lava. More important, a radical change in climate occurred since that time.[52] It is hard to imagine that at the same time the fig and the magnolia thrived in northern Greenland. Tropical coral reefs fringed the entire Arctic coast of North America, and water lilies and cypress flourished on the island of Spitsbergen that lies in the Arctic region between northern Norway and Greenland. No satisfactory explanation has yet been offered for these phenomena, especially when conventional thinking today accepts no catastrophic events in the past nor any change in earth's past history from what is occurring now.

Plants in Prehistory

When the Plants Were Cursed

Francis Schaeffer calls attention to Genesis 3:18 where thorns and thistles suddenly appear.[53] Where all had been good previously, at least some plant life changed. Thistle refers to luxuriously-growing but useless plants. According to Schaeffer, the phrase suggests a mutation. The plants had been one kind of thing; God spoke, and they began to bring forth something else. It seems reasonable to assume that similar events took place in the animal world. First there were only plant-eating species. Then carnivores and poisonous creatures came as a result of the curse.

Tracing Plant History

For the scientist who attempts to relate the history of plants to the theory of evolution, the facts are dismal. Not a single group of modern plants has been traced back to a common ancestor. According to Darwin, the rapid development of all higher plants "within recent geological times" was an abominable mystery. Darwin conceded that this mystery was the "most obvious and gravest objection" that could be urged against his theory.[54] We must add that Darwin said this also about other frustrating mysteries that did not fit his theory.

It is generally held on supposed fossil evidence that the first plants came long after the first animals. Despite this belief, however, it seems that plants really came before animals on the basis of simple logic. The life of animals depends ultimately on plants. Genesis 1:30 speaks of the time before any carnivore feasted on another species, that is, before the fall:

> To every beast of the earth, and to every bird of the heavens, and to every thing that creeps upon the earth, wherein there is life, I have given every green herb for food.

This passage gives the sequence of events: plants first, followed by animal life.

Mysteries of Growth

In 1974 Soviet scientists reported strange growth effects from a silicone mixture on both plants and animals. If true, this illustrates the fact that there is much remaining to be discovered about the effects of trace elements, environment, and other elements on growth. The solution caused the hair of mice to grow so long it could be braided. It had striking growth effects on human hair and fingernails, and on fur, antlers, and hooves of animals. Research is continuing with the effects of the solution on certain crops and trees.[55] Indirect support for the Soviet research comes from two similar stories. Plants grown in earth enriched with moon soil are reported to grow to unusual size, but little has yet been published about this unusual effect. Another interesting report came out of Missouri. Long ago a giant meteor struck the earth 14 miles north of Lebanon in that state. This site, known as the Decaturville Crater, was known as a geological curiosity since the 1880s, but it was not identified as a meteorite crater until 1967. The original crater near Lebanon, Missouri, was 500 feet deep and 3.5 miles in diameter. Descendants of the Osage Indians who lived in Missouri passed down tales about the place where minerals had strange curative and growth stimulating properties. It is believed that the superior size and strength of the Osage Indians came from the plants and animals nurtured in this soil and water.

Long-lived Seeds

The astonishing durability of seeds receives notice from time to time. A report from Argentina stated that Cauna Compacia plants were grown to full and normal height from seeds found in a 600-year-old tomb. Other examples are known from seeds much older that still grew when planted.[56]

In the Arctic's frozen muck, burrows of lemmings have been found containing seeds of lupine. Some of these seeds, despite the fact that they were dated by C14 at about 10,000 years old, were planted. These grew into plants that are the same as the present wild plants. As lemming burrows are poorly ventilated and damp with mold, only a sudden event could have buried and preserved the seeds.

Microspores

In India, scientists reported microspores and wood fragments from the Middle and Upper Cambrian formations in Kashmir, Spiti, and other locations. Several scores of microspore types were recorded. In 1949 Naumova, a Russian scientist, reported many microspores from the Lower Cambrian blue clay of Estonia, Latvia, and Lithuania. The samples were believed to be free of contamination. Microspores from vascular (woody) plants were also reported from Cambrian deposits in Siberia.[57]

The mystery of Peat

Measuring the thickness of peat beds was used in the past for dating purposes. The scientist, de Perthes, taught that peat grows at the rate of a fifth of an inch per century. Later, Roman roads were found in Scotland covered with eight feet of peat. By this formula the roads were a mind-boggling 48,000 years old. More attuned to the real world, local farmers over a century ago stated that peat may grow at the rate of 2.5 inches per year or about one foot in five years. Standard texts in geology now indicate that peat may grow at the rate of one foot in thirty years. Yet extinct elephants covered by a few inches of peat are commonly reported as millions of years old by scientists in the media.[58]

Land Plants in Wrong Places

Early reports from the shale deposits exposed in road cuts at Cleveland speak of mysteries or anomalies. For example, land plant fossils have been found in the marine deposits far away from any known land at the time the shales were deposited.[59]

Banana. Wind, drifting debris on the seas, and birds are properly credited with spreading the seeds of many plants to other areas. When we come to the spread of the banana plant around the tropical world, however, we must look for another explanation. Bananas do not produce seeds and they spread only from sucklings from the plant. A curious exception to this was noted in 1885. One species of the banana plant was said to seed in the Andaman Islands in the Indian Ocean.

According to John Cohane, the banana was the cultivated species carried the farthest and to the greatest number of places by ancient men. The name *banana* comes from Africa, and the same word was also the native name in Guiana on the coast of South America, before the coming of European explorers. The banana was carried to India, to China, to the Malay Archipelago, to Java and Indonesia, to remote islands of the South Pacific, and to the west coast of Africa. Further evidence of the importance and early spread of the banana lies in the fact that the same distinct varieties of the banana bore distinct names in very different Asiatic languages, such as Sanskrit, Chinese, and in Malay.

Bottle gourd. This useful plant is a native of Africa and grows only there without the aid of man. We can follow the trail of gourds left by ancient man in remote times. Some authorities hold that the bottle gourd is the oldest domesticated plant taken from the Old World to the New World. The dates offered are highly conjectural. The bottle gourd is found in Asia to the Indo-China coast, and westward from the Amazon River in South America. An Atlantic route is indicated because American gourds are closer to the African variety than the Asian. The gourd has been cultivated by man in Africa since about 40,000 BC according to the conventional chronology. It is believed that it was brought to South America around 14,000 BC and carried up the Amazon to the Andes by 11,000 BC. It is still in active use in Peru today.[60] Intensive agriculture appeared earlier in the tropics of the Amazon than on the coast of Peru. The bottle gourd was thought to have arrived in Middle America as early as 7000 to 5500 BC at the Ocampo caves, and somewhat later to Tehuacan Valley (4900 to 3500 BC). It arrived on the Peru coast at least by 3000 BC.

The bottle gourd was used for its fruit, for utensils and containers, and more important, as a water skin during sea voyages. Remains of the gourd were found in prehistoric graves in South Pacific islands. The island name for the gourd *kimi* is the same name given to the gourd in old Peru. In Hawaii the Polynesians cut charts of ocean areas on bottle gourd shells for use in navigation.

Buttercup. About one-seventh of the earth's land surface is muck in the Arctic ranging in depth from a few feet up to more than a thousand feet and consisting of sand, silt, and earth. In this muck are incredible numbers of both extinct and modern animal species of different climates. In one instant of time these animals were mashed, splintered, buried, and quick-frozen. This catastrophic moment is perfectly caught by the example of fresh buttercups in the mouth of a frozen elephant. Before it could chew and swallow, all life in a vast area of the earth ended in a sudden catastrophic event.[61]

Cotton. How long would it take for a coastal seabed to rise up 80 feet above sea level? It would not be unusual for such estimates to run into many millions of years. Darwin found exactly this situation on the mainland near Lima and the nearby island of San Lorenzo in Peru. The evidence was clear enough. Seaweed and marine shells were found at that altitude, but they gave no indication of great age. Nearby was the dried channel of a large river where a ridge or line of hills had uplifted directly across the stream bed. The district once had been very fertile, and the area was covered with ruins and bore the marks of ancient cultivation. What astonished Darwin was that mixed together with the seaweed and marine shells were pieces of cotton thread and plaited rush. In historic times the seabed had suddenly risen, and a civilization had come to an abrupt end. Cotton thread was an undeniable clue.[62]

Peruvian archaeology is gloriously confused. There are all kinds of conjecture about the early cultures that lived there thousands of years before the Incas. It is not possible to sort out which authorities are reliable and which are not. Lanning comments on the so-called Encanto Complex on the Peru coast that thrived around 3600–2500 BC. Sea foods were the basis of life. Interestingly at this early date, cotton and gourds were cultivated here. Twined cotton textiles and cotton fishing nets were found at two sites of this complex.

Gerald Hawkins[63] states what many authors have noted, and here is where the mystery of cotton deepens. The cotton of ancient Peru was genetically related to that of Egypt. In a work edited by Carroll Riley, it was suggested that Old World reached the New World by 5000 BC and perhaps as early as 5800 BC, for example, at Tehuacan, Mexico. Cotton seed cannot survive in sea water, so the remote possibility is granted that the seeds were carried over the ocean by birds. This, however, seems a bit too incredible. At any rate, Old World cotton was cultivated in abundance in the Americas in an age before pottery or maize (corn) was known.

Cotton fabrics dated about 3000 BC were found at Mohenjo-Daro of the Indus Valley civilization, in the Zambesi valley (not dated), at the Huaca Prieta site in Peru about 2500 BC, and in the Tehuacan Valley of Mexico 3000 BC and perhaps earlier. The cotton found in the Cape Verde Islands means that it was brought there by some form of transatlantic travel, and cotton found in the Marquesas Islands involved some kind of transpacific contact. All of these finds long predate the coming of European explorers to these lands. The Cape Verde Islands were discovered in 1460. There were no signs of ancient inhabitants there, but wild cotton genetically linked to the New World was discovered. One conclusion is inescapable. In prehistoric times, men brought Old World linted diploid cotton to the New World, where it was hybridized with New World lintless species to produce luxuriantly linted. It is hardly necessary to add that a high degree of sophistication was involved in this agricultural achievement.

Cotton was important in prehistoric times. Maria Ambrosini notes that the same variety of cotton was being cultivated both in Peru and in the Indus Valley in 2500 BC.[64] Kenneth Macgowan states that looms with the same eleven work points have been found in the Old and in the New Worlds, which again points to transoceanic contacts at a very early date. He also observed that species of cotton native to Eurasia and Africa have been found in Middle America.[65] Cyrus Gordon states that American cotton is a hybrid of wild American species and Old World species, for example, Egyptian, and therefore the Old World species had to be brought over by ship.[66] Looms of the Old World and the New World were remarkably similar and weaving patterns and methods were strikingly alike. This was observed in the mummy shrouds of Egypt and Peru. A rare fragment of lace in the museum at Harvard was found in Ceremonial Cave in the Hueco Mountains in Texas.

Edward Lanning puzzles over the special problems of cultivated cotton in the New World.[67] It was grown in southern Mexico by 3400 BC, and in central Peru by 3600 BC. Different species were developed through the hybridization of two or more wild varieties. The Mexican hybrid species differs from that developed in Peru. The heddle loom spread across ancient Peru somewhere between 1800 and 1500 BC. Lanning makes a special point of saying that the heddle loom and other cultural changes were "gradual in their development." Statements such as this litter the literature of prehistory. How does one go out and "gradually" develop a heddle loom? No answer is offered, no evidence is presented.

John Cohane and others observe that Peruvian cotton and other plants were carried in the remote past over long stretches of the ocean to various Polynesian islands. A *Time* article speaks of agricultural villages going strong on the Peruvian coast south of Lima about 4000 BC, but does not mention cotton until the coming of the Chavin people to that area about 700 BC.[68] Erich von Daniken wonders why the Incas cultivated cotton in Peru in 3000 BC, though they did not know or possess the loom. This is not exactly a brilliant statement. The Inca did not gain power and become known as a people in Peru until the fifteenth century AD, about 4,500 years later than von Daniken places them. No one knows where the people were in 3000 BC who later were called Inca. Furthermore, there is still much to be learned about ancient Peru, and it is not prudent to say whether or not the ancients there had the loom in the early stages of cotton cultivation.

Barley, boars, and beer. The pig, or boar, and barley were both originally called in English by the same name, *bar*. These two products, boar and barley, were spread throughout the world by sea in very, very ancient times.

Barley may have been a native plant of southwest Asia, but southeast Asia and Ethiopia are also considered possibilities from which the grain spread around the world. The oldest known clay tablet in Babylonia, thought to date about 6000 BC, shows a priest preparing barley for brewing beer. Barley came to Switzerland about 3000 BC, and to Britain and China by about 2000 BC. Borneo was a very early prehistoric center, and the first name for this region was the island of Barley. Barley was considered sacred among the ancients, and we have a relic of this belief in more modern times in the traditional role of monks tending grain and brewing beer. Pigs and barley figure in Homer's *Odyssey*. Circe, from which comes our word *cereal*, trapped the companions of Ulysses. Circe mixed barley meal and penned the men up in pig sties.

Barley beer dates back to the earliest times in Sumer and is mentioned in Egypt as early as the Fourth Dynasty. Beer was the usual drink of the German tribes when the first Roman contacts were made with them. Beer was known to the Gauls. The Kaffir races of South Africa from the earliest times made a kind of beer from millet, and similar drinks were common in Nubia, Ethiopia, and other parts of Africa. Tribes in Russia used barley and rye, and the Chinese and Japanese used rice for intoxicating beverages. Beer was known in Neolithic times at Catal Huyuk, Turkey, and at Jericho. For lack of a better source, the Picts, an ancient tribe in Scotland made beer out of heather. The dubious art of brewing intoxicants spread world-wide at a very early time.[69]

Grain. The consonant pattern of the word *grain*, GRN, is the same as our word *corn*, CRN. They are the same word in origin, coming from the Latin word *granum*. The familiar grains include wheat, barley, rice, maize or corn, millet, and rye. In large measure, these miracle plants, so insignificant-looking in their wild state, have sustained the world from the earliest times. The word *corn* often means the kind of cereal that is the leading crop in a district. In England, corn means wheat, while in Scotland and Ireland, corn is oats. In the United States corn is corn, but also Indian corn or maize.[70]

According to John Cohane the name of the oldest fertility goddess in the world among the ancients is derived from Eve, the mother of all living, who was deified throughout the ancient world. Wheat often figured prominently in ancient fertility rites, and Cohane finds the name of Eve in the word *wheat*, derived from another form of the name Ava.

Anthropologists refer to an explosion of innovation in prehistory which they date to approximately 8000 BC. The great innovators had an amazing botanical knowledge that enabled them to transform wild wheat grass into cultivated wheat by a shrewd selection of seed. Evidence of early botanical wisdom has been found at Jarmo (Iraq), in Turkey, and at Jericho. No satisfactory explanation has ever been given by evolutionists for the stroke of genius that produced sophisticated agriculture. It is indeed beyond imagination to picture rock-chipping, half-human creatures suddenly create the foundation of modern civilization. There are better answers.

The knowledge of agriculture spread quickly as though all were taught by the Old Ones before they spread over the earth. Carbonized wheat kernels found in eastern Iraq, dated about 7,000 years old, were very similar to present-day wheat. Catal Huyuk in Turkey is presumed to date back to 8,000 years ago. The people here cultivated barley, wheat, and peas. The finest kind of bread-making wheat was found in a fine red and black jar in a Sumerian house at Kish in Mesopotamia and was dated to about 3500 BC. Grain seeds dated to about 4000 BC were found in the excavation of a sophisticated site in Macedonia. The first known appearance of rice was believed to be in eastern India dated to about 2000 BC. A recent discovery in Thailand suggests an even earlier date there, and the practice of terracing slopes goes back to earliest agriculture.[71]

There is much mystery and conjecture about the origin of corn, though there is general agreement that the plant is native to tropical America. It is unknown in the native state, but some hold the view that maize originated from a common Mexican fodder grass known as Teosinte. Small grains of an unknown variety have been found in the ancient tombs of Peru. The cultivation of corn in Peru predates the sudden rise of the sea bed. Darwin, for example, found heads of maize embedded in old shores which are now 85 feet above the present sea shore. The spread of corn from America predates the age of European exploration in the Americas. On the basis of linguistic evidence, maize was brought by the Arabs into Spain in the thirteenth century, and this presupposes earlier contact with America. Maize was cultivated in South Africa before Columbus. A drawing of maize was found in a Chinese work on natural history dated 1562, which is too early to be influenced by European explorers in America.[72]

According to Carroll Riley, carbonized grains of maize found in India cannot be later than AD 1435. Maize pollen has been found in various sites in India, for example, in Kashmir, which date back to the thirteenth and fourteenth centuries AD and even earlier elsewhere in Asia. Nigel Jeffreys cites evidence that points to the introduction of maize from America to Asia Minor. From this point maize spread eastward to Persia, to China, and to Europe. The Quechua Indians say that maize was first grown near Lake Titicaca. Due to its high altitude today, corn will no longer grow there.

Honey. Catal Huyuk, that old, old city partially excavated in southern Turkey is unique so far in showing many links between the remote hunters of the Paleolithic, the culture of the Neolithic, and more modern times. Some day it may provide a clear key that conventional chronology needs drastic revision. We now are learning that the Paleolithic hunters overlapped in time with those living in Neolithic cities just as we can show today people still living in a so-called Stone Age. (It is interesting that the lowest levels excavated at Catal Huyuk give indications that honey was known and used.)[73]

Nopal cactus. Plant and animal life team up in Mexico to produce a potent red dye, for example, near Cuilapan in the state of Oaxaca. The microscopic cochineal plant lice grow only on nopal cactus. Some authorities have found a curious circumstance in India. In southern Tibet, Lahore, Kabul, Nepal, and Bengal, cochineal insects were brought in long ago. The insects thrive on cactus there that is similar to nopal, but on no other. It is thought that in ancient times the insects were carried over to India to produce the much-admired dye in that area.[74]

Peanut. The peanut is another New World plant that has been widely distributed around the world since the time of the European explorers in America. In another hint of ancient contacts between Old and New Worlds, peanut remains were found in Chekiang Province, China, and were dated between 2100–1800 BC.[75]

Pineapple. The pineapple is accepted as native to tropical America. The Inca appreciated its taste, and pineapples were carved on Inca wooden cups. The first mention of the fruit in English literature occurred when John Evelyn in his *Diary* wrote that he had tasted a pineapple from Barbados at the table of King Charles II. Attempts were made in 1712 to cultivate the fruit in England. But there is unmistakable evidence that the pineapple had come to Europe much earlier. This means that the ancients traveled more widely than we customarily think. The pineapple appears in a mural excavated at Pompeii, and the identification is considered certain. Pompeii was buried by the eruption of Mount Vesuvius in AD 79, over 1,600 years before the pineapple was first brought to England.[76]

Reed. Hardly anything seems more insignificant than the reed. We have word pictures to convey this idea: the bruised reed, and the reed shaken in the wind. Yet the reed played a remarkable role in prehistory. We shall stretch the term a bit to include rushes and vegetable fibers. Swamps still exist in parts of Egypt, but the papyrus from which we have the word *paper* is no longer found there. This plant still flourishes elsewhere. Sebastian Englert shows many links between South America and remote Easter Island, and among these is the totora reed that grows in the crater lakes of Easter Island.[77] This is the same species of reed that grows off the shore of Lake Titicaca in Bolivia and Peru. Plantings of this useful species were undoubtedly carried on sea voyages from South America to Easter Island in ancient times. On Easter Island, burials have been found where the body was wrapped in mats of the totora reed. The Incas of Peru followed a similar burial practice. Across the Atlantic and off the coast of Africa, the same kind of burials was found among the Guanches of the Canary Islands. These mysterious and now extinct people deserve a separate discussion.

Instead of brown paper bags, we can assume the ancients used plaited rush containers to hold food. As noted earlier, Darwin found plaited rush in uplifted sea beds near Lima, Peru. At the ancient mining center of Bomvu Ridge in Swaziland mattings of grass and vegetable fibers were made for beds. These were C14 dated at more than 50,000 years, a figure we need not take seriously.

There is evidence of ancient shoemakers plying their trade at an early date. They used vegetable fibers to make intricately woven sandals. Fifty miles southeast of Bachelor Butte near Fort Rock, Oregon, a cache of 100 sandals was discovered. They were dated by means of C14 as 9,000 years old. In another source on the same remarkable discovery we learn that there were 600 sandals, or 300 pairs. In a more reliable report, Wormington informs us that 75–100 well-made sandals of shredded sagebrush bark were found in Fort Rock Cave. Obviously somebody can't count. All the sandals were charred from the Newberry Crater eruption that is dated no earlier than about 300 BC. Luther Cressman of the University of Oregon made a similar discovery of 200 pairs of woven fiber sandals in Lamos Cave in eastern Nevada, and these were also dated over 9,000 years old. They were skillfully made by an artisan and look like modern beach sandals.[78]

The reed was used by the ancients for making homes. This was true of western Indo-Europeans. The German word *Wand*, meaning wall, and the related word *winden*, refer to constructing walls out of intertwined reeds or twigs.[79]

I. W. Cornwall marvels at the continuity between ancient and modern times. Around 4000 BC the Sumerians used long reeds tied in bundles for house frames, including arches. The frames were covered with reed matting. Drawings of this kind of home were found on clay tablets a few decades ago in excavations at Al'Ubad on the Lower Euphrates river. The remarkable thing is that the Ma'dan people, also called Marsh Arabs, build precisely this kind of home today, following an old tradition.

But we can marvel still more. The same reed structure appears in an intriguing wall painting at Catal Huyuk, one of the oldest sophisticated cities of the world in southern Turkey. The house is unmistakable. It is built with four gables separated by five pillars. Again, the same reed structure is found in predynastic Egyptian art. The same kind of hut is also built today in the marshy regions around the lake of Eber, northwest of Konya, Turkey. Despite invasion and catastrophe over many centuries, the blueprint of this structure has been carried down in memory to the present.

Thor Heyerdahl convincingly demonstrated that the reed boat could travel over enormous distances on the ocean. The ingenious reed boat of the Nile was similar to that used today on Lake Titicaca in South America. Gerald Hawkins believes he has found the site where papyrus rafts were launched from Africa for the long voyage to South America in ancient times. Shimish, which lies beyond Gibraltar, is on the northeast coast of Morocco. It was originally called Maqon Smese, the City of the Sun. The site was ancient when the Romans came and built Lixus on its ruins. Recently huge megalithic walls were uncovered under the ruins of Lixus. The stone work is equal to that of Egypt or Peru. According to Hawkins, the site is a critical one in the world-wide pattern of astro-megaliths. He notes one more thing: The papyrus raft was used on the river at Shimish, just as in Egypt and on Lake Titicaca.[80]

Among the Indians of the coast of Venezuela there are traditions of people who arrived on their shores in reed rafts. We know that the reed boat was used in ancient Sumer. Again it is interesting that boats, about 15 feet long and made of bundles of reeds, are used today by the fishermen of Bahrain Island in the Persian Gulf. They are buoyant but not watertight, and are therefore technically rafts. They are similar in design to the reed boat used on the Nile earlier than 2000 BC. Early explorers in central California found that the Indians shaped bundles of reeds into boat form and used them for water travel. Moellhausen, an early traveler to the West, found to his surprise that Indians on the Colorado River floated along on little rafts made of bundles of rushes. These were the only watercraft he observed in the Colorado River valley. Many ancient canals have been noted in the desert of southwestern Arizona near the Salt and Gila rivers. One ditch was 25 miles long. In this area archaeologists found remains of reed rafts. The trail of the reed around the world is an impressive one and tells an eloquent tale of ancient world-wide travel and a common source of technology.

Sweet potato. Peruvian sweet potatoes were carried in the remote past over long stretches of ocean to various Polynesian islands. When the first European explorers came to Pacific islands, they found extensive plantings of sweet potatoes on Easter Island, Hawaii, New Zealand, and the rest of Polynesia. The natives had legends that the plant was brought to them. In the islands the sweet potato is called kumara, the same name given to this plant in old Peru. Pearl Buck suggests that in the course of extensive voyaging in the Pacific, the Polynesians may well have reached the coast of the Americas. They could have returned to the islands with plantings of the sweet potato and other plants.[81]

Carroll Riley adds another dimension. The unique occurrence of the Peruvian word for sweet potato was found in a Chinese document dating to about AD 300, along with a rather vague description of it. The plant eventually spread to Melanesia, Micronesia, and Indonesia.

Tobacco. The use of tobacco has an unexpected history.[82] On July 12, 1970, the *New York Times* carried a Soviet report regarding Asians discovering America. As evidence, ancient Tibetan maps were cited. These maps refer to a green land lying far across the Eastern Sea (Pacific Ocean), which could only be America. There is evidence that American tobacco was used among Asians around 1000 BC. Furthermore, the purely American word *tobacco* penetrated a number of Oriental languages far back in antiquity. We have already noted that New World tobacco was in use many centuries earlier in Egypt.

Wine. Wine has been known from the earliest times. Noah must be credited with the first mention of wine in historical records. It is interesting, but no surprise, that wine residues were reported in ancient jars in the highlands not far south of Mount Ararat, and are thought to date back to 5400 BC. The report also stated: "Based on current archeological information and historical considerations, a single origin for the domesticated grapevine in some northern mountainous region of the Near East makes the most sense."[83]

Wine was a familiar drink in Catal Huyuk, Turkey. The site of Photolivos in the Macedonian province of Drama was recently excavated. A very sophisticated village was uncovered where copper was extracted from ore. Pottery was well-preserved, and one ceramic piece resembles the camel, a hint of domestication some thousands of years before the official date for this development. The site is dated to about 5000 BC, long before the Old Kingdom in Egypt began. Among the finds were carbonized grape pits, tentatively dated at 4000 BC. Wine was a part of their culture.

Concluding Thoughts—the Message

Language and Literary Clues

We have repeatedly found indications of ancient and sophisticated travel over much of the earth and across all the oceans. John Cohane, for example, is satisfied with the evidence that long before the Phoenicians, and even before the time of the Egyptians, certain key names and words were taken to all parts of the world from the eastern Mediterranean by land and water.[84] In addition to many geographical names, names of animals, birds, fish, flowers, and trees lie only half concealed in the spoken and written languages of widely separated peoples. Such key words, blended in many combinations, are found all over the world. Almost everywhere in the world there are deep and persistent myths that the original settlers or the early visitors arrived by ship from overseas. There are strong indications of two great dispersions of people in ancient history. Key words in the first group are found in all parts of the world. Another group of key words is found widely dispersed, but limited to the Mediterranean, Europe, Africa, parts of Asia, West Indies, Brazil, the Gulf Coast of Central America, and east coast of North America, Japan, Philippines, Australia, and New Zealand.

No End of Mysteries

We must not convey the impression that creationists are agreed on every detail about the early world, even though the biblical framework is fully accepted by them. On many questions there is simply inadequate evidence, or we do not know how to read the evidence that lies before our eyes. It is freely granted that many questions may never be satisfactorily answered, nor is there any real need to do so, other than to satisfy our curiosity about the past. Many valuable and interesting insights have been gained from a study of scientific facts, as opposed to conjecture. This is a valuable undertaking which shows that the Christian who takes Genesis seriously certainly has nothing to fear from scientific data. We must never confuse explanation imprisoned by evolutionary theory with the observations we can make in the world around us.

Evolutionist Charles Reed summed up his life's work in a massive volume on the origins of agriculture.[85] After working through the volume, I wrote to the author/editor. I pointed out to him that every fact reported in the volume fits exceptionally well into a biblical framework. Our only points of disagreement were the labored and unconvincing interpretations that showed a loyalty to evolution.

When all is said and done, everything in this chapter that we were able to explore and observe about plant life around the world fits well within a biblical framework—facts, mysteries, and all.

For Discussion and Reflection

1. How can plants tell us important things about ancient history, land bridges, sea travel, radical changes in climate, catastrophic events?

2. When Noah planted a vineyard (Genesis 9:20) after the ark landed, does this hint that he and his family brought seeds and perhaps seedlings into the ark for planting in the new world?

3. Jesus taught important spiritual lessons to His hearers using familiar plants. What were some of the messages when Jesus spoke of good seed, bad seed, good plants, evil plants, good trees, useless or evil trees, etc.?

4. What is the lesson of the kernel of wheat that falls to the earth and dies (John 12:24)?

5. What was Jonah's experience with a plant (4:6–11)?

6. When God finished creation, He declared that everything was very good (Genesis 1:31). How do you explain the cactus and poisonous plants? See Genesis 3:17–18.

7. See Genesis 1:28–29. What was the role of plants in creation? Note that in God's creation there were no carnivores. They became one of the consequences of the fall into sin.

8. Note that in this chapter, as well as all the other chapters, Christians have nothing to fear from scientific discoveries. Is interpretation by evolutionists the same as scientific discovery? If not, how do they differ?

9. Darwin observed that the failure to trace back any plant group to its ancestor was an abominable mystery and a grave failure of his theory. It seems fair to say that evolution explains nothing and predicts nothing. Does a young, created world provide a good framework for our world?

PART FOUR

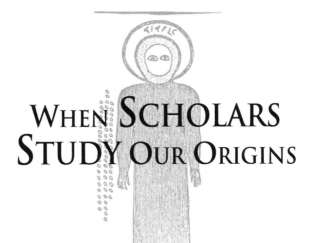

WHEN SCHOLARS
STUDY OUR ORIGINS

CHAPTER THIRTEEN

REFLECTIONS ON DARWINISM

JOINING THE CAUSE

In the continuing debate between creation and evolution, a university student provided this insight: "We are determined to believe evolution, not because it is true nor that we believe there is any evidence for it, but because it has become the symbol of our liberalism." [1]

That seems to say it all. As this chapter unfolds, we need to quote Einstein again: "It is the theory which decides what we can observe." Wherever evolution seems to look good, it is because much contrary evidence is not permitted to be considered.

How the Feather Evolved

Who can resist this explanation of how feathers evolved? Randy Wysong gives this account of Gerhard Heilmann's 1926 work about the origin of birds:

> The pressure of the air, acting like a stimulus, produces chiefly longish scales developing along the posterior edge of the forearm and the side edges of the flattened tail. By the friction of air, the outer edges of the scales become frayed, the frayings gradually changing into still longer horny processes, which in course of time become more and more featherlike, until the perfect feather is produced. [2]

The above is total nonsense, of course. However, it reflects the enthusiasm of evolutionists who fervently believe they can explain anything and everything about life. Loren Eiseley gives us a candid picture of the mind frame of the evolutionists. They replaced God with time and chance.

Science scoffed at the belief in creation because modern people should not believe in myths and miracles. Then nothing turned out to be provable in evolutionary theory, so evolutionists had to invent their own myths and miracles.[3]

The above reminds us of early efforts by evolutionists to evolve a tailless mouse. In the experiment, scientists lopped off the tails of many generations of mice, assuming that eventually mice would be born without a tail. The experiment failed.

Erosion of the Bible

Man has never been at a loss for finding reasons to reject the Word of God. Charles Darwin, whose *The Origin of Species* appeared in 1859, was a natural consequence of the Enlightenment of the previous century. Rationalistic and supposed scientific approaches then were applied to religious, political, social, and economic issues, and thus the influences of Wellhausen, Marx, Spencer, Freud, and Dewey were felt. Darwinism was one of the elements in such eroding influences as the denial of all absolutes, downgrading the Bible into a set of secular writings, and placing a focus on change and process. The church and the Bible seemed all but swept aside by evolution in the great enthusiasm for what was believed to be the great unifying principle for all human knowledge. Sweeping predictions were made that a large number of fields of knowledge would be greatly modified if not revolutionized by the application of Darwin's theory: astronomy, geology, biology, anthropology, philosophy, ethics, religion, history of social institutions, sociology, psychology, education, history, political science, economics, and others.

Quick and Easy Non-Science

In the rosy dawn of evolutionary thinking, there were quick and easy answers to show how everything had evolved, and each time fellow evolutionists applauded the marvelous new revelations. No one thought to ask whether this was really science. For example, it was claimed that the family evolved according to the following pattern: sexual promiscuity, various states of matriarchal family, progression through stages of patriarchal family, and the small conjugal family of today. In economics mankind was alleged to have gone through these stages: food collecting, hunting, cattle breeding, agriculture, and industry. Technology was supposed to have followed the stages of wood, stone, bronze, and iron. In the legal field development was conceived to have taken the following steps: from common to individual property, from status to contract. In the realm of religion the evolutionary development was portrayed as having been: magic, animism, totemism, personal deities, and from polytheism to monotheism. Traces of this kind of thinking remain in textbooks, but evolutionists generally refer to such naive pronouncements with embarrassment.[4]

The Peppered Moth

When evolutionists are asked for their best proof of evolution, they still often point to the peppered moth in England. It always had two varieties, white and black. Before the Industrial Revolution the white was more numerous. After smokestacks blackened the countryside, whites were often picked off by predators and blacks became more plentiful—a wonderful example of evolution, they said. Every biology textbook showcased pictures of these moths exposed on tree trunks. The late Stephen Gould of Harvard taunted those who believe in creation with the moth example.[5] Creationists did not take the matter seriously because nothing more than variation was involved, and the moths remained moths. Recent new research confirmed that neither kind of moth spends time on exposed tree trunks. In fact, the famous textbook pictures were faked by an Oxford University professor. Moths were glued to tree trunks to show what was believed to be evolution in action. Nothing about the story was true. In yet another stirring defense of evolution the two shades of the peppered moth are still featured as a prize exhibit, though the author has the grace to say this is a "disputed" example of evolution in action.

A Revolt Against Christian Beliefs

Darwinism was the banner under which many battles in many areas were to be fought. Professions of faith in Darwin were less a conviction in some cases than a signal of solidarity in the minds of the adherents against outmoded religious beliefs and other superstitions. Evolutionist Loren Eiseley declared that evolution "is essentially a revolt against the Christian conception of time as limited and containing historic direction with supernatural intervention constantly imminent."[6]

We often need to translate the jargon concealing the myth. Eiseley here says that there is no God or God is irrelevant, and we live on a very old earth that gradually evolved into what we see today.

Blind Loyalty to a Cause

There are some peculiar and interesting instances of total loyalty to evolution, yet anger and frustration among evolutionists about futile attempts to apply evolution in certain ways. Because of the factor of loyalty to a cause that overshadows actual practical applications, we can speak of the great influence of Darwinism and yet stand amazed at persistent but vain attempts to apply it to various areas of knowledge. Allegiance, not practical application or valid explanation, is the name of the game among evolutionists.

A Radical Shift

Even evolutionists were forced to realize that the fossil record provided no support for slow gradual change, Darwin's most basic teaching. Change from one species to another over time is known as macroevolution, but the fossil record provided no examples. Leading evolutionary experts from around the world met in 1980 in Chicago to grapple with the problem.[7] Starkly put, the problem was to dump Darwin, yet remain 100 percent loyal to Darwinism. Various schemes over a period of some decades had been proposed with such names as cladistics, hopeful monsters, saltation, punctuated equilibrium, etc. Basically these had a focus on sudden change, so sudden that there was no time for the fossil record to show such changes. Instead of embarrassment over what the fossil record did not show, the new position in a way was to show that the proof for evolution lay with what the fossil record did not possess. With this view, missing links, that is, transitional forms, were no longer necessary. This was fortunate for the evolutionists because about 150 years of searching had produced no such forms. The driving force for the new belief, called the "fact" of evolution, was Harvard's Stephen Gould. Proving things by what is not there has never been a respectable position, but evolutionists were forced into it. Despite the new doctrine by the leading experts, textbooks have changed little. They are still steeped in gradualism, and students are not permitted to ask honest questions about it.

Competent Critics Speak Out

One does not need to go to the writings of creationists to point out fallacies and contradictions of evolution. We give only a few of many, many examples. Charles Darwin stated that if it could be demonstrated that any complex organ, such as the eye, existed that could not possibly have been formed by numerous, successive, slight modifications, his theory would absolutely break down. Michael Behe, biochemist and author of *Darwin's Black Box* states: "Applying Darwin's test to the ultra-complex world of molecular machinery and cellular systems that have been discovered over the past 40 years, we can say that Darwin's theory has absolutely broken down."[8] Evolutionist Herbert Nilsson, distinguished botanist, said:

> My attempts to demonstrate evolution by experiment carried on for more than 40 years, have completely failed. At least, I should hardly be accused of having started from a preconceived anti-evolutionary standpoint.... It may be firmly maintained that it is not even possible to make a caricature of an evolution out of paleo-biological facts.... The deficiencies are real, they will never be filled.[9]

A basic teaching of evolution is that nothing happened in the past that is not going on today. Luther Sunderland asked many prominent museum officials whether they knew of any example of fossil deposition going on today on the bottom of any lake, sea, or ocean. No one could give an example.[10] This is a devastating admission.

Phillip Johnson, a tenured law professor at the University of California at Berkeley, has created a great impact with his series of books, the first of which is *Darwin on Trial*. One of his specialties is "analyzing the logic of arguments and identifying the assumptions that lie behind those arguments."[11] In this must-read amazing book, with many examples, Johnson shows that evolution is simply a belief system posing as science, and very poor science at that.

Mixing the Vague and Precise Theories

A half century ago Anthony Standen explained evolution theory this way. There are really two theories of evolution, the vague theory and the precise theory. The vague theory seems to prove that all forms of life are connected in some way. But to show in what way they are connected a precise theory is needed, namely that all forms of life on the earth today came from some original form of life by a series of changes which, at every point, were natural and explainable by science. By mixing up the vague theory with the precise theory, they give the impression that both have been proved, whereas the precise theory has never been proved at all—it is accepted as a faith.[12] Here is the dilemma for the evolutionists: It is easy for them to imagine how man was evolved from an amoeba (vague theory), but even today they cannot form a plausible guess as to how two closely related herbs were evolved, either one from the other, or both from a common form (precise theory). As a grand and glorious viewpoint, evolution is as easy as can be, but in getting down to any actual details, difficulties begin. One must observe that this is faint praise for the merits of the "fact" of evolutionary theory. It is one thing to say the oak is our cousin, but quite another matter to begin to spell out how this all came to be.

Strange Orthodoxies in Evolution

Another acknowledged spokesman for evolution speaks of orthodoxies in evolution that have been found wanting: Reliance on mutation as the one important source of change; natural selection ideas; the principle that evolution could never reverse itself; and the value of studies of fruit flies. Others can easily be added; for example, Standen states that the many genealogical trees showing how man developed from simpler forms of life have fallen to pieces. Yet the inadequate orthodoxies cited by William Howells and others keep appearing in the texts right up to the present without a blush and without any hint that they are less than true.

What one evolutionist discards as untrue, inadequate, or simplistic, other evolutionists seize upon as convincing proofs of the doctrine. For example, Ronald Duncan says that the present impasse in evolutionary thinking, productive of so many fallacies, is due chiefly to the interpretation of biological fact in terms of out-of-date physical theory. What physicists abandoned in the nineteenth century, evolutionists eagerly seized and misapplied in the next century. He points out the absurdity of going from non-life to life, unless we take leave of science and find refuge in scientific myth.

Carl Sagan tells us that the information contained in a simple cell, invisible to the eye, is equivalent to about a hundred million pages of the *Encyclopedia Britannica*. Yet we are to believe that a sterile speck of dirt or drop of sea water somehow evolved into this awesome miracle, and that it is irrelevant to ask how this miracle came to be.

How Evolutionists View Creationists

The evolutionist is certain that creationists are anti-scientific, as though being opposed to evolution is the same thing as being opposed to science. The argument commonly offered is that if one accepts water as being made up of oxygen and hydrogen, then one should also in good conscience accept evolution because both are "science." The issue is much deeper. Evolution rests on faith. Creation rests on faith. It never was an issue of science versus religion. Evolutionist O.H. Matthews, who wrote a preface to a university edition of Darwin, *The Origin of Species*, made this clear statement:

> [B]iology is in the peculiar position of being a science founded on an unproved theory. . . . Belief in the theory of evolution is thus exactly parallel to belief in special creation—both are concepts which believers know to be true but neither, up to the present, has been capable of proof.[13]

Man the Machine

In 1969 McGraw-Hill promoted the book *Mechanical Man* by Dean Wooldridge, which was said to draw on recent discoveries in the fields of biophysics, biochemistry, neurophysiology, and electrophysiology.[14] The author concludes that the origin of life and all human behavior "are entirely the consequence of the normal operation of the ordinary laws of physics in inanimate chemical matter." It would be difficult to imagine a lower and more incorrect view of humans.

George Gaylord Simpson has been credited with producing the best general work on the meaning of evolution prior to 1950. The difficulty of the evolutionist in speaking with relevance to religion, to the humanities, to the social sciences, and to the process and purpose of education may be seen by the premises on which evolution rests. As we see below, Simpson believes that whatever is good, right, ethical, and valuable must somehow be derived from the evolved universe, so there is no need for a God to enter the picture. Christians sharply disagree.

Simpson, for example, states the following:

> Man is the result of a purposeless and materialistic process that did not have him in mind. The discovery that the universe apart from man or before his coming lacks and lacked any purpose or plan has the inevitable

corollary that the workings of the universe cannot provide any automatic, universal, eternal, or absolute ethical criteria of right and wrong.[15]

The Influence of Darwinism on Religion

No one can doubt the great influence Darwinism has had on our culture. The appearance of Darwin's book in 1859 caused a great stir within the churches.

Initially there was opposition by churchmen, but many soon found ways to accommodate the theory of evolution to the Bible. This triumph for evolution illustrated to no small degree the intimidating power of theory presented as science. Thomas Huxley led the attack:

> You (clergy) tell your congregations that the world was made 6,000 years ago in six days, and that all living animals were made within that period ... I am bound to say, I do not believe these statements you make; and I am further bound to say that I cannot call up to mind amongst men of science and research, and truthful men, one who believes those things, but, on the other hand, who does not believe the exact contrary.[16]

Raymond Surburg's survey a half century ago of the eroding effect of Darwinism on religion may be summarized as follows, and is as true today as penned decades ago:

> For the vast majority who soon came to accept evolution, embracing the theory meant progress and acceptance of change. As new scientific concepts were discovered and developed, the belief was that religion to be relevant must change with science. Many welcomed the fact that at last reason could take an honored place within religious beliefs. In harmony with beliefs about evolution the Bible was reinterpreted to show how the idea of one God gradually developed over a long period of time. Similarly doctrines gradually developed into their present form. St. Paul's doctrines evolved out of Jesus' teachings. It was of course impossible to hold on to literal meanings of the Bible in order to accommodate evolution as the scientific explanation of how the creation really happened. Biblical accounts became valid only insofar as they taught a lesson of some kind. Since the Bible was viewed as a human document like all other so-called sacred books, Christianity in order to be honest had to recognize the validity of all other religions. Thus the idea of sending missionaries became invalid and the focus was placed on hospitals, schools, and agricultural assistance. Since science has no room for an afterlife, concerns were limited to this life with the Gospel becoming the social gospel—the Golden Rule.[17]

Many Christians, however, believed they had cause to reject evolution, and they continued to accept in faith the creation story just as it was written. They could see that to reject or revise the creation story would be only the first link of a long chain. If the creation story were abandoned, the fall of man would fall with it. God's covenant with man culminating in the coming of Jesus Christ would lose its meaning. Natural selection was unacceptable as a substitute for being the children of God.

At issue were some very fundamental points: The uniqueness of Jesus Christ and the atonement; unchanging truth; inspiration; the infallible Word; and interpretation of the text—what is to be taken literally. Those Christians did not want a Bible where nothing really happened the way it was described, and where someone had to come along constantly and tell them what God really meant to say with this or that story. Evolutionists were determined to rid the church of what they saw as outworn mythologies, such as the old argument about design in nature showing God as architect, whether the world was really created in six days, whether Adam and Eve were real persons, did the serpent actually speak, did the flood really happen, and the like.

The Influence of Darwin on the Humanities

The effect of Darwinism on philosophy and history deserves examination. Other areas of the humanities, such as literature, music, and the other arts can be related to Darwin little if at all unless one attempted to build a bridge back to Darwin through Freud, who welcomed Darwinism as a means of stimulating a decline of religious belief. In literature, works by George B. Shaw and many others reflect evolution, but it is in history and in philosophy where Darwinism appears as a significant influence.

A generation ago Raymond Surburg summarized the influence of Darwinism on philosophy in this way:

> Various evolutionary theories and concepts had been developed long before Darwin. The ancient Greeks, plus Lucretius, Kant, and Laplace, advocated some form of evolution. By the time of Darwin, however, a great deal of data on animals and plants had accumulated as well as information on the geological structure of the earth. This information seemed to make possible the production of a reasonable and comprehensive theory of organic evolution, in addition to the theories of cosmic evolution and physical evolution that had been advanced previously by Hegel and Kant. At this juncture Darwin's contribution was made. At one stroke it appeared that man, nature, and the universe itself could be traced back to the very beginning.[18]

These new concepts of the evolutionary process eventually affected ideas about the nature of reality, the nature of knowledge, and the nature of morals. In the philosophical fields of ontology, epistemology, and ethics, the influence of

Darwinian thinking was very pronounced. In the second half of the nineteenth century Herbert Spencer tried to work out a synthetic philosophy based on the first principles taken from biology, psychology, sociology, and ethics.

Spencer's eagerness to apply the theory of biological evolution to all of life, including ethics, however, showed disastrous consequences. No one would support his formulation today, despite the fact that his words flow very logically out of Darwinism.

> The poverty of the incapable . . . starvation of the idle and those shoul-
> derings aside of the weak by the strong . . . are the decrees of a large, far-
> seeing benevolence.[19]

Henri Bergson (1859–1941), French philosopher, wrote *Creative Evolution* in 1907 in which he accepted evolution as fact. In America Darwinism was a factor in the development of a new school of philosophy, pragmatism, with which Charles Pierce and William James exercised a great influence upon twentieth century thinking. Pragmatists minimized the distinction between thought and action, asserting that thinking was a part of a process of interaction between the organism and the environment. Thus the mind was a part of the total organism and therefore subject to development and change in the same way as the organism itself, thus reflecting Darwin's theory of natural selection.

Darwin and the Belief in Naturalism

Darwinism influenced the system of thought known as naturalism, the belief that everything can be explained in terms of natural causes and laws. Naturalism assumes that there is no spiritual or supernatural element involved in any event. Thus the ultimate explanation for every question was to be found in nature. Man is both a product of nature and continuous with nature. It is evident that naturalism is an element within many philosophies such as positivism, pragmatism, and socialism. Naturalism laundered God out of everything.

Pierre Teilhard de Chardin (1881–1955) illustrates an attempt in the twentieth century to apply Darwinism to both religion and philosophy. Teilhard, anthropologist and Jesuit priest, was obsessed with the necessity of using evolution to breathe new life into the church, and his rather mystical writings on the subject became something of a sensation in the 1960s, but neither scientists nor theologians could warm up to his grandly eloquent but eccentric scheme. His biographers, Mary and Ellen Lucas, granted him cult status among admirers in France and the United States, but placed his works "quite outside the mainstream of modern thought." In the 1970s British historian Hugh Trevor-Roper dismissed him as one of the "great charlatans of modern letters."[20]

The Influence of Darwinism on the Writing of History

This is found in both obvious and subtle applications. When historians treat the ancient world, they commonly present some kind of chronology that may begin with the birth of the universe or with the supposed date of the earliest man-like creature. The student is then taken step by step down to the sudden but unexplained appearance of urban civilization in Sumer or in Egypt. The chronologies follow evolutionistic assumptions and the conjectural nature of the dates offered is seldom hinted at. As all writers copy the same dates for the same supposed events in the dim past, the sheer repetition of the dates gives a ring of "truth" about them, at least to the true believers.

It is instructive to observe how historians who accept evolution as a fact chafe in their futile attempts to apply it to ancient history. Stewart Easton stresses the almost total ignorance about prehistoric man and the experts violently disagree about the little that is thought to be known. Speaking of evolution, Easton comments:

> In this age, on principle, we are inclined to prefer even the most far-fetched of material explanation to the possibility of any kind of divine guidance or intervention, or the fulfillment of any divine purpose. Chance and probability appear to us so much more scientific, and therefore more credible, than a super-human power and wisdom which could direct the course of evolution.[21]

Another evolutionist, Giorgio de Santillano states:

> The simple idea of evolution, which it is no longer thought necessary to examine, spreads like a tent over all those ages that lead from primitivism into civilization. Gradually, we are told, step by step, men produced the arts and crafts, this and that, until they emerged into the light of history . . . Those soporific words "gradually" and "step by step", repeated incessantly, are aimed at covering an ignorance which is both vast and surprising. One should like to inquire: which steps? But one is lulled, overwhelmed and stupefied by the gradualness of it all, which is at best a platitude, only good for pacifying the mind, since no one is willing to imagine that civilization appeared in a thunderclap, . . . The lazy word "evolution" had blinded us to the real complexities of the past.[22]

In *The Roots of Civilization*, archaeologist Marshack is equally irate:

> Searching through the historical record for the origins of the evolved civilizations, I was disturbed by the series of 'suddenlies.' Science, that is, formal science, had begun suddenly with the Greeks; in a less philosophically coherent way, bits of near-science, mathematics and astronomy, had appeared suddenly among the Mesopotamians, the Egyptians, the early Chinese, and much later, in the Americas; civilization itself had appar-

ently begun suddenly in the great arc of the Fertile Crescent in the Middle East; writing . . . had apparently begun suddenly with the cuneiform of Mesopotamia and the hieroglyphs of Egypt; agriculture . . . had apparently begun suddenly . . . the calendar had begun suddenly . . . .[23]

The Influence of Darwin on the Social Sciences

The effect of Darwinism on sociology and psychology will be examined here in some detail with passing attention also given to other social sciences where some effect of Darwinism may be seen. Johannes Stark and Conrad Waddington[24] note that in the second half of the twentieth century the social sciences were predominantly Darwinian, but that shifting Darwinian modes of thinking from nature to society and culture created difficulties. Karl Deutsch and others developed an admittedly subjective list of sixty-two great breakthroughs in the social sciences for the period of 1900 to 1965. We can see or at least infer some possible influence of Darwinism in ten of them, for example, psychoanalysis and depth psychology (Freud, Jung, Adler); gradual social transformation (Shaw, Wells); pragmatic and behavioral psychology (Dewey); conditioned reflexes (Pavlov); sociology (Parsons, Wilson); structural linguistics (Jakobson); culture and personality and comparative child rearing (Benedict, Mead); operant conditioning and learning (Skinner); and structuralism in anthropology (Levi-Strauss). Perhaps there are several more. For the all-embracing claims made for the importance of Darwinism, ten of sixty-two breakthroughs do not seem impressive. Even in the ten cases cited, the Darwinian influence is sometimes marginal at best.

Darwin and Sociology

Darwinism was a strong influence in the early development of the field of sociology. Evolution was applied to explain social behavior, to trace the origin and development of social structure and social institutions. In 1876 and following years Herbert Spencer may be said to have founded the field of sociology, grounded in evolutionary thought. The views of Spencer found ready acceptance, more so in America than in his native England.

In the past decades there have been other efforts to apply evolution to the field. Ideas of social evolution seemed dead by the mid-twentieth century, but new life came into the concept in the form of Edward Wilson's sociobiology, at least for a time.[25] It was a renewed but bizarre attempt to find the biological basis of such social behaviors as aggression, genocide, male dominance, military discipline, spite, altruism, ethical behavior, love, and division of labor. Wilson's argument rests on supposed similarities between human and animal behavior. Having been burned before, however, sociologists are in the main wary of an approach that seems too much like what has already been discarded.

With so much emphasis today on caring, helping relationships in the community, social work as an application of the field of sociology cannot look to Darwin for a helping ethic. Darwin brutally stated disapprovingly that civilized men do their utmost:

> [T]o check the process of elimination; we build asylums for the imbecile, the maimed and the sick; we institute poor-laws; and our medical men exert their utmost skill to save the life of everyone to the last moment. ... Thus the weak members of civilized societies propagate their kind. No one who has attended to the breeding of domestic animals will doubt that this must be highly injurious to the race of man.[26]

Such talk became grist for militarists, nationalists, and racists, as we know only too well.

As we have seen, text authors appear to accept biological evolution, but it seems fair to state that in some cases more space is devoted to pointing out that it has not been successfully applied than to any virtues of applying evolution to the field of sociology.

Psychology and Darwin

Psychology has been significantly affected by Darwinism. Darwin, for example, suggested that many human expressions of emotion are merely continuations of actions useful in the animal world, for example, the sneer is a continuation of the animal's preparation to bite. A lengthy comparison of the mental powers of man and the lower animals was made by Darwin, who believed that animals showed evidence of imitation, curiosity, imagination, and even of reason. Darwin's approach was extended by Herbert Spencer, who wrote one of the first books specifically on psychology, and by George Romanes, who developed the field of comparative psychology, that is, animal studies intended to discover insights into human behavior.

William James accepted the idea of evolution in his *Principles of Psychology* and described mental life as a biological function of adjustment between impressions made upon the body and the reactions of the body upon the world. He also claimed that mental activity was always accompanied by bodily changes and that ideas lead to action.[27]

George Stanley Hall (1904) won acceptance for a time in applying recapitulation theory to psychology, that in the development of the individual before and after birth are re-enacted all the steps of evolution, beginning with the remotest primitive ancestral organism. This quaint notion has been long since abandoned by evolutionists, but it is not unusual to find elements of this naive and totally unsupported idea in textbooks today.

Freud and the Neo-Freudians have exerted the greatest influence in the twentieth century in some areas of psychology such as the study of personality and psychotherapy. Freud's influence has extended beyond psychology into biography, literature, ethics, history, religion, and the arts, though it must be said that much of these expressions rest on a pop-psychology level. While no direct tie with Darwinism is apparent, Freud's concepts of sex and aggression as the driving life forces for man is in harmony with concepts of evolution.

John Watson's behaviorism is perhaps the most clear cut example of evolutionary principles applied to psychology. Watson argued that all behavior could be reduced to pure physiology. Human and animal behavior were of the same stripe. Learning in man and in animal consisted only of acquiring appropriate stimulus-response bonds. Skinner, the most influential psychologist in the last half of the twentieth century, developed operant conditioning that rests on the same evolutionary assumptions and principles.

Evolution is accepted by most behavioral scientists as a biological fact. The assumption is made that the learning process is basically the same across species lines, and thus animal studies are considered relevant to human learning by most psychologists. Thus we find much of what we think we know in learning theory based on applications of Darwinian assumptions. Some areas of psychology, however, are not so influenced, because no way has been found to do so. It should be noted that there is marked controversy within the field of psychology on the relevance of animal studies. Some argue that apart from any biblical considerations, some human functions closely resemble some animal functions with the extent of such correspondence unknown. Further, the scientist may accept the supernatural and spiritual values as spheres not open to study with the tools available to the scientist. Yet it must be said that many scientists believe in a world with no miracles and nothing supernatural.

It took Charles Judd of Columbia University 125 pages of his text in educational psychology in 1939 to build his case for the theory of organic evolution as a necessary foundation for the understanding of psychological concepts.[28] A generation later many texts mention neither Darwin nor evolution, though as noted above, the application of animal studies to human behavior normally (but not necessarily) assumes a continuity between man and other animals. In most cases, however, one can assume that the de-emphasis of evolution in psychology texts does not stem from a lack of commitment to the doctrine of evolution, but rather that, as in the case of sociology, nothing of value has been found. Even in the case of animal studies, researchers may prefer animals to humans for reasons of ethics, cost, and for better control of experimental conditions, rather than for any particular belief in evolutionary principles.

Some psychologists severely criticize animal studies in psychology, because the studies treat tasks so simple and artificial that they have no counterpart in real life. For some time there has been a strong movement within psychology, called humanistic psychology, which stresses the fact that man is qualitatively more than an animal, but this does not mean that humanistic psychologists have any particular objection to the doctrine of evolution as long as it is not applied in areas of their direct interest.

Evolutionary theory can be traced in other social sciences. Marxism predates Darwinism and preached revolution rather than a laissez faire evolution, but sometimes Marxism found it useful to borrow such terms as "natural selection" for certain purposes. Walter Bagehot (1826–1877) went further than Spencer in applying Darwin to politics. Thus Darwinian concepts opened the door for racism and the non-compassionate use of accumulated wealth to be sanctioned and made respectable in the name of "science." Later Darwinism applied to politics helped produce, along with many other factors, an even more far-reaching conclusion: the Nazi movement in Germany. Ernst Haeckel, who championed evolution in Germany, was honored by the Nazis as a precursor of national socialism. Himmler stated the law of nature must take its course in the survival of the fittest. The result was the gas chambers. Hitler pressed the idea that Christianity and its notion of charity should be "replaced by the ethic of strength over weakness."[29]

The Effect of Darwinism on Education

With all the fervor of a new messiah, Spencer derived the "perfect" set of objectives for education from evolutionary principles. His classic statement follows:

> Thus to the question with which we set out—What knowledge is of more worth?—the uniform reply is—science. This is the verdict on all the counts. For direct self-preservation, or the maintenance of life and health, the all-important knowledge is—science. For that indirect self-preservation which we call gaining a livelihood, the knowledge of greatest value is—science. For the due discharge of parental function, the proper guidance is to be found only in—science. For that interpretation of national life, past and present, without which the citizen cannot rightly regulate his conduct, the indispensable key is—science. Alike for the most perfect production and highest enjoyment of art in all its forms, the needful preparation is still—science. And for purposes of discipline—intellectual, moral, religious—the most efficient study is, once more—science. The question which at first seemed so perplexed, has become, in the course of our inquiry, comparatively simple.[30]

It scarcely needs to be said that for Spencer, the heart of science was evolution. Spencer looked forward to the day when religion would become completely discarded to be replaced by science rooted in Darwinism. Statements of objectives of education developed since the time of Spencer clearly show their derivation from his original version above. Christian values are filtered out, and the stated objectives apply equally well to worms, animals and people.

The strongest influence of twentieth century education in America has unquestionably been John Dewey. In his classic work, *Democracy and Education* he stated:

> The philosophic significance of the doctrine of evolution lies precisely in its emphasis upon continuity of simpler and more complex organic forms until we reach man.[31]

Dewey's movement in education was called experimentalism or instrumentalism. Dewey, following the ideas of Rousseau, held that the child is naturally good. The child is encouraged toward independent thinking and to pursue physical freedom in a changing world. While one version of Dewey's philosophy of education, progressivism, was discarded in part from the schools in the 1950s, it is clear that many of his basic ideas are firmly planted in the American educational system today.

Darwin and the Philosophy of Education

Pragmatism was brought to its peak by John Dewey. In an earlier work, Dewey showed the influence of Darwin on philosophy. In it he asserted:

> The *Origin of Species* introduced a mode of thinking that in the end was bound to transform the logic of knowledge, and hence the treatment of morals, politics, and religion.[32]

The purpose of philosophy, according to Dewey, was to help find a way into a better order for our society. All theoretical problems must be conceived of in practical terms; they must arise in life situations, political, social, ethical, or scientific. Dewey denied all absolutes and advocated the experimental or scientific method as the only valid way for determining knowledge.

One can readily see that the influence of John Dewey has extended to the schools, the courts, the laboratories, the labor movement, and the politics of the nation. His thinking has had a profound effect upon the philosophy of science, art, religion, sociology, economics, law, psychology, and education. It is especially in the field of education theory and practice that Dewey's influence was felt.

It seems reasonable to say that all major approaches to public education today take evolution for granted as can be seen in some of the assumptions made. The stimulus-response view of learning fostered by Skinner rests in large part on animal experimentation, and on the denial of any kind of inner life. For Skinner, education simply means a form of training in no way qualitatively different from shaping the behavior of any other "animal." For him there was no such thing as good and evil; there are only desirable and undesirable behaviors that can be added to or subtracted from by further conditioning.

The cognitive-discovery view of learning rejects the approach of Skinner, but evolution is accepted as a fact here also. Children again are good by nature and are capable of discovering what is of value around themselves. Similarly the humanistic approach to education offers choices to the child who is naturally good.

Jerome Bruner, a strong force in shaping modern educational thinking, illustrated how central evolutionism is to his thinking in his model curriculum for the teaching of social studies at the elementary level, *Man, A Course of Study (MACOS)*. A unit on baboons is described as follows:

> Infant rearing, food gathering, defense against predators, intergroup relationships and communication are studied as background against which to examine human social behavior. Baboon behavior raises some interesting questions about the functions of dominance, aggression, sharing and reciprocity, territoriality and exchange, and various interpersonal relationships with small groups, human and non-human alike. As the only other group-adapted primates, baboons offer an unusually provocative contrast for examining the child-rearing practices and social behavior of man.[33]

Perhaps this example shows as well as any what the real issue is for parents who must decide between an evolution-based or a Christian-based education for their children. None of the major systems of educational philosophy offered today sees the child in a biblical light. Thus there are serious deficiencies in public education as well as actual dangers to the child. At best, public education is incomplete education.

Concluding Comment

Despite the conspicuous failures among Darwinists to apply evolutionary principles to the fields of knowledge examined here, we must never underestimate the overpowering influence this doctrine has had in our culture. It is the supreme irony that evolutionists, unable to convince people with evidence for evolution, have become increasingly militant, resorting to threats, blacklisting, firings, ridicule, and intimidation. Unique among theories, and in the absence of evidence for evolution, students, as well as adults, are not permitted to ask questions or ask for supporting evidence. An article in *Nature* advocated censorship of any who asked questions about evolution.[34]

The November 2004 issue of *National Geographic* bears the title: "Was Darwin Wrong?" The feature article gives us the answer. "What a silly question! The evidence is overwhelming and increasing daily." This book gives us the opposite response. We are "underwhelmed" for the many reasons and examples spelled out throughout this book.

It is therefore no surprise when we learn that "many biologists have one set of beliefs at work, their office beliefs, and another set, their real beliefs, which they can speak openly about only among friends."[35] What a devastating judgment this is on Darwinism! As some have observed, if in the future Christians are burned at the stake, evolutionists will light the matches. Yet with all the pressures to accept evolution, it is really astonishing to realize how irrelevant that doctrine is in daily life. No wonder that after 150 years of pressures and attempted brainwashing, the majority of the population does not buy into this pseudoscience.

Many Christians have read, studied, and evaluated evolution only to find nothing of value. Indeed it is all too clear that the evolution belief system is designed not to explain the world, but clearly to attack and erode the faith and values of Christians. In another context, St. Paul (2 Corinthians 4:2 TLB) furnishes us with a closing statement that fits our beliefs precisely:

> We do not try to trick people into believing—we are not interested in fooling anyone. We never try to get anyone to believe that the Bible teaches what it doesn't. All such shameful methods we forego. We stand in the presence of God as we speak and so we tell the truth, as all who know us will agree.

For Discussion and Reflection

1. When I visited Westminster Abbey in London, I saw that Charles Darwin and James Ussher are both buried there. This is also where many kings and queens were crowned and eventually buried. Why do you suppose this honor was given to Darwin?

2. Darwin's book *The Origin of Species* was initially welcomed by many clergy in England but sharply criticized by many scientists of that period. Why do you suppose these reactions took place, when one might have expected just the opposite responses?

3. When I read *The Origin of Species*, I discovered that Darwin never defined the word *species*, nor did he discuss the origin of any species. Do you think I was expecting too much from this work? Now, 150 years later, are you able to find the origin of any species explained in any textbook?

4. Why are vast amounts of time such an important concept for Darwin's theory?

5. In your experience in schools, was evolution treated as theory or as fact? Did any teacher permit discussion of that issue?

6. What is your reaction to the first paragraph of this chapter where a student expressed why she believed in evolution?

7. Can you think of any field of knowledge that has been revolutionized in a positive direction by the application of evolution?

8. Reread the quotation by Easton regarding the rejection by evolutionists of any kind of divine guidance or intervention or purpose in the world. Everything must have a physical explanation. If you have experienced any years of public education, is this what you experienced too?

9. We sometimes hear the argument that most or all of the sciences would be undermined and even destroyed if the theory of evolution is rejected. Have you ever heard an actual example of how any field of study would be undermined if God the Creator was believed?

10. Read once more the final paragraph of this chapter, which is taken from 2 Corinthians 4:2 (TLB). How well does this pa ssage express your own view of the doctrine of evolution?

CHAPTER FOURTEEN

VIGNETTES OF ANCIENT ASTRONOMY

THE UNIVERSE UP THERE

No one can fail to be awed by the absolutely marvelous photos beamed back to the earth from such triumphs of technology as the Hubble space telescope. These sights are a perfect illustration of the psalmist who exclaimed that the heavens declare the glory of God (Psalm 19).

This chapter is not a short course in astronomy. That is the task for astronomers. Danny Faulkner and Don DeYoung carefully spell out what is involved in developing a creationist astronomy, no simple task.[1] However, in this brief introduction we shall express some of our assumptions and attitudes about the science of astronomy. We illustrate these with a few items about astronomy in the news in recent years that show the extremely speculative nature of some important aspects of astronomy.

The purpose of this chapter is to show how deeply immersed ancient people were about the heavens, how they applied astronomical knowledge to daily life, the degeneration of some aspects of astronomy into evils, such as in astrology, and some of the evidence for strange catastrophic events in the past from the skies.

A Notorious Quotation

Evolutionists sometimes accuse Christians of restricting the free exercise of scientific endeavor, and sometimes this has been true. It is not difficult to find examples of blindness also on the part of scientists in the past, and evolutionists throughout their history. About 200 years ago the Academy of Sciences of France declared: "In our enlightened age there can still be people so superstitious as to believe stones fall from the sky."[2] Eyewitness accounts of meteor falls were of no avail at that time because "science" had spoken. In this chapter we shall examine whether that mentality still exists within the field of astronomy.

Is Astronomy a Science?

Every astronomer in the world would be shocked at such a question. There can be no doubt that billions have been spent to make the study of the universe a truly sophisticated pursuit. Then why is it commonly stated that astronomy is the most speculative of all the sciences? As we shall illustrate below, there is a great gulf between what is observed and how the observations are interpreted. We have no quarrel with data gathering, nor with speculation properly labeled as such. It is only when speculation is presented as fact that we must emphasize that hunches fitting a particular bias are not the same as truth. The Christian, for example, realizes that when astronomers speak of billions of years and millions of light years, they base such notions on assumptions that are far from verified.

Ernest Brown, past president of the American Astronomical Society, confessed that many of the beliefs regarding the solar system cosmogony, dynamics, and stability that he had held throughout his life were illusions, mere articles of faith, adhered to for non-rational reasons, and impossible of legitimate presentation as the logical consequences of observations and valid calculations.[3] Now these are shocking statements and should be reread whenever one reads sweeping statements about the nature and age of the universe. Things have not changed.

ASTRONOMY IN THE NEWS

The Big Bang?

A most remarkable drama occurred about the Big Bang some years ago and received very little attention after the first big splash in the news media. The Big Bang phenomenon has been taught as fact for years despite much contrary evidence.

Act 1 of this mighty Big Bang drama was announced in April 1992 at the meeting of the prestigious American Physical Society.[4] A discovery was made that was described in the following manner:

- Explains how stars and galaxies evolved
- Evidence for the birth of the universe
- One of the major discoveries of the century, in fact, of all time
- Unbelievably important: its significance could not be overstated
- They had found the Holy Grail of cosmology
- Solved a major mystery and deserved the Nobel prize
- The discovery is like looking at God

This was indeed heady stuff, and so one must inquire just what was discovered? With more than 300 million measurements, the astronomers discovered huge ripples of matter near the edge of the universe. How did they know this? The measurements were all of temperature with sensors pointed in different directions from the earth. In averaging the 300 million measurements, astronomers found a temperature difference in different parts of the universe. How big was the difference? One report stated 30/millionths of one degree; another said 10/millionths of one degree.

That study cost $400 million dollars and 28 years of work by many scientists. Then came an embarrassed silence. Apparently there is no instrument in the world that can measure such an infinitesimal difference. Presumably a gnat flying across a sensor 100 miles distant would create a greater temperature difference. All that had evidently happened in the study was averaging an almost infinite number of meaningless measurement errors.

All was not lost, however. Act 2 of the Big Bang drama was announced nine months later at the annual meeting of the even more prestigious American Astronomical Society. The great new discovery was described in these terms:

- Strong new support . . . that the universe began some 15 billion years ago with a Big Bang
- Precise measurements of remnant energy from the Big Bang gave results exactly as the theory predicted
- It was the toughest test yet of the theory
- The powerful new evidence is verifying the textbooks

Again we must ask just what was found? We now learn that the 300 million measurements taken at a cost of $400 million dollars were just preliminary results prematurely released to the media a few months previously. The new study, hundreds of millions of measurements later, is 30 times more precise. What was found? Nothing—that is, no temperature differences at all, which is exactly what the first study should have concluded.

But read these studies again! Two opposite results are reported, and both claim to be exactly what the theory required. It is one thing to present such things to captive students in a classroom, and to equally captive colleagues who do not dare to raise an eyebrow, but one must be forgiven for thinking this is another version of the wonderful story of the emperor who wore no clothes. Even the premier British science journal *Nature* proposed that the results be ignored. William Corliss, no friend of creationism, was aghast to see that distinguished astronomers were claiming that opposite results each proved the same thing about Big Bang. Thus Big Bang was "fact" despite what research revealed![5]

Kohoutek (1973)

Noted astronomers excitedly called Kohoutek the comet of the century, and it was to become the most intensely scrutinized celestial object in the history of astronomy. Astronomers promised that the display would be 50 times as spectacular as the awesome appearance of Halley's Comet in 1910. These studies would help to solve the great mysteries of how our solar system was formed, what changes will occur in the next few billion years, and how life began. How could any foundation resist financing such important studies? People around the world flocked to the best vantage points, and cruise ships were booked to capacity for those who wanted an even better view away from city lights. Now we know what happened. Nothing much. The predictions were so grossly exaggerated that one astronomer had an egg thrown at him when he mentioned Kohoutek at a lecture, and another astronomer had to testify before a congressional committee to explain the comet's failure to follow the predictions.[6]

Despite getting burned with Kohoutek, hope springs eternal, and special tours organized by astronomers to view Halley's comet were sold out three years in advance. People paid more than $10,000 each to sail with Carl Sagan between New Zealand and Australia for the choicest view of this expected dazzling display. Again, the scene was spectacularly less than advertised.

Another comet, Hyakutake, appeared in 1996 and promised to rival Halley's spectacular appearance in 1910, which was indeed an awesome sight.[7] Those who still believed in such predictions strained their eyes and necks and barely could make out a little fuzz ball in the sky, again far less than predicted.

In 1910 noted astronomers had it all wrong then, too, about Halley's Comet. Here are three samples of the dire prophecies:

1. About 4:25 a.m. our planet will be enveloped in a deadly cloud of poisonous gases and cosmic dust that makes up the tail of the comet. Hydrogen, carbon, nitrogen, hydrogen cyanide, and potassium cyanide will turn the globe into a monstrous gas chamber. No one will escape, or at most a fortunate few in the areas around the poles, who may not be directly struck by this terrible fate which is approaching us out of space. There will be a catastrophe.

2. The sun will become dark, glaring lightning will illuminate the pitch-black sky, monstrous fiery masses will plunge from heaven. The eruption of the chained volcanic fires will alter the face of the earth.

3. The axis of the earth will be displaced. The bodies of water in the oceans will leave their beds and break over the continents. A hundred thousand human beings will meet a terrible end in this new deluge, and all traces of our civilization will be obliterated in a single night.

As you might suspect, the world is still here, and nothing of the sort happened. We must remember that these are not crackpots from the lunatic fringe speaking. These were noted twentieth century astronomers.

Three Up—Three Down

Hindsight, of course, is a great gift, but the nagging question remains. If the best astronomers err so badly on the easy matters of relatively nearby objects, how can they be taken seriously about much more fundamental mysteries and questions? We have no quarrel at all with the science of astronomy. Speculation posing as fact that is then used to undermine the Bible is quite another thing, and that is where our strong objection lies.

There is more. We must not confuse amazing technology with speculation about what has been viewed in the skies. There have been two recent asteroid scares.[8] An asteroid named 1997 XF11 caused screaming headlines around the world. Astronomers predicted a disastrous collision with the earth. Only one day later evidence emerged that there would be no such impact. *Discover* tagged this scary but false announcement as one of the twenty worst scientific debacles of the past century. In late 2000 astronomers warned the world another object, 2000 SG344, might smash into the earth in 2030. A short time later other astronomers showed that no such danger existed as predicted.

Astronomers who believe in evolution would have us believe that they can prove the age of the earth and the universe in terms of billions and billions of years. Yet we have seen over and over above how shaky their pronouncements are with far easier questions. Sensible people are not impressed.

When Astronomy Was Part of Daily Life

Daily Life Long Ago

It seems that the average citizen today feels quite accomplished if he or she can look up into the night sky and actually pick out the seven stars of the Big Dipper, that is, The Great Bear. Things were very different in ancient times. The stars then dominated life to an extent difficult to understand in our time.

The farmers and shepherds of long ago could name the stars and constellations, knew when stars rose and set, watched for the first appearance of fixed stars annually to fix their calendar, and used the stars as a compass in travel. The sky face was a clock; the stars were almanacs. They hunted, fished, sowed, and reaped by the stars. In remote African jungles in the ancient past men used the stars to guide them and taught every boy how to lead the group back to their camp. There are tribes today that still teach this ancient wisdom.

Ancient astronomers knew the cycles of the sun, moon, and planets and could predict eclipses. Already in Dynasties I and II in Egypt, sophisticated star sighting instruments were in use. Richard Burton outlined astonishing evidence that astronomers of antiquity were well acquainted with the ground lens, the mirror, the telescope, and other sophisticated ways of studying the heavens.[9] The Old Ones mastered astronomical knowledge for two reasons: to know the time of the year for specific purposes, and to know their whereabouts anywhere on the earth when traveling so they could always find their way back whether on land or sea.

The Alphabet

It seems like quite a stretch to relate the alphabet to the stars.[10] Joseph Seiss and others have observed that all the known ancient alphabets had the same number of letters, including seven vowels, and all began and ended with the same letter sounds. Even from a casual examination we see several odd things about ancient alphabets. First, all seem to contain unnecessary letters or letters that duplicate others. Second, the vowels are placed in a rather odd pattern among the consonants. Gustavus Seyffarth believed that the seven vowels represented the seven planets, while the twenty-five letters of the old alphabet were placed into the twelve signs of the zodiac, the first and twenty-fifth occupying the same space. Thus the vowels show the actual placement of the seven planets supposedly in the year 3447 BC. Seyffarth held that this date marked the end of the flood. One needs to arrange the consonants in a circle divided into twelve sectors representing the constellations. The vowels, representing the planets, were inserted among the consonants to show the location of each at a particular point in time. We must confess that we would like to have had a simple, clear diagram of what these scholars believed to be true about the alphabet.

Along the same lines, Richard Burton, that incredible scholar of the ancient world, also related the older alphabet to the twenty-eight mansions of the moon, linking the form of each letter and their order to the constellations involved. Dr. Kelley, in his massive study of alphabets around the world, concluded that the lunar zodiac was the common source of more than 200 phonetic sounds. This is a persuasive analysis.

Among remarkable things about the alphabet, it appeared full and complete from the start; it was not something that gradually evolved out of earlier picture symbols, as some believe. The fact that Hebrew, Greek, and other ancient languages gave numerical as well as phonetic value to each letter seems to support the use of the alphabet as described above.

All the evils we associate with pagan worship of the heavens and astrology came much later.

The Stars in Our Language

The central place that the skies held in the life of ancient people and the power people believed that the stars had over their life is reflected in our language. English is a language borrowed from many sources, so star terms are more clearly evident in other languages, yet we may be very surprised at the origin of some common English words.[11]

Tell me, do you <u>consider</u> matters carefully (that is, observe the stars carefully)? Do you <u>contemplate</u> (that is, scan the heavens to look for signs in the stars)? Have you ever suffered a <u>disaster</u> (evil from the stars)? What do you <u>desire</u> (being without a star)? Have you ever been <u>moonstruck</u> or were you called a <u>lunatic</u> (evil from the moon)? Is your personality <u>jovial</u>, <u>saturnine</u>, <u>martial</u>, or <u>mercurial</u> (determined by Jupiter, Saturn, Mars, Mercury)? And <u>venereal</u>, associated with lust, is derived from Venus.

The Chinese number seven is written like our 7 upside down and backwards, derived from the constellation Little Dipper or Bear, with its seven stars just like our 7, best written with a stroke across the handle to connect the two lesser stars. Blaze, blast, blitz (German for lightning) may have been derived from the brilliant light of Bel, the sun god, still celebrated with May-Day fires and dances. Did you as a child honor Bel or Baal by dancing around a maypole at school? And circus refers to the sacred ring consecrated to the sun. Asterisk (*), of course, is a little star. Perhaps you know someone named Esther or Astrid or an astronaut or astronomer, names derived from the word *star*.

The Stars in Our Calendar

In a way the division of the year into twelve months is purely arbitrary.[12] After all, the word *month* is derived from "moon." We all know there are thirteen moon months a year, yet the division of the year into twelve months is universal. The answer lies in its relationship with the zodiac, a word derived from zoo or zoology, in this case meaning "little animals." The twelve signs of the zodiac are found wherever there are traces of civilization. From the most ancient times the sky was mapped into twelve very arbitrary constellations of stars, and we accept the thought that originally they had served a useful purpose for people before they became perverted into evil. One need not believe in astrology as a fortune-telling device to think that there may be embedded in such lore some information important for daily life in ancient times. "[L]et them [the stars and planets] serve as signs to mark seasons and days and years" (Genesis 1:14). This is an accurate statement.

In the course of the year the sun seems to trace an ever-changing path against the stars, traveling about 1/12 of the great circle of stars each month. This path is called the zodiac. There is evidence to show that originally after creation, the year was exactly 360 days and the lunar month 30 days (as in Genesis 7–8), as opposed to the strange mix we have today of months 28, 29, 30, or 31 days long. Thus the Sumerian culture handed down and exported the division of the circle into 360 parts and their calendar of 360 days, and this is the origin of the degrees of the circle.[13]

Over the centuries it took great efforts to determine the true length of the year, with painful corrections applied from time to time. More than vanity was involved when our calendar was changed to honor Emperors Julius and Augustus. It is really bizarre that our ninth month, September, means seventh, and October through December are also gloriously misnamed. We must note that our Julian calendar in use since 46 BC is not as precise in the length of a year as the Mayan calendar.[14] Another relic from the ancient past is still in place in Scotland. Quarterly rent payments are due in February, May, August, and November, and each date falls on an ancient pagan festival day, such as the fires of Beltane (Venus) on August 1.

From the time of ancient Egypt the days of the week were named after the sun, moon, and five planets, namely Mars, Mercury, Jupiter, Venus and Saturn, as in the French: *Dimanche, Lundi, Mardi, Mercredi, Jeudi, Vendredi*, and *Samedi*. Our English names for these days of the week are the exact equivalent of the French and many other languages. The strongest efforts of the church to change the names to Christian terminology failed.

Called the rage of the Arab world, the popular game of backgammon closely reflects the calendar. The 30 disks stand for days of the month; black and white for night and day. Each board has four sections (seasons), 24 positions for the hours of the day and night; 12 months of the year in each player's position; The point of the game is to see who can complete the year more quickly. It is thought that this game was played over 5,000 years ago in Ur of the Chaldees.

The game of checkers dates from the so-called Stone Age and illustrates how intricate counting was done at Stonehenge and at other stone circles for astronomical purposes. The counters, representing dawn and dusk, followed fixed rules for moving, jumping or not jumping, to determine cycles, eclipses, annual festivals, and the like. The fifty-six Aubrey holes at Stonehenge figured somehow in such elaborate counting, but its precise use is long forgotten. Likewise the vast complex of standing stones in Brittany was used for calendar purposes. The name of the largest alignment is Menec, meaning counting or reckoning, perhaps derived from the Phoenicians long in the past. Adding "Al", Arabic for "the", we find our word *almanac*, the calendar book.

The Stars in Ancient Navigation

In the South Pacific, where amazing ancient navigational skills are to be seen today, people still use the stars for brilliant navigational feats. This is more than a little hint of the superb skills of ancient mariners centuries before the time of Abraham. As Thor Heyerdahl experienced firsthand, these great navigators could take their bearing by sun in the day, and be guided by the stars at night. They possessed an astonishing knowledge of the heavens.[15] They knew the earth was round long before Europeans had this knowledge. They knew the concepts of equator and the Tropics of Cancer and Capricorn. With twigs and shells they made detailed maps showing the locations of islands. They knew the planets and where in the sky different stars rose and set again. Along with their astounding knowledge of stars, they could read the ocean currents, and the barely perceptible ocean swells. The behavior of sea birds far distant from land told the ancient mariners where they could expect to find land. In this fashion they discovered and colonized islands over an area of 15 million square miles.

In the remote past, mariners could determine longitude (not rediscovered until in the eighteenth century). They conducted worldwide exploration and interconnection to gain access to raw materials by land and by sea.

When one speaks of the four corners of the earth or four pillars, or the ends of the world, does this display ignorance or an otherwise primitive grasp of the skies? Far from it. In ancient writings these all refer to the position of the sun at the times of the equinoxes and solstices, the points that mark the beginning and end of our seasons. How would a scientist express these important concepts more scientifically? Here is a modern example: Sunrise and sunset are very inaccurate terms, yet the best of our meteorologists today have not found a better way to express these concepts.

Astronomy in Stone

Before 2000 BC almost every corner of the world had been visited by people who possessed amazing technical skills. They erected vast astronomical instruments, circles of erect pillars, pyramids, underground tunnels, gigantic stone platforms, straight tracks and alignments marked by notched mountain slopes, stones, mounds, and earthworks that boggle the mind. They are found on remote uninhabited islands as well as on the continents and major islands. The intensive study of the stars, daily life, travel, sea navigation, calendar, and religion were all closely knit together.[16]

Alexander Thom claimed these ancient solar observatories were capable of defining solstices, equinoxes, and predicting eclipses. Hundreds of such stone circles and stone alignments in Britain and elsewhere clearly show lunar and solar significant events. On remote Tonga Island a massive stone lintel atop three huge stone pillars has two incised lines that indicate winter and summer solstices. Computer analyses and other studies in recent years have not had much success in discovering other meaningful alignments. However, much ancient literature and other evidence indicate that dramatic events—catastrophes—occurred in the skies which then may have played havoc with the assumptions made by modern scientists about the skies in ancient times.

Whether it is Peru or India or England or elsewhere, there are signs of sudden abandonment of megaliths such as Stonehenge, a strong indication of catastrophic events, a topic we shall examine more closely later in this chapter. Francis Hitching, in his studies of the great megaliths around the world, states how extraordinary it is that the megaliths in the Tibetan highlands are so similar in pattern to those in Carnac, France, to illustrate that the same technology and patterns are found thousands of miles apart around the world, a powerful indication that this all stemmed from one source, and with the same obsession about the skies. Stonehenge (perhaps derived from Saxon "hanging stones") is a dramatic example of stone circles found throughout the world. There are strongly held beliefs that such circles had both ritual and astronomical purposes.

In America as well as in the old world there are so-called solar serpent mound formations. The Ohio Serpent Mound, with almost a mile of sinuous curves, is perhaps the most dramatic of them all. This serpent and others elsewhere in the world shows the solar serpent in the act of swallowing the cosmic egg. The serpent patterns too are believed to have astronomical significance, but there is much mystery about them.

Hundreds of strange stone spheres, arranged in little understood formations or patterns, were discovered in the jungles of Costa Rica, Guatemala, and Mexico, ranging in size from a few inches to 8 feet in diameter, and up to 16 tons. They are perfectly shaped and many are finely polished. They are described in various sources to be of volcanic rock, or granite, or of a very hard type of coquina. As there are no quarries within many miles of the sites, they must have been rolled or rafted into position, incredible technical feats. There is much mystery about them, but some believe they may picture arrangements of planets, constellations or other astronomical patterns.

The ancient world is filled with mysteries, but it would be difficult to find anything to compare with the Nasca lines in Peru and similar formations in Bolivia and Chile. There are many dramatic lines and designs in the desert, straight as an arrow for many miles over plains, hills, and gullies. Best guess is that lights were used to stake out such incredibly precise lines—many hundreds of such lines. The outlines were formed centuries ago by removing the darker colored top layer of reddish and blackened pebbles, thus exposing a lighter colored layer of sand and clay. In addition to lines and abstract shapes there are numerous precisely drawn figures of such creatures as lizard, spider, flower, monkey, condor, and unknown creatures. Undoubtedly scale models were constructed and then multiplied many times in the desert version. In fact, 6×6 feet plots have been discovered where designs had been carved in miniature. Thus, from a small model drawing, the lizard was drawn accurately 600 feet long. Some solar and lunar alignments were discovered here, but other astronomical discoveries are hard to come by.

Yet the huge figures and extensive lines seem to make sense only for astronomical purposes, and some consider it a gigantic astronomical calendar. The fact that the grandeur of this mighty display may be appreciated fully only from the air has led to bizarre theories by the so-called lunatic fringe, but condemning wild theories does not lessen the mystery.

Cupmarks

Ancient obsession with the stars is illustrated once again by strange cup-like depressions in rocks, sometimes with rings around them, and often connected with others by lines. These are not to be confused with somewhat similar "nutholders" in the rocks used by Indians to hold nuts for harvesting.

The cupmarks go back to very ancient times. They are found on dolmens and passage tombs in Palestine, Corsica, Italy, Spain, France, Germany, Scandinavia, Great Britain, Ireland. They are found in Asia, Africa, America, Peru, Polynesia. Cliffs in China and around the Gobi Desert are dotted with them.[17]

What is the connection of many cupmarks with astronomy? In France, for example, cupmarks made up the shape of Ursa Major (Big Dipper), Ursa Minor, and Pleiades. Evan Hadingham provides an interesting and more detailed answer to the question.[18] In studies in Britain, it was found that cupstones formed accurate patterns, but in mirror image, of constellations and first magnitude stars. Thus we see that small pools of water were formed to mirror areas of the night skies for careful study, and these little pools in turn were the model from which cupmark designs of stars were made in stone. We may speculate that these and other cupmark patterns were used in part for planning and recording widespread travel/navigation.

The Stars in the Bible

As noted above, the Bible tells us of the real purpose of the stars, and how this purpose was later perverted:

> And God said, "Let there be lights in the expanse of the sky to separate the day from the night, and let them serve as signs to mark seasons and days and years, and let them be lights in the expanse of the sky to give light on the earth." And it was so. God made two great lights—the greater light to govern the day and the lesser light to govern the night. He also made the stars. God set them in the expanse of the sky to give light on the earth, to govern the day and the night, and to separate light from darkness. And God saw that it was good. And there was evening, and there was morning—the fourth day (Genesis 1:14–19).

The psalmist adds (104:19 TEV): "You created the moon to mark the months; the sun knows the time to set."

The Magi from the east knew the Scriptures and they knew something of signs in the stars when they asked, "Where is the one who has been born king of the Jews? We saw his star in the east and have come to worship him" (Matthew 2:1–2).

The Hebrew word for *sign* is used eighty times in the Old Testament and regularly indicates a religious symbol, a mark with spiritual significance.[19]

The earth received its light from a source other than the sun before Day 4, a miracle we leave in God's hands. The Northern Lights, however, are an example of light from a non-solar source. The Bible describes other miraculous astronomical events that no one can explain: the sun standing still (Joshua 10:12–14); the shadow moving backward on the dial (2 Kings 20:10–11); the star at Bethlehem (Matthew 2:9); the failing of the light at the crucifixion (Matthew 27:45). We may say, however, that God rules the heavens. God is not subject to human limitations.

God said in Genesis 1 that there would be signs in the skies, but in Jeremiah 10:2 we read, do not "be terrified by signs in the sky, though the nations are terrified by them." We can see how the skies were used to orient people to the cardinal points of the compass in ancient times, and it was assumed that one would begin by facing the rising sun in the east, as we see in Job 23:8–9. Older Bible translations use the words *forward, backward,* the *left hand* and the *right hand,* but the New International Version shows us the obvious intent: *east, west, north, south.* Heathen temples and other structures provided a clear path down the center from the entrance for the rays of the sun at appointed times to shine on a fixed point inside

With Israel, however, the glory of the Lord was within the most holy place of both tabernacle and temple. Is it then not remarkable that the number of pillars in front of the tabernacle was odd, while the number placed between the holy and most holy was even, thus preventing the light of the rising sun from entering (Exodus 36:36, 38). In the temple, as best we can understand the description, the entrance to the holy place consisted of two doorways with a central post between them, while the entrance to the most holy place was one opening, thus again effectively blocking the sun's rays from entering the temple (Ezekiel 41:2–3). These worship centers were thus dramatic witnesses against heathen worship of the sun, moon, or stars. The appropriate use of the stars for navigation and night travel is clearly shown in such passages as Job 38:32 that mentions the constellation of the Bear, and Acts 27:20 on the use of sun and stars for navigation.

Ancient people divided the skies into three categories: non-setting stars; rising and setting stars; and the stars hidden under the horizon in the south. Job (9:9) mentions all three types, which seems remarkable for a desert dweller who lived before the time of Moses: the Bear, then non-setting in that latitude; Orion and the

Pleiades, rising and setting stars; and the (hidden) chambers of the south. And, of course, the ancients had amazing knowledge of the planets and the constellations. Job (26:7) knew that God had suspended the earth in free space.

There are interesting but unverifiable legends about Enoch, Abraham, and Moses and their supposed deep knowledge of astronomical matters. For example, the Book of Jubilees states that Enoch was among the angels of God six jubilees (6 × 50 years). Enoch was the first among the children of men who learned writing, science, wisdom, and who interpreted the signs of heaven according to the order of the months and wrote them in a book because the sons of men know the time of the year, subdivided according to the order of months. Josephus, using sources we do not have today, said that Abraham taught Egyptians arithmetic and astronomy. Josephus referred to Berosus, a priest at Babylon about 250 BC, who mentioned Abraham without giving his name: "In the 10th generation after the Flood, there was among the Chaldeans a man, righteous, and great and skillful in celestial science.[20]

The black stone of the Ka'ba in Mecca became a holy thing and later this meteorite was brought into the Islamic faith as an object of reverence. We may compare this with Acts 19:35:

> Men of Ephesus, who is there who does not know that the city of the Ephesians is temple keeper of the great Artemis, and of the sacred stone that fell from the sky?

Worship of celestial bodies was sharply forbidden throughout the Bible. Moses (Deuteronomy 4:19) repeatedly warned Israel:

> And when you look up to the sky and see the sun, the moon and the stars—all the heavenly array—do not be enticed into bowing down to them and worshiping things the Lord your God has apportioned to all the nations under heaven.

Job (31:26–28) described the allure:

> If I have regarded the sun in its radiance or the moon moving in splendor, so that my heart was secretly enticed and my hand offered them a kiss of homage, then these also would be sins to be judged for I would have been unfaithful to God on high.

We learn from Stephen (Acts 7:41–43) that the golden calf made by Aaron became the focus of the worship of the heavenly bodies and all the evils that heathen worship brought with it. God condemned Babylon to a terrible destruction because of their star worship, and this is mirrored in many similar passages directed against Jews and other nations (Isaiah 47:13–14)

Let your astrologers come forward, those stargazers who make predictions month by month, let them save you from what is coming upon you. Surely they are like stubble; the fire will burn them up.

Both the clergy and the scientific community strongly challenge the pretentious claims of astrological charlatans, yet 1,200 American newspapers meekly buckle under pressure and print admittedly absurd horoscopes daily, and these are consulted by at least 50 million Americans. About 185,000 full- and part-time astrologers are busy ministering to the gullible. The power of astrology is stronger in much of Asia and elsewhere today than in the United States.

P. E. Cleator describes the obsession in the past and much of it holds true today:

Such was the importance attached to (consulting the stars) that no monarch thought of engaging in war, or of setting out on a journey, or of disposing of an unwanted wife, without first being assured of a favorable outcome, any more than one of his subjects considered starting up in business, or of moving to a new address, or of cheating his best friend, without first ascertaining the attitude of the high gods (stars) to the proposal.[21]

The Gospel in the Stars?

Because of some very mysterious oddities in the makeup of the zodiac, some have explored the possibility that the stars once carried an astonishing visual message. David was inspired to say (Psalm19:1–4):

The heavens declare the glory of God; the skies proclaim the work of his hands. Day after day they pour forth speech; night after night they display knowledge. There is no speech or language where their voice is not heard. Their voice goes out into all the earth, their words to the ends of the world.

The zodiacal system has been in use around the world since the earliest times, and all have used the same twelve major signs for the constellations involved. Each sign of the zodiac is accompanied by three additional constellations called decans. Independent invention is out of the question because the stars in each constellation bear no resemblance to its name. As with many other developments described in this book, the zodiac spread from one original source.

Joseph Seiss and Ethelbert Bullinger, originally published in 1882 and 1893, have written interesting books on aspects of ancient astronomy. Their focus is on how an original godly message in the zodiac degenerated into heathen worship and the general disrepute of astrology today. It is because of the evil reputation of star worship and astrology that few today discuss the possible original message of the stars for fear of being misunderstood. Thus Henry Morris speaks cautiously about the "imaginative" reconstruction of the possible story of God's plan and covenant as told in the zodiac and decans. To give just a bit of the flavor of such a reconstruction, however, Bullinger is one of several who have outlined the possible story of the stars:

- The prophecy of the promised seed of the virgin (Virgo: the virgin)

- The Redeemer's atoning work; the price deficient balanced by the price which covers (Libra: the scales)

- The Redeemer's conflict; the scorpion seeking to wound, but itself trodden under foot (Scorpio: the scorpion)

- The Redeemer's triumph; the two-natured conqueror going forth (Sagittarius: the archer)

- The blessings procured for the redeemed; the goat of atonement slain for the redeemed (Capricorn: the goat).

- The blessings ensured for the redeemed; the living waters of blessing poured forth for the redeemed (Aquarius: the water-bearer)

- Blessings held in abeyance; the redeemed blessed though bound (Pisces: the fishes)

- Blessings consummated and enjoyed; the Lamb that was slain, prepared for the victory (Aries: the lamb)

- Messiah, the coming judge of all the earth; Messiah coming to rule (Taurus: the bull)

- Messiah's reign as Prince of Peace; the twofold nature of the King (Gemini: the twins)

- The possessions of Messiah's redeemed; the possessions held fast (Cancer: the crab)

God had good purposes for his creation of the stars, and they were to declare his glory. If Seiss and Bullinger are right, what better way to declare God's glory in nature than to tell the wonderful story of salvation in the heavens. There is no condemnation of constellations in the Bible—only of idolatry and degrading superstition.[22]

EVIL FROM THE SKIES?

The Skies Also Cursed

We have gone to great lengths in other writing to show that, as a result of sin, God cursed the earth, the animal kingdom, the plant kingdom, and the earth itself.[23] The curse, as we shall see, also involved the heavens or the skies (Romans

8:22): "We know that the whole creation has been groaning as in the pains of childbirth right up to the present time." The evidence is remarkable and dramatic.

In ancient times people were preoccupied with the stars, not for recreation but as something that permeated every part of life. The ancient religions generally involved worship of the heavenly bodies. Josephus wrote that Abraham was alone in his belief that one should worship one God, the Creator. The ancients knew of great irregularities in the heavens, accompanied by gigantic upheavals on the earth. Thus they tracked the heavenly bodies with great concern and fear. Details of specific catastrophes are pictured in ancient literature and mythology, and they are utterly frightening: unexplained darkness, earthquakes, falling fire, falling stones, widespread volcanism, terrible noises in the sky, reversals of the earth's magnetic field in historic times, tumult, lightning, and overpowering winds.[24] A Golden Age came to an abrupt and shocking end. Animals and people fled into caves to avoid the horrors.

No one could tell the dramatic story of the battle of the stars better and with more precise astronomical details than Ovid in *Metamorphoses*. It is easy to believe that he was relating eye-witness accounts handed down to him by the ancients.

The Mayas described the disaster of 3113 BC, which could well have been the time of Babel when God "splattered" the people far and wide across the whole earth: Discord in the heavens caused universal disorder on the earth. In great convulsions the sea erupted onto the continents. Climates changed suddenly; ice settled over lush vegetation, while green meadows and forests were transformed into deserts. The people fled from a torrent of meteors. Crazed survivors sacrificed victims endlessly to appease the gods. Appeasing the gods through fire-worship and the sacrifice of children were horrors practiced in biblical times, and a major cause for God to command the total destruction of the heathen who practiced this horror. It is a shock to learn that a remnant of the old fire-worship and sacrifice of children persisted well into the nineteenth century. For example, in Scotland, Lady Baird mentioned that on her estate it was the practice of peasants on May Day to gather round a fire and throw their children across from one to another through the fire. Similar descriptions and other elements of the old pagan worship of Baal, if not of Moloch, were noted in other areas of England and Ireland. The peasants no longer knew why they continued the practices, but the tradition was compelling. As Georgio de Santillana said so well, myth is language for the perpetuation of a vast and complex body of astronomical knowledge, and science someday alone will explain it!

The Bible tells of evil from the skies, and therefore we know the real source of this punishment on evil people. Violent natural catastrophes expressed God's righteous wrath on wickedness (Psalm 97:3–7):

Fire goes before him and consumes his foes on every side. His lightning lights up the world; the earth sees and trembles. The mountains melt like wax before the Lord, before the Lord of all the earth. The heavens proclaim his righteousness, and all the peoples see his glory. All who worship images are put to shame, those who boast in idols—worship him, all you gods!

One has only to read a book, such as D. S. Allan and J. B. Delair's 1995 *When the Earth Nearly Died*, to see compelling evidence of a catastrophic world change. The book makes a strong case for the disastrous near approach of an object from outside the solar system. We do not agree with the chronology in this book, but the evidence for stupendous disasters on earth in ancient times is very persuasive.

The idea of welcoming a comet today is one that would have sounded strange to our ancestors, who regarded these visitors with terror as most ill-omened and precursors of plague, famine, war, and other catastrophes. Such beliefs can still be found today. Comets were mysterious, and they seemed to violate majestic law and order.

While the problem is glossed over, comets present a great difficulty for conventional theory, for at each encounter with the sun the comet loses some of its material. If the universe indeed is very very old, comets should all have perished billions of years ago. It seems more reasonable to believe that comets are another of many evidences for a young earth as well as a reminder of catastrophic events in the heavens in the past.

Part of America's folklore is Mrs. O'Leary's cow that kicked over a lantern, thus starting the disastrous Chicago fire of October 8, 1871, destroying about 18,000 buildings and leaving 100,000 homeless. For decades natural causes were believed responsible for this horror. We now know more. At the same moment two more terrible fires flared up violently hundreds of miles away. A small comet composed of three parts and made up of frozen gases, water, and clusters of gravel and sand struck on either side of Lake Michigan, near Green Bay, Wisconsin, and near Manistee, Michigan. People heard a terrible roar like a tornado crashing through the forests. Then its trailing apex struck moments later in Chicago. The sky seemed to have burst into clouds and whirlwinds of flame. In the cities great stone buildings and blocks of pavement instantly melted and disappeared. Great masses of glass, china, mortar, brick, and earthenware were cemented together in bizarre shapes. In ancient myths we read of similar catastrophes. Now we know that they were eyewitness accounts of terrible events.

Fear of the Sky

These awful memories were handed down from generation to generation. Alexander the Great asked Adriatic Celts what they feared most. They replied that they feared no one, but they did fear that the sky might one day fall upon them. This is the truth embedded in the old nursery tale of Chicken Little, otherwise known as Chicken Licken or Henny-Penny. Whether it was Mexico or England or Babylonia, astronomy preoccupied ancient people. They watched the planets because they were afraid another disaster would occur. They were a society of frightened sky watchers.[25]

A vivid example of such fear is found in the royal correspondence of the Assyrian Empire. Assyrian ambassadors in various countries spent their nights carefully watching the sky for unusual or threatening signs. The royal records include regular and reassuring reports to the emperor from the ambassadors that all was well in the heavens.

We have discussed elsewhere that the dragon in art and literature was an actual memory of the dinosaur. Those who studied the heavens now saw the shapes of winged serpents and dragons in the skies.

A Sign of Catastrophe

One of the less obvious signs of the catastrophes described above is the orientation of ancient temples. C. Ernest Wright is one of the archaeologists that documented changes in temple orientations over a period of centuries.[26] Temples in

A boulder of Kettlepoint calcite at Ann Arbor, Michigan, cushioned inside a bed of clay, was discovered during construction 110 miles from its source in Ontario. This distinctive rock, also known as Stink Stone, emits a strong odor of petroleum when struck on a hard surface. Besides the mystery of its origin, we marvel at the mystery of how it was carried by flood or ice from that distance.

ancient times were very carefully aligned with the cardinal directions or with an important star or constellation. After destruction of a temple we would expect it to be rebuilt on the same foundation as before. Wright found a significant shift in the orientation of the great temple at Shechem when it was rebuilt, and he calls it peculiar, inexplicable, and an unsolved mystery. He was ready to believe that something had caused the earth's axis to change. Only a force from outside the earth could cause such a catastrophic change.

Tektites and Meteorites

Billions of tons of little glassy black objects are scattered around the earth and in the oceans, but concentrated in some areas, such as Texas, Georgia, Germany, and Australia. A geologist who studied tektites for 17 years summed up what he knew: "They are strange beasties!" Long thought to be splashings from the moon, that possibility seems to be ruled out because of their chemical makeup. Best guess currently is that they are splashings from large meteorite impact sites, though this is difficult to verify.[27]

There are many native legends about tektites falling from the sky, and local names for tektites show that natives knew their correct origin. But evolutionists discard all that and maintain that tektite fields vary in age from 700,000 to 35 million years, relying on potassium/argon (K/Ar) dating. Amazingly, tektites lie on the surface of the earth and on ocean bottoms, indicating a very recent fall, and more realistic estimates are in the range of 3,000 to 6,000 years.

In 1975 the largest known meteorite was thought to be the Grootfontein in southwest Africa, weighing about 60 tons. Recently China claimed the honor for a granite-covered meteorite in northeastern China near Shenyang. Their geologists gave the measurements as 600×250×300 feet, with an estimated weight of 2 million tons. This, of course, requires verification.

Perhaps something even larger struck the Sudbury, Ontario, area long ago, creating the vast mining complex there, including one of the largest deposits of nickel in the world.

Bays and Blemes

An incredibly large mass of meteors grazed the tops of the Blue Ridge Mountains in the distant past, then struck the coastal areas of North and South Carolina, creating more than 3,000 oblong-shaped depressions best viewed from the air. Some estimates go as high as half a million parallel craters, locally called bays.[28]

Thousands of lakes in a remote northern Alaska region form a similar pattern. They were first discovered in the 1940s. Other areas around the world are under investigation as well. The catastrophic consequences of such strikes as described above were dreadful.

Space photography in the past several decades has revealed startling and gigantic scars on the earth, called astroblemes, similar to the impact sites we observe in photos of Mars and the moon. Although well over a hundred have now been identified, and more are reported each year, here are a few dramatic items that illustrate earth's turbulent past.

A mighty crater, 150 miles in diameter, was located near the South Pole under a mile of ice, and first identified in 1975. Sweden currently boasts the largest bleme in Europe, a crater 27 miles across. A crater 55 miles in diameter was discovered in the 1970s in Wisconsin, and twin craters were spotted in Siberia, the larger measuring 40 miles in diameter. At Ishim, Kazakhstan, scientists found a stupendous crater 217 miles in diameter. Such a strike was powerful enough to obliterate an entire country the size of France. The south part of Florida, including the Everglades, is one gigantic crater not fully verified until 1985! A great cosmic body crashed in Argentina at a location called the Field of Heaven. The most interesting one of all, perhaps, is found in Ohio. The Great Serpent Mound, known all over the world, is located precisely in the center of a great astrobleme crater, not verified by geologists until 1960. Were the ancient people there appeasing angry gods?

Comment

What is even more startling than these gigantic blasts on the earth with stupendous consequences for the geologic processes and life on earth, is that at least through the mid 1970s not one geology text to our knowledge discussed these impacts. One can only conclude that the evolutionists who write the texts wanted nothing to do with such catastrophic matters, because that supported the belief in a young earth.[29] We believe these terrible events took place chiefly in the cursed pre-flood world and at the time of the flood.

ANOMALIES AND CONTROVERSIES

Mysteries Everywhere

There are many tantalizing mysteries about the ancient world, especially about the skies, too many to make even a partial list of them.

Despairing of developing a convincing theory of how life arose from non-life on earth, eminent scientists Francis Crick and Leslie Orgel suggested in 1974 that a spaceship brought tiny organisms to earth from a distant planet.[30] From this all life on earth evolved. Interest in such a possibility is alive and well today. Meteorites are currently under careful examination for signs of fossil life or for some of the building blocks that make up life. Contrary to what textbooks written by evolutionists say, this bizarre theory shows how bankrupt evolution is in its attempt to explain how life arose from non-life.

We can only wonder about the number 432,000 that occurs in so many contexts of myths and cycles. The number occurs in connection with ancient astronomy, but no one has been able to penetrate the mystery of what it means. While the number of seconds in 5 days is 432,000, that fact carries no meaning for us. The most ancient collection of Hindu sacred verses consists of 432,000 syllables. This work was carefully crafted to honor somehow that mysterious quantity.

A Matter of Arithmetic

There has been a cry of outrage for a long time that many children do not learn to do the simplest math in our schools. When world renowned scholars, however, make horrible, inexcusable errors in simple math, we are stunned to see that there is no outcry at all.

Over two decades ago the Club of Rome was so famed in its economic forecasts that the media referred to it in hushed tones of reverence. No scholar could rise higher in eminence than membership in that think tank. Based on its projections of pending ecological disaster, a team of scientists at the renowned university, M.I.T., forecast global horror and collapse by the end of the next century. When another scientist independently reviewed the study, he found a typo that multiplied the horror tenfold. When this error of simple arithmetic was corrected, the whole prediction of collapse fell apart, and no crisis could be forecast. Not only was the error stonewalled, but the preeminent American journal *Science* refused to publish the correction, which then was accepted in the British journal *Nature*. The scientists were so renowned that they could not admit their error.[31]

How Old?

Many factors are thought to be clues on the age and structure of the universe. It is a rare year in which we do not read about a new discovery that seems to contradict previous thinking. For instance, a physicist announced in 1986 that there is only one millionth of the amount of invisible gas and dust previously thought likely in the vast regions between clusters of galaxies.[32] If correct, this evidence would have a radical effect on thinking about the age and nature of the universe.

In one sense, however, such announcements are good, and they show the progress that this science may be making. But we must not forget that taking impressive photos of the heavens is not the same thing as interpreting the many mysteries that surround the earth. Today's speculation is replaced tomorrow with a new speculation, and this is not necessarily progress at all.

In the early 1980s the universe was confidently described as 20 billion years old. Then *Time* told the story of a Harvard astronomer who did what good scientists do.[33] He carefully reviewed the work that went into making this estimate about the age of the universe. In so doing, he found to his astonishment that an arithmetic error of 10 billion years had been made in the calculation. Rather than admit such an embarrassment, the age of the universe for many years afterward was reported to be 10–20 billion years old.

Let's examine briefly the tale of the age of the universe, treating just the past few decades. Edwin Hubble argued that the more distant a galaxy is from earth, the faster it is moving away and developed a constant to show the relationship between assumed distance and assumed speed. Around 1929 his constant yielded an age of 2 billion years for the universe. Geologists, using their own set of assumptions, were taken aback and pointed out that some rocks on earth were much older than that. Astronomers embracing another set of assumptions were positive that some stars were 12 to 16 billion years old, and could hardly have hung around all by themselves for 10–14 billion years to welcome the beginning of the universe.

There was another problem. Astronomers had worked out a theory to explain star color from red to yellow to white-hot to blue-white over a period of around 20 million years, they thought. Again geologists sharply objected and demanded at least a hundred times as much to fit their belief in gradualism and to give evolution time to go from non-life to what we observe today.

The solution was to change the value of the constant to make the universe old enough to silence the critics. Bitterness remains, however, and "old universe" astronomers today are not civil toward "young universe" astronomers.

By the early 1970s the alleged Big Bang was dated (in billions of years, of course) by leading astronomers as follows: several, 4.3, 5, 7, 10, and 18. Astronomer Allan Sandage "refined and improved" the distances, thereby changing the Hubble constant and announced the age of the universe as 15–20 billion years. The 10 billion year error described above once more left some stars supposedly 6 billion years older than the universe itself. A California astronomer adjusted the Hubble constant once again and came up with a universe 12 billion years old. In 1985 Sandage announced "new data" and upped the universe once again to 18–20 billion years. His chief rival, Dr. Wendy Freedman, cited studies in support of a universe 10–13 billion years old. Astronomer Thomas Matthews sought to bring peace to the feuding camps and announced a new model in 1993 that gave an age of 15 billion years for the universe. Other astronomers kept insisting that some stars were 16 billion years old. As of the late 1990s we read about chaos in the study of the universe, astronomers too angry to speak to rival camps, bitter reactions, bizarre studies, and theory busting. Freedman has held to a universe 9–12 billion years old, while Sandage has come down considerably to 12–15 billion years. Sandage thinks Freedman's group is making serious errors, while

Freedman suggests that Sandage's group would have better results if they would dump some of their "ratty data."

If any of the above discussion sounds like science to the reader, kindly explain to the author. Both factions believe in an old universe, and billions are tossed around like confetti. All the ages suggested are based on many unproven assumptions, and there is nothing compelling about them. Our belief in a young earth is founded on the Bible, and there is nothing in the way of evidence to shake that belief. Speculation is not evidence, and as we have seen, astronomers disagree violently among themselves. To all of this we must note a growing amount of literature that presents evidence that there never was a Big Bang.

Color of the Stars

Basic accepted theory about the age of stars holds that the colors and temperatures of stars should remain constant for great spans of time measured in multi-millions of years. Yet Greek and Roman writers described certain prominent stars as reddish or fiery red, but those stars bear radically different color descriptions since that time.[34] Another basic "law" is that double or triple stars must be the same age because a star cannot annex another star. Yet there are clear examples of double stars of strikingly different colors, thus supposedly impossible hundreds of millions of years apart in age.

The Red Shift

Every schoolchild is taught the story of Big Bang as the moment of creation of the universe, and the strongest support for this theory is known as red shift. The importance of this idea, called Hubble's Law, is the supposed relationship between distance and the amount of red shift. Color is supposed to change or shift much as we observe with the sound changes of the whistle of an approaching or departing train. Hubble is revered for what is thought to be the most important astronomical discovery of the century, as we note with the naming of the Hubble telescope. Now enters the villain, Halton Arp, a premier astronomer and the world's leading authority on the mysterious quasars, who documented numerous and clear examples where the red shift is not behaving as the "law" requires.[35] There is clear and very unwelcome evidence that red shift has nothing at all to do with distance. Arp's reward for this important discovery, and for using the scientific method to evaluate a theory, was to be banished from the astronomic observatories in America. He was denied further use of the great telescopes to continue his studies, and now lives in Europe where he has received a friendlier reception.

We see the ugly specter of important knowledge suppressed or ignored, because it does not fit the accepted theory. Is this unusual? One of the great intellects of the twentieth century, Alexis Carrel, observed already in 1935 the following:

Certain matters are banished from the field of scientific research and refused the right of making themselves known. Important facts may be completely ignored.... (Scientists) willingly believe that facts that cannot be explained by current theory do not exist.

Red shift is so fundamental to the belief in the Big Bang that evolutionists do not dare examine the evidence that undermines that belief.

Sun and Moon

Ancient man was preoccupied with the magical moon from the earliest beginnings.[36] Even the rock markings of the Neanderthals were astronomical in nature. In recent decades scientists have discovered that carved lines, strings of dashes, rows of dots, and other symbols often recorded moon phases and other cycles. Such markings were found in many so-called ice age caves of Spain and France, as well as throughout Europe, Asia, and Africa.

What led the Incas and other peoples around the world to fear that the sun would leave its course in the sky? In each town of the Inca empire a special post was erected and ceremoniously tied to the sun to prevent it from straying.

The lure for spending billions on space research includes much emphasis on solving the great mysteries of the universe, especially its age. Typically a space venture results in adding "a vast store of fresh knowledge" about the moon or the planet studied. Such statements almost invariably add the thought that the venture leaves "unanswered most of the puzzling questions about the planet and evolution." We must say that issues about an old universe or in support of evolution have not fared well. Our closest astral neighbor, the moon, is not without mysteries of its own. An early shock was that dust on the moon was measured at one billion years older than the rocks they rested on. This reminds us that staunch promoters of evolution such as Isaac Asimov calculated the amount of meteoric dust on the moon over 4.6 billion years, and Asimov warned that he pictured our first space ship sinking majestically out of sight in the dust of the moon. The actual depth was measured in terms of a few inches, indicating that once again the evolutionary time scale did not treat the real universe.

Dating laboratories couldn't wait to date moon rocks when they were brought back. The amount of argon in the samples was unexpected. Some "unacceptable" dates resulted, ranging from 2 million to 28 billion years. Fantastical solutions were offered. In some unwelcome cases the samples were ruled as contaminated, and the dates were discarded. Another solution was to apply "correction factors." For example, if a sample was dated at 8 billion years and the desired date was 4 billion, the wrong date was multiplied by a correction factor of .5 thus giving the desired age. How often does one see the above information when the age of the moon is discussed?

One early analysis of moon rocks revealed that the moon and earth have very different chemical compositions. This undermined the many theories that the moon was originally ejected from the earth.

There is much literature about how the ancients studied and depended on the cycles of the moons, and we may all remember old Indian stories that used the phrase "many moons ago." Ancients in Mesopotamia discovered by observation that the moon forms a cycle every 18 years and 11 days (or 223 lunations) to which they gave the name *saros*. But saros had a second meaning, that is, 3,600 of anything. When an inscription was found that listed the ten kings that lived before the flood, their total reign was said to be 120 saros. Which meaning should be taken? Scholars, eager to show the mythical character of such accounts naturally multiplied the 120 by 3,600 to give a total of 432,000 years, an absurdity often found in accounts dealing with ancient myths. If, however, we multiply 120 saros by the much more logical amount, we find that the ten kings (now read "patriarchs") lived a total of 2,164 years, a total astonishingly similar to the Vatican Septuagint total of 2,242 years from Adam to Noah. But, as we have noted elsewhere in this book, humanists cannot bear the thought of anything actually supporting the Bible as an historical record. It is not that unusual to find a word with two very different meanings. *Billion* in England, for example, is an amount a thousand times greater than in the United States.

Sirius and the Sothic Cycle

One of the most sacred beliefs about the chronology of the ancient world centers on the dog-star Sirius and a time cycle called the Sothic Cycle. Seven inscriptions treating the rise of Sirius on the horizon at sunrise (called heliacal risings of Sirius) have seemingly locked in Egyptian chronology, which in turn determines key points of the chronology of many of the surrounding ancient civilizations, including the Bible lands.[37] But as some scholars have observed, astronomers know nothing of ancient history, and historians know nothing of astronomy, hence a curious tale emerges.

Egyptian sources refer to the rising of Sothis (or Sirius) at the same time as the sun rose. Astronomers determined that this occurred only once in 1,460 years, called the Sothic period, cycle, or year. As the sunrise appearance of Sirius migrated through the calendar months, one could calculate the precise year BC of an inscription by calculating what fraction of 1,460 years the calendar date was. All that was needed was to know the year when the cycle began. Egyptologists calculated that one Sothic cycle had begun ca. 1320 BC, and from that they needed only to add or subtract 1,460 to determine other beginning dates.

One papyrus inscription read "You ought to know that the rising of Sothis takes place on the 16th of the 8th month." This inscription interlocked with other papyri, for example, one that noted the date of a new moon and the 31st year of the reign of a king later thought to be Sesostris III. All of this put together yielded, it was presumed, an unshakable date of 1872 BC. Dynasty lists and lengths of reigns were believed to provide a quite secure picture of Egyptian chronology. Inscriptions and artifacts such as scarabs in the strata of ancient ruins dated various historical periods. Evan Hadingham is typical of hundreds of writers on ancient history who showed what Sirius meant in unscrambling ancient chronology, or so they thought:

> Fortunately, a secure chronology was quickly established for ancient Egypt since its king lists stretched back to the 1st dynasty, shortly before 3000 B.C. A calendar note found on a papyrus scroll mentioned the date when the dog-star, Sirius, appeared to rise in the sun's path above the horizon. Astronomers can independently pinpoint this date to 1872 B.C. . . . We could perhaps say that most of the prehistory of the eastern Mediterranean is now firmly 'anchored' in historical time, all by comparison with those vital king lists.[38]

We might add that the scholarly world does not treat kindly anyone who questions what are called "astronomically fixed" dates. They are considered science at its best. If the 1872 BC date falls, the entire structure of ancient dating falls with it. It does not take much searching, however, to find grave problems with such dating. Here are just a few of the problems, in addition to gaps, frequent overlaps, and fictions of the Egyptian king lists:

1. Some astronomers identify Sothis with a different star than Sirius, which, if true, would upset the entire dating scheme.

2. The cycle of 1,460 years is based on a year of 365.25 days. When the true length of the year is used, the length of the cycle is about 1,507 years, a serious error when 1,460 is deemed correct, especially as great precision is claimed for the method. For starters, the so-called Sothic cycle is very bad and careless arithmetic.

3. Both ancient writers and modern astronomers disagree on starting dates for a given Sothic Period. For example, authorities differ by as much as 400 years on the cycle many believe began in AD 140. Strangely, astronomers have not settled such discrepancies, despite of claims by astronomers that they can roll back the skies in their observatories to any time in the past, again with great precision.

4. There are serious problems in defining the actual rising of Sirius because the rays of the sun would effectively block out the light of Sirius if it rose the same time as the sun.

5. The Egyptians did not leave a single trace of the fixed calendar required by the supposed Sothic cycle. Several mentions of Sothis in inscriptions do nothing to establish such a scheme.

6. There is no evidence that the seasons gradually shifted throughout the year in the 1,460 year cycle as assumed.

Donovan Courville listed eight basic points that must be established to verify the Sothic system and concluded that not one of the points has been established.[39] We must say that it appears that Sothic dating is clung to because nothing else is known that could fix dates in ancient Egyptian history. This, by the way, shows that the radiocarbon dating process (C14) has not been successful in sorting out Egyptian chronology. Is it not reasonable to suppose that if both C14 and Sothic cycles were valid measures, each would validate the other?

Bode's Law

In 1772 astronomer Johann Bode discovered a peculiar arithmetic law that seemed to govern the distance of the planets from the sun.[40] It cannot be proven and there are some problems with it, yet its value nevertheless has been inestimable. It helped in the discovery of Neptune in 1846. With a minor adjustment to the formula, there is no place for Venus. Thus one can argue that Venus was a newcomer among the planets.

The most interesting thing about Bode's Law is that it presumably shows where a missing planet used to be. In this blank space astronomers found a series of 90 small planet chunks (asteroids) and many thousands of other fragments of what may once have been another planet. It is fascinating to learn that ancient myths treat this missing planet, indicating that the terrible event occurred in historic times.

Venus

When Pioneer Venus 2 descended to that planet December 9, 1978, there was consternation among astronomers, one of whom stated that they now had to rethink all theories about how the planets had been formed.

What was the problem? First, Venus was abnormally hot, an idea ridiculed since Velikovsky proposed it three decades earlier. He suggested there was truth to the myth of the birth of Venus which maintained that Venus was a latecomer among the planets. The greenhouse effect offered as explanation scarcely begins to account for the unexpectedly high surface temperature of the planet. Before 1956 astronomers said Venus was cool, but within 20 years it rose to 900°F. Although Velikovsky is the only one to offer a viable explanation, astronomers have never recognized his prediction, because he was an outsider. Second, the amount of argon-36 on the planet is up to 300 times more than expected. Without getting technical, it is in sharp contrast to the amount on the earth, once called the sister planet of Venus. The amount also hints of the extreme youth of the planet, a suggestion orthodox astronomers will never accept. Another strange finding, hinting strongly of its catastrophic past, is that the atmosphere rotates at many times the rotational velocity of the planet. Venus always presents the same face to the earth, which means, strangely, that its rotation is controlled not by the sun, but by the earth. Further, its rotation is in the opposite direction of all but one other planet. This all tells of a powerful electromagnetic event in the past between the two planets.[41]

Another shock was that Venus is pockmarked with rugged craters up to 100 miles in diameter. This was baffling because the atmosphere was a sandblasting kind that should quickly grind down any surface scars in short order. We ask again, could Venus be a newcomer?

Ancient tablets of Venus observations seem to show an erratic behavior far different from its peaceful behavior today. There are many ancient descriptions of Venus that horns grew out of her head, thus the worship of the bull with horns. At a later time the horns disappeared and bull worship was replaced by the sacred cow. In India the Brahmans neither worshiped the cow nor refused to eat its meat. Their treatment of the cow as sacred is believed by many to go back to Venus worship when the planet was behaving dangerously.

In both the Old World and the New, the ancients were acutely aware of the 52-year cycle of Venus. There was intense fear near the end of each cycle that something would once more go wrong in the heavens. People believed it was a time of great danger. There were mass destructions of belongings, ritual mutilations applied to everything, human sacrifice to appease the gods. There are vivid accounts of the horrors and destructions in the past when Venus did not behave.

As usual, uniformitarian scientists would not permit Venus to misbehave and therefore they "corrected" the ancient tablets to read the way they were supposed to read. Hugh Rose relates the shocking procedure! And this is called science?

Note the ease with which Elizabeth Langdon, J. K. Fotheringham, and Robert Schoch can take away a month here, add a month there, take away 20 days, and add 20 days there, all for the sake of reconciling the text with modern observations of what Venus appears to be doing. This is what I called playing the uniformitarian game. They don't want any 5-month invisibilities, so when the ancient texts report one, they rewrite those texts so that it isn't there anymore and so that what is there will be in accord with modern observations. Then, after the surgery has been completed, they find to no one's surprise that "the observation, thus restored, is excellent."

Currently, except for endlessly and mindlessly repeating the conjecture about the greenhouse effect, we hear little about further study of Venus because conventional theory has nothing convincing in the way of explanation. Among astronomers, Venus is the planet with no answers, only questions.

Mars Misbehaves

The planet Mars also held a place of great fear and terror in certain ancient periods when some scientists believe it followed an erratic path for a time. The result for the earth was gigantic ocean tidal floods, earthquakes, renewed volcanism, and radical lightning strikes. One interesting evidence that Mars had moved closer to the earth for a time is that the ancients as far back as the Sumerians knew very well about the two steeds (satellites) of Mars. Homer and Virgil spoke of the red planet, Mars, the god of war, and described the two steeds that drew his chariot through the heavens, named Phobos (terror) and Deimos (rout). These satellites were only rediscovered by means of the telescope by Asaph Hall in 1877.[42]

Ancient literature, especially the myths, is full of stories about the god of war, Mars, and his actions.

Other Planets

There are eerie glimpses of how much ancient people knew about the heavens.[43] The Shilluk people of South Africa called Uranus "three stars." In 1781 Uranus was discovered with its two chief satellites, that is, the planet and its two great moons, confirming the mysterious tribal wisdom. There is the ancient legend about the planet Uranus eating and then disgorging his children. Modern telescopes show that Uranus regularly covers its moons, which then again become visible. How could the ancients have seen all this without the aid of the telescope?

Jupiter, another fascinating mystery planet, was in the news in early 1982 for a most peculiar reason. Two young scientists, now instantly faded into obscurity, published the book *The Jupiter Effect*.[44] The nine planets would all be in a straight line on the same side of the sun in early 1982. This would set off a complex chain of events that would trigger a disastrous earthquake on March 10, 1982, and the most likely site was the Los Angeles area. A number of scientists attacked the prediction as inaccurate, incomplete, biased, and speculative. As we have come to expect, the doomsayers guessed wrong, and 1982 came and went. Los Angeles is still there. Other similar forecasts were made in the 1990s and were just as wrong, e.g., Author R. Noone announced doom for the earth would be May 5, 2000.

CONCLUSION

Concluding Comment

Attempting to treat the ancient and the present state of astronomy is much like the gnat who resolved to swallow the ocean. Yet we hope the reader has gained a measure of understanding and awe about the heavens that declare the majesty of God. We see something else as well, that the stars played a vital role in the daily life of our remote ancestors, an intimacy now long lost. We observe also that the wonder of the stars soon became perverted into many forms of heathen worship. This evil has come down to our day largely in the form of astrology.

We see something of the overwhelming mystery about the heavens, and the futile attempts of many astronomers to fit all this into the bankrupt theory of evolution. We see the weakness of science when specialists are punished severely for discovering the wrong things—evidence that contradicts the beliefs of evolutionists. In the classic book *The Encyclopaedia of Ignorance*, written by some of the world's most prominent scientists, we see a ray of hope for the scientific method in our day.[45] These great specialists were eager to confess their ignorance of many of the most basic beliefs about the heavens. Others confessed in the same manner regarding other basic areas treated by science, such as evolution, human fossils, and the mystery of the human mind. When arrogant specialists in positions of power shut off all dissent, real learning screeches to a halt. We are pleased, however, to see that there are exceptions in the scientific community. We have illustrated the evil of closed minds here and there in this chapter. The faulty assumptions or beliefs of the evolutionists turn off knowledge, as we have illustrated. There is a vast difference between assumptions and evidence.

For Discussion and Reflection

1. Psalm 19:1–6 is joyful praise of God for the wonder and beauty of the heavens that declare the glory of God. What are some of your feelings and expressions when you see and ponder the sky at night?

2. Genesis 1:14–18 tells us of some good uses of the stars. What are some of the examples of this you found in the chapter? Are any of these new to you?

3. When astronomers who believe the universe is 20 billion years old will not talk to astronomers who believe it is only 12 billion years old, what does this suggest to you about facts versus interpretation? Could both groups be very wrong?

4. Astronomers use very powerful telescopes and provide the world with wonderful photos of stars and galaxies. Yet astronomy is called the most speculative of all the sciences. Have you noted examples of speculation when reading this chapter? Why can't astronomers be more precise?

5. When you read articles or books on astronomy and of new discoveries, how much effort is displayed to distinguish between fact and assumption and interpretation?

6. Satan is skilled at taking something very good and making it into something very evil. Can you give examples of any evils involving astronomy?

CHAPTER FIFTEEN

AN IRREVERENT REVIEW OF PREHISTORY

THE LURE OF THE PAST

Searching Out the Past

The view of history all depends, of course, on one's assumptions and presuppositions. Much, if not all, the evidence used in books on prehistory will fit more than one possible theory, at least on a superficial level. A more penetrating analysis may reduce the range of possibilities. Yet there can be only one true story of the ancient past. Ancient history does not interest everyone. Still, the distant past holds a powerful lure, and there are many attempts to unravel the mysteries, or to reveal—generally for the very first time—the whole truth about man's historic past. When people by the millions around the world buy the books of authors such as Erich von Daniken that claim to reveal man's true past, we see some indication of how curious many are about prehistory. Standard college texts, and the media in general, tell one kind of story about man's past, but on the fringes of the academic community there has been no lack of attempts to present the wild and wonderful "genuine" truth.

We hasten to add that *prehistory* is an impossible word, but it is in common use. There can be no history before history. It is normally understood, however, to refer to history that took place before written records of some kind existed.

A Novel Approach

The approach of conventional texts is well enough known to require little comment here. It is the story of cavemen in chapter 1, followed by the story of Egypt or Greece in chapter 2. In this book, however, we take a serious look at the lepers outside the city walls, at books that are not likely to be found on reading lists for university courses on prehistory. The following discussion should not be considered as book reviews, but rather as an evaluation of representative books that treat or touch on prehistory in some way to see what, if anything, each might contribute toward possible insights into man's past. The focus is on the kind of evidence offered—not on the theory proposed. It must be emphasized that a great deal of improbable "evidence" has been rejected out of hand as too improbable. It is freely acknowledged that one walks a precarious path through this kind of literature by accepting one statement, but rejecting the next one.

Why not do the accepted thing and use only the works of recognized scholars? There are several answers. First, apparently no one has ever taken a systematic and thoughtful look at "fringe" books to see what evidence is actually offered. This in itself is sufficient justification for the venture. Second, such offbeat books contain some very interesting material that ought to be considered in theory construction, if the facts offered can be supported by other parallel evidence. An internationally renowned scholar, the late Cyrus Gordon, noted "valuable material can be gleaned from old books written by uncritical authors who were out to prove a theory. The pickings in them may be lean, but they are nevertheless there."[1]

In starting our journey, we shall adopt a face-value technique by assuming that a fact is a fact, whatever its source. If several of a class of facts can be verified from more reliable sources, the point is made without further verification. When an error or a fraud is uncovered, it will be exposed for what it is. Enough of these occur both in conventional and in unconventional works that they deserve full and separate treatment—an enticing project for the next chapter of this book.

What Triggered These Books?

In searching the literature, we find an amazing fact. There have been many books over the years challenging conventional views of the past. Although there has been a sprinkling of offbeat books every decade, the 1970s seemed to spawn more books about prehistory than all other decades combined. Why should that be? We can point to some factors that seemed to be involved. America went into shock on October 4, 1957, when the Russians launched Sputnik I, the first earth-orbiting satellite. The space age began that day and forced all eyes to the heavens. It took America another twelve years to regain the lead in space science. The first space walk on the moon took place in 1969. The late 1960s were also the years of the hippie movement with wholesale rejection by millions of traditional values. "Space gods" seemed to fill some of the void for a time, but with lack of substance, interest in such strange books has waned. Our interest in this chapter is purely on the evidence offered, not for the wild theories proposed.

After the following review of a number of books that in some way treat prehistory, we shall pause to consider what these sources may have to offer toward an understanding of man's prehistory. There will be some surprises.

THE LITERATURE OF THE PAST

The von Daniken Syndrome

In this section we examine authors who have found that ancient man was far more sophisticated than one would have expected. Conventional evolutionary views of gradual development have been considered to be inadequate explanations. The writers, of whom Erich von Daniken has been the best known for more than three decades, appeal to someone or something in outer space as explanation for man's sudden sophisticated development. Other writers who have offered critiques of this view are included. The enduring popularity of von Daniken and similar writers may be attributed to widespread interest in space travel, the belief in flying saucers, and acceptance by more millions of astrology. Such interests and beliefs may, in the minds of some, bridge the gap between ambiguous evidence and a proposed theory. For years I had thought von Daniken had passed into obscurity. Then, to my surprise, *Archaeology* featured the new von Daniken theme park in Switzerland.[2] The article stated that this has to be the most bizarre archaeological experience on the planet. Obviously, people still love to be fooled and will pay handsomely to experience it.

Following, I list representative authors, titles of their books, followed by comment and evaluation.

- **von Daniken, Erich.** *Chariots of the Gods?.* New York: Bantam, 1969; *Gods from Outer Space.* New York: Bantam, 1973; *The Gold of the Gods.* New York: Putnam's Sons, 1973; *In Search of Ancient Gods.* New York: Bantam, 1973; *Signs of the Gods?.* New York: Berkley Books, 1980; *The Eyes of the Sphinx.* New York: Berkley Books, 1996; *Odyssey of the Gods.* Boston: Elemeny Books, 2000.

To evaluate von Daniken's writings, one must examine carefully his unstated assumptions and premises, and these are summed up below. These five statements would be acceptable to millions of intelligent, educated people around the world today. If the statements are actually valid, von Daniken comes out as a real winner, the champion of long suppressed truth. It is no surprise then that he has enjoyed a large following for some years.

1. Erich von Daniken accepts conventional dating by evolutionists of the ancient past as used by geologists, historians, and archaeologists.

2. He believes in a patient, quiet universe where he can confidently reconstruct the past on the basis of natural processes operating now, but he allows for catastrophic events of a rather mild and localized sort, not destructive enough to affect dating of the past.

3. Evolving man, patiently chipping away at stone artifacts with little change millennia after millennia, is simply not far enough along to create the sophisticated technologies that appeared suddenly all over the earth.

4. We know that many ancient writings indeed seem totally obsessed with vivid, mysterious, and wonderful happenings in the sky; and also note in those writings that disaster came from the sky more than once. The conventional explanation is that these are childish myths and legends that are not to be taken seriously, but this explanation simply does not satisfy many intelligent people.

5. There are an astonishing number of mysterious structures, carvings, and highly sophisticated technologies going far back into the past for which there is no convincing explanation.

Without detailing the reasons at this point, it is the major premise of this book that the first three assumptions are in error, and that the fourth statement is roughly correct, but for entirely different reasons from those expressed by von Daniken and other writers. The fifth statement is factual. In due time each of these assertions will be thoroughly tested.

To his credit, von Daniken has collected a large amount of interesting material from other writers and from sites he has visited. Some of the evidence offered, however, is very ambiguous for interpretive purposes. For example, von Daniken will describe a helmet or headdress from an ancient carving as a space helmet, or he will compare an abstract design with a modern electronic circuit. These are interpretations, not facts. No arguments can be won on the basis of these kinds of data. Erich von Daniken's theoretical position consists of several vague speculations that beings from outer space visited the earth many thousands of years ago, taught a very primitive mankind some of the arts of civilization, left many mysterious signs of their presence around the world, and then left as mysteriously as they had arrived. They may be keeping an eye on things now, and they may return.

In his first best seller *Chariots of the Gods* von Daniken notes evidences such as the following: the accurate mapping of the earth before the Antarctic was ice covered; the strange geometric lines of Nazca, Peru, huge animal effigies, peculiar rock carvings and paintings all over the world, ancient calendars and advanced astronomical knowledge, huge megaliths (stone structures), catastrophic events, rock vitrifications (rock melted from intense heat of unknown origin), other surprising evidence of the high sophistication of the most ancient cultures, and transoceanic cultural contacts.

Gods from Outer Space is a reissue of an older volume titled *Return to the Stars* (1968) on the same theme as above. The same kinds of evidence are presented, centering on giants, huge dressed stones, tremendous heat and poisonous vapors (related in old epics), ancient ruins, rock engravings and cave paintings, effigies, and ancient astronomical alignments.

Gold of the Gods claims to explore a vast network of caves in Ecuador filled with gold and writings that go back in time to a prehistoric era of the gods. Not long after publication, however, von Daniken admitted that the story was pure fabrication. No such place existed, but in the book von Daniken exclaims: "What tricks scholars will use to displace this fabulous gold treasure of inestimable archaeological and historic value!" The reader should be aware of how von Daniken used evidence. For example, he illustrates (p. 26–27) an Australian cave painting, and he assumes that the figure is a god. There are sixty-two small circles included in the work. He muses about this prehistoric message transmitter. The simple, correct interpretation of this painting, however, has been a matter of public record since 1901, and of course has nothing at all to do with his thesis.[3] The message on the head of the figure in old Japanese (before the twelfth century AD) is "The number of the hopeless ones is 62 shipwrecked Japanese." The book otherwise contains a small number of descriptions of other mysteries in various locations around the world, but each of these topics is treated better elsewhere.

In the book *In Search of Ancient Gods* von Daniken still sees signs of ancient gods wherever he looks. Rock drawings and paintings are important to him, and he has located ceramic reproductions of insect-headed beings. Gigantic markings have now been located in the deserts of Chile somewhat comparable to those previously reported in Peru. The reader cannot help but feel that von Daniken is running low on material.

In *Signs of the Gods* Erich von Daniken muses on a variety of topics such as ancient king lists, the impressive but relatively modern ruins at Zimbabwe in Rhodesia, ancient symbols, and many other evidences of sophistication in ancient times. A good deal of space is devoted to the truly extraordinary unsolved puzzles in stone on the island of Malta, including the large number of unexplained parallel deep grooves on the island, some of which continue on into the sea bed, and the amazing ancient stone structures there. The fact that there are no plausible explanations for many of the structures and artifacts he discusses is not to say that anything gives support to his quaint theory.

His more recent books, *The Eyes of the Sphinx* and *Odyssey of the Gods*, contain speculative musings about the sphinx and the ancient Greeks, but nothing of value for our search.

Critics of Erich von Daniken are evaluated in the following:

Wilson, Clifford. *Crash Go the Chariots.* New York: Lancer Books, 1972; *The Chariots Still Crash.* New York: New American Library, 1975; *Gods in Chariots and Other*

Fantasies. San Diego: Creation-Life Publishers, 1975.

In the book *Crash Go the Chariots* Wilson cites a number of points where von Daniken is obviously in error, a task not very difficult to do. While one can agree with Wilson's conclusions that the Bible is a faithful but incomplete account of some aspects of man's ancient history, Wilson does not offer much in the way of support. Better answers are at hand than the ones provided.

The Chariots Still Crash contains little not already better said elsewhere. Wilson does, however, cite a number of interesting accounts from ancient literature that may refer to such biblical events as the ten pre-flood patriarchs, and Babel and the confusion of languages. The obvious purpose of the book is further refutation of von Daniken's works. Wilson attempts to relate evidence of man's early sophistication within a biblical framework of history.

Gods in Chariots and Other Fantasies contains a number of references to errors in von Daniken's works. The book, however, seems hastily put together, and better critiques are available.

Thiering, Barry, and Edgar Castle. *Some Trust in Chariots.* New York: Popular Library, 1972.

Thiering's welcome book is like a purgative after reading too many unrelated and unevaluated statements alleged to be facts. Competent persons have examined Erich von Daniken's writings and have found them to be full of misrepresentation and error. This little book is probably the best response to the nonsense that von Daniken has been writing.

Story, Ronald. *The Space-Gods Revealed.* New York: Harper & Row, 1976.

This critique of von Daniken is worth reading, and it is at its best in careful evaluations of the evidences von Daniken puts forth in support of gods visiting the earth in ancient times. The point is made over and over that ancient sophistication is not support for von Daniken's theories. The book freely acknowledges such remarkable and unexplained mysteries as the Piri Reis map. This in turn illustrates that there is much about the ancient world for which conventional explanation in terms of primitive-to-modern development is just as clearly wrong as von Daniken.

Having examined von Daniken and his critics, we now view a group of authors, some of whom preceded von Daniken in crediting gods from outer space for sophistication in the ancient world, and others who imitated him.

Bergier, Jacques. *Extraterrestrial Visitations from Prehistoric Times to the Present.* New York: New American Library, 1973.

Bergier attempts to support a case for extraterrestrial visitors and settlers to the earth in the past. He does not cite any evidence that does not also appear in many, many other similar sources. The principal point made is that ancient man was too sophisticated to be accounted for by evolutionary theory. He seems unaware that one can accept his point without accepting his solution.

Charroux, Robert. *Legacy of the Gods.* New York: Berkley Medallion Books, 1965; *One Hundred Thousand Years of Man's Unknown History.* New York: Berkley Medallion Books, 1970; *Forgotten Worlds.* New York: Popular Library, 1971; *Masters of the World.* New York: Berkley Medallion Books, 1974; *The Gods Unknown.* New York: Berkley Medallion Books, 1979; *The Mysterious Past.* New York: Berkley Medallion Books, 1979.

Charroux, a French author, preceded Erich von Daniken in expressing an outer-space theory. In *Legacy of the Gods* he follows a catastrophic model of early earth history. He has brought together many interesting items from the past, such as man's early use of metals, descriptions of catastrophes in ancient literature, evidence of writing in alphabetic script long before the date commonly accepted today, and evidence of giants, whatever one might make of that.

In *One Hundred Thousand Years of Man's Unknown History* Charroux's uncritical acceptance of conventional dating inadvertently renders a service to a different view of prehistory from that offered in texts today. His thesis is another variation on von Daniken—the conspiracy to conceal the real truth about our Superior Ancestors from outer space. He cites interesting evidence of advanced culture in

Erich von Daniken used this cave painting as evidence for gods from outer space. If he had searched the literature, he would have found this translation of medieval Japanese: "The number of the hopeless ones is 62 shipwrecked Japanese." (Photo from the *Victoria Institute.*)

ancient times that is diametrically opposed to the theory of primitive- to-modern development of man. Among these are alphabetical characters 10–15 thousand years old (in his chronology), ancient steel nails and manufactured glass that are much too old, a human jawbone in Tertiary strata (anywhere up to 70 million years old supposedly), modern clothing depicted 15–20 thousand years ago, very old but sophisticated astronomical charts, a decimal system long before its known history, and advanced writing in Peru that may be older than Egypt. While his chronology is naive, some of his data are both useful and fascinating.

Forgotten Worlds promises to provide even more authentic and startling facts than one can find in *Chariots of the Gods*! Robert Charroux confesses to some minor errors in his earlier writings. If the reader can survive a great deal of the fanciful and mystical on far out topics, he is rewarded with interesting information about the ancient past, much of which can be used in a fresh approach to prehistory. The author enjoys making nasty remarks about conventional archaeologists and prehistorians. Apparently he is reacting to critics of his earlier writings. *The Gods Unknown* was written in the same year as *Chariots of the Gods* and promises to offer exciting new revelations. Mountains of mystical writing are mingled with a few molehills of fact, yet there are some items of interest. The use of the wheel in Middle America is discussed rather convincingly. The fantastic astronomical lines of Nazca are mused about, mythological writings are examined in their ambiguous glory, mounds and ancient megalithic stone work are shown to relate around the world, and the wrong races are found in the wrong places, e.g., fair-haired, blue-eyed ancient people in Peru. The author also discusses giants, ancient alignments on the land, sophisticated ancient astronomical knowledge, and writing that has been found to be much older than that found in Sumer or in Egypt.

In *The Mysterious Past* Charroux has collected a large number of interesting facts about the past. Of special interest is his detailed description of the many glazed forts and walls in Europe, where apparently blasts of heat partially melted the stone. The reader must sort through a large mass of mystical and unintelligible writing to find a few items of interest.

Masters of the World contains a variety of interesting facts useful in the study of prehistory. Among these are a list of the great meteorite craters of the earth, and a list of meteorites worshiped around the world in various temples. He has brought together many curious facts to show the great sophistication of ancient people.

Drake, W. Raymond. *Gods and Spacemen in the Ancient West.* New York: New American Library, 1974; *Gods and Spacemen in the Ancient Past.* New York: New American Library, 1974.

The first title is billed as the ultimate proof of super-beings from space, and it is claimed to be more convincing than *Chariots of the Gods*. The author promises to lift the fog shrouding our hidden history. Drake makes giant leaps between interesting scraps of information and his preconceived notions. Among the evidences cited are details about planets known to the ancients that are not visible to the eye, like the moons of Uranus and the rings of Saturn. The possibility of a relationship between the explosion of a supernova and mass extinctions on earth is suggested. Other items are strange footprints, worked metal in old strata, mythology as coded history, rock paintings and carvings, and a very old map that shows California as an island. This is sparse going and indicates nothing more than the fact that the ancient ones were more sophisticated than is generally believed. The second title contains additional useful material drawn from a wide variety of ancient literature. He accepts uncritically vast ages in man's prehistory.

Kolosimo, Peter. *Not of This World.* New York: Bantam, 1971; *Timeless Earth.* New York: Bantam, 1973.

This is another apparent attempt to ride on Erich von Daniken's coattails (his assumptions and hypotheses are the same) due to the worldwide interest in "gods" from outer space. While Kolosimo has borrowed heavily from other writers on the fringes of prehistory, *Not of this World* contains useful data not found elsewhere that may apply to other theses on prehistory. He presents evidence of strange rock constructions, including those in the Pacific area and in Siberia, most unusual astronomical knowledge possessed by primitive tribes, themes of ancient rock carvings, man-made artifacts in strata too old for man, ancient calendrical knowledge, catastrophes in historic times, a magnetic pole reversal during Etruscan times, transoceanic links with the Americas, desert ruins dating to a time before the land became desert, giants, fossil footprints, vitrified ruins caused by intense heat of unknown origin, prehistoric irrigation systems around the world, myths regarding catastrophes that may have historic content, and deeply buried artifacts accidentally excavated.

In *Timeless Earth* Kolosimo has brought together an additional fascinating and extensive collection of oddities from around the world. He accepts conventional dating, but assumes that man evolved to a much higher state of civilization in the past only to relapse into chaos and barbarism. While the author is not always careful with his facts, the book is very useful for the study of various aspects of prehistory. His coverage of prehistory in South America is especially interesting.

Landsburg, Alan, and Sally Landsburg. *In Search of Ancient Mysteries.* New York: Bantam, 1974.

The Landsburgs wander a bit aimlessly about in this world to make two points in Erich von Daniken style. Advanced technologies existed long before they should have in the world as conventionally viewed. As man presumably had not yet developed, the assumption is made that beings from outer space colonized the earth. A considerable number of interesting evidences, almost all of which are found in other similar sources, illustrate the surprising sophistication in very ancient times. The authors note that people tend to believe what they want to believe, and that it is difficult for them to face isolated unexplained facts if they conflict with their mental picture of the world, and that our view of ancient history may well seem ridiculous to people in the future. To this we can agree. Ancient technology requires a better rationale and explanation than what is found in texts today. As usual with this sort of book, there is a vast gap between the evidence cited and any theory of beings from another planet.

Mooney, Richard E. *Colony: Earth.* Greenwich, CT: Fawcett Publishers, Inc., 1974.

Mooney has brought together an incredible smorgasbord of facts that do not fit into a conventional pattern of ancient history. His conclusion is that the earth was colonized from somewhere else in the universe, but he does not attempt to specify where. The book is well worth reading for the information brought forward in support of his views. He gives examples of man's remarkable sophistication in the past, and shows that sudden developments around the world can hardly be attributed to some kind of evolutionary development. He also stresses the fact that so-called primitive peoples today are only the wreckage of more highly developed societies in the past.

Norman, Eric. *Gods, Demons and Ufo's.* New York: Lancer Books, 1970.

The title of the book shows clearly Norman's orientation. He has noted a number of strange finds of metal in places apparently too old for them to occur. He has collected accounts of the artifacts and graves of giants and of tiny people, which, if true, are intriguing mysteries from the distant past. Among mysteries are huge stone structures around the world and enormous figures made in the deserts of southwestern United States. There is no relationship between his facts and his bias.

Sendy, Jean. *Those Gods Who Made Heaven & Earth.* New York: Berkley Publishing Co., 1972; *The Coming of the Gods.* New York: Berkley Publishing Co., 1973.

With assumptions similar to those of von Daniken, Sendy presents what he terms a novel of the Bible showing that men from outer space reached the earth about 23,580 years ago. In *Those Gods Who Made Heaven & Earth* Sendy is interested in various conventional chronologies, the possibility of catastrophes, the remarkable fact that all the great civilizations appeared abruptly with a high level of sophistication within a relatively narrow East-West strip of the earth, the earlier "explosions of innovation", e.g., technologies, domestication of plants and animals, and the sophisticated astronomical knowledge of the ancient ones. The second title follows a similar theme but is almost totally free of any supporting data for the author's views.

Tomas, Andrew. *We Are Not the First.* New York: Bantam, 1971; *On the Shores of Endless Worlds.* New York: Bantam, 1974.

As the titles indicate, Tomas' theses are the same as von Daniken's—superbeings who once walked the earth. A review in *Faith and Thought* notes that *We are Not the First* is irresponsible and bizarre.[4] Much documentation is poor and suspect, such as claims of everlasting lamps still lit that have been found in ancient tombs. Some of the evidence given is, however, useful, which makes the book of value apart from the theory he espouses. Among the kinds of information he furnishes, much of which is not found in other books, are the following:

He notes many examples of the sophistication of the most ancient cultures, all of which appeared abruptly within a short time span. He asks the crucial question: Who were their teachers? The question is especially appropriate because there is no evidence of less sophistication before the Old Kingdoms. Marked deterioration of culture as time went on—another very crucial question—is noted in the old cultures. Tomas cites ancient texts and other evidence that show advanced astronomical knowledge among the ancients. He cites the discovery of artifacts that boggle the mind if the dating for them is accepted. He describes very old footprints, cave and rock drawings and other paintings of long extinct animals, old megaliths and sculpted stones, some of enormous size, worked metal in formations too old for them, transoceanic cultural contacts, strange vitreous areas of the earth once subjected to enormous heat, an ancient sophisticated calendar system, detailed maps of many regions of the earth dated from before the Greeks that are of remarkable accuracy, including the Antarctic and Greenland before they became ice-covered, electroplating done by 2000 BC, and, in general, evidence for an ancient Golden Age—not a savage past. Catastrophic events from ancient literature are also cited.

In the 1974 volume Tomas offers additional support for his view. He focuses on the planets, strange footprints, catastrophic creation of the Sahara Desert and unusual rock paintings in Australia. He has uncritically accepted von Daniken's tale of vast man-made underground caverns in South America—a story that von Daniken has retracted.

Summing Up

We may say that in sampling the kind of literature Erich von Daniken has popularized the past two decades, we may draw several conclusions: The writers have located a large pool of interesting facts, though documentation is often poor and it is sometimes difficult to distinguish fact from fancy. Although the topic of gods from outer space seems to attract millions of people around the world, we must in all honesty say that not one of the hundreds of interesting items noted and summarized above has any necessary connection with the proposed theory. And this includes a sizable number of mysteries for which no one at this time has a satisfactory explanation. The point of our discussion is that interesting and useful facts may indeed be found in some very peculiar sources, and these facts may be set apart from any theory proposed.

Atlantis and Mu

There is much interest in and discovery of drowned lands around the world, and archaeologists predict interesting and important discoveries in coming years. Yet it takes a certain amount of courage to scan the books that treat Atlantis and Mu. The legendary island of Atlantis, lying somewhere beyond the Pillars of Hercules, is said to have sunk beneath the sea, according to Plato in the *Timaeus*. If Atlantis did indeed exist at one time, we cannot be certain that the Pillars of Hercules necessarily referred to Gibraltar, and therefore many locations for the island have been suggested with considerable fervency. The even more legendary land of Mu, somewhere out in the Pacific, is another matter. Dictionaries that will identify Atlantis as a legendary land have to draw the line somewhere. Many will not give even legendary status to Mu. The reviewer of this kind of literature will understandably be very cautious. However, we must be reminded that we are seeking supposed facts—not evaluating the likelihood of whether or not Atlantis or Mu ever existed. We take the view that facts may be considered apart from the theories to which we may find them attached. With that explanation we now proceed to review some admittedly odd literature. The following authors and titles of their works are discussed:

Donnelly, Ignatius. *Atlantis: The Antediluvian World.* Edited by Egerton Sykes. New York: Gramercy, 1964; *Ragnarok: The Age of Fire and Gravel.* Blauvelt, NY: Multimedia Publ. Corp., 1971.

Books on Atlantis crowd the marketplace. Donnelly's books are the exception to many. One may reject his notion of Atlantis and yet find a wealth of data and citations from ancient writers for other uses. Donnelly ranges over the whole world to point out convincing examples of very ancient transoceanic contacts. Elephant mounds in America, surprisingly ancient use of metals, mummies in America, Africans in ancient America, inscriptions, and evidence of ancient catastrophic events all furnish the ingredients for a new look at ancient history. As the books were written between 1882–1883, we see errors in his work, but overall Donnelly is fascinating and useful.

Ley, Willy. *Another Look at Atlantis.* New York: Ballantine Books, 1969.

This book is very sparse going for anything of value to the study of prehistory. One item of interest is evidence of the giant sloth in Argentina living in very recent times versus the conventional view of extinction many thousands of years ago.

Michell, John. *The View Over Atlantis.* New York: Ballantine Books, 1969.

On many counts this book is a remarkable contribution to the study of prehistory. Evidence is convincing that a prehistoric sophisticated civilization spread all over the world, but the identity of the civilization is unknown. There are astonishing alignments in England and in many other areas of the earth, a good many of which were recently discovered from aerial photos. Such lines, miles long, set down in prehistoric times, are in use today, e.g., roads, boundaries. Among other evidence, he notes the value to archaeologists of ancient place names, ancient sophisticated astronomy, the "dragon" sites, significance of color to the ancients, ancient stone and earth structure and effigies, ancient mapping, an event of catastrophic heat in historic times, and a "reason" for human sacrifices.

Tomas, Andrew. *The Home of the Gods.* New York: Berkley, 1972.

In this work, Tomas has left the fold of Erich von Daniken and has embraced Atlantis as the answer to many mysteries around the world. Apart from his theory, there are many interesting items that deserve careful study and possible application to a study of prehistory.

van der Veer, M. H., and P. Moerman. *Hidden Worlds.* New York: Bantam, 1971.

M. H. van der Veer has collected much interesting information about the past. He is content to live with an evolutionary model of the earth's history. However, he believes that one civilization far in the past reached an extremely sophisticated level but was then destroyed. He thinks that survivors from this civilization then became the teachers of all the early civilizations we now know in our past. He believes that when we solve the identity of the first great civilization, we will also have solved the mystery of Atlantis.

Mavor, James. *Voyage to Atlantis.* New York: G. P. Putnam's Sons, 1969.

The book focuses on one of the most exciting finds of the century—the excavations on the island of Thera that just might some day answer centuries of dispute about Atlantis. There is a possibility that this island was the cultural and religious center of Atlantis, really the Minoan culture. The book in a way is premature because excavations are still in an early stage, but welcome, nevertheless. Mavor's attempt at ancient chronology seems weak because he depends on conventional dating systems that are supported only by the prestige of the originator and by a few ambiguities that hold out various possibilities of interpretation.

Luce, J. V. *The End of Atlantis.* Frogmore, St. Albans, Herts: Paladin, 1969.

Although Luce follows a conventional chronology for the Greeks, including a Dark Age that is becoming more and more vulnerable to question, the book is a thoughtful attempt to link Atlantis with Thera. Following the destruction of the Cretan civilization with the explosion of the island of Thera, power and control of this part of the Mediterranean passed on to the early Greeks. Luce has made a careful examination of ancient Greek literature to support his thesis.

Ferro, Robert, and Michael Grumley. *Atlantis.* New York: Bell Publishing Company, 1970.

This is another book on a familiar theme that consists of about one page of useful data with all the rest of no interest or purpose except to serve as a filler. The authors recognize the probable validity of the theory that Atlantis is to be identified with Crete and Thera. This book tells what is known so far of possible undersea ruins by the islands of Bimini in the Bahamas—a supposed discovery that has become quickly discredited.

Ebon, Martin. *Atlantis: The New Evidence.* New York: New American Library, 1977.

Ebon is another modern author who accepts the explosion of the island of Thera as the end of the fabled civilization of Atlantis. Ancient travel to the Americas is assumed. We have no reason to believe that this book should be any kind of major reference for prehistory. One item of considerable interest, however, is the growing recognition of the fact that huge meteorites struck the earth in the past. The evidence of these massive strikes, called astroblemes, is drawing much interest. The book has a useful list of actual and possible astroblemes.

Berlitz, Charles. *The Bermuda Triangle.* New York: Avon, 1974.

Berlitz is intrigued by many reports of ruins being found under the ocean in the Bahamas, off the coast of Peru, and in many other places. While the title of the book suggests most of the content, a number of curious facts quite unrelated to his title are sprinkled throughout the book. Berlitz is interested in the possibility that Atlantis may be found in the shallow waters off the Bahamas—a forlorn hope.

Earll, Tony. *Mu Revealed.* New York: Paperback Library, 1970.

Nothing in the way of evidence gives any support to the title of this book. A few gleanings, poorly or not at all documented, may be found, e.g., rock carvings, prehistoric ruins, artifacts, and practices like mummification with possible relationships to the East Mediterranean, and fossil tracks of man and animal.

Summing Up

As we may observe in the comments above, the same kinds of evidence used by Erich von Daniken and his followers to bolster up the concept of gods from outer space are used by other writers in support of Atlantis and Mu concepts. Of course, some of the above authors would argue that both views are really dealing with the same matter. Again, we see in the above material that there is no necessary connection between the facts offered and the theories they supposedly support.

Potpourri of Other Theories—Other Directions

Potpourri, defined as a combination of incongruous elements, is a good description of what is to follow. Many writers have been struck by the discovery of strange mementos of the distant past, and these have been attached to all kinds of explanations. In this section I hope to make the reader aware of human ingenuity of many stripes—some convincing, some outlandish. While I cannot resist a comment here and there about theories offered, the principal purpose is to identify once more the kinds of apparent facts seized upon to support a given position. Here we meet catastrophists, hollow earth fans, pushers of ancient transoceanic travel, people obsessed with mysterious pyramids, and defenders and attackers of ancient chronology schemes. These sources defy a neat ordering, so the reader is invited to wander through swamp and thicket to marvel at some unique literary creations of his fellowmen. Here are some representative authors and their works:

Berlitz, Charles. *Mysteries From Forgotten Worlds.* New York: Dell, 1971.

Berlitz has done a great deal of musing about the strange world of ancient times, and has written a book that is well worth reading. This book is many cuts above the average on the mysteries of the past, and much interesting data are provided not easily found elsewhere. In fact, Berlitz has almost drowned himself in data collection and finds it difficult to organize himself. His special interest is the many remarkable remains of drowned coastal and ocean areas around the world that show signs of former habitation. Berlitz accepts conventional dating and ends up with one or more ancient, very advanced civilizations that spread all around the world, including the Antarctic before it became ice-covered. Long before Sumer and Egypt began, a worldwide catastrophe destroyed the world of the Old Ones. He is interested in the days when the oceans were lower, islands were larger and more numerous, and before climatic changes and seismic upheavals modified whole sections of the world and destroyed or scattered its population. Berlitz, with considerable justification, takes a few good-humored jabs at conventional views of the ancient world, e.g., the Bering Land Bridge that everybody accepted except the Indians themselves. He also quotes a fellow sufferer who stated that if a splendid sunken city were found at the bottom of the Atlantic, scientists would label it as a sunken Greek shipment of building materials.

Bernard, Raymond. *The Hollow Earth.* Revised edition. Clarksburg, WV: Saucerian Books, undated.

Enormous mileage is derived from an ambiguous statement made by Rear Admiral Richard E. Byrd in 1947: "I'd like to see that land beyond the Pole." For the faltering in faith about a hollow earth, a diagram is offered to show how men can sail a ship over the lip of the earth at either pole without realizing the earth is hollow. The whole wonderful story would emerge for the public, according to Bernard, except for strict censorship imposed by unnamed agencies in Washington. Hence we have only a hint of the real truth: The earth is hollow. The poles are phantoms. In the interior of the earth are vast continents, oceans, mountains, and rivers, peopled by races unknown to us. The book is unstained by any evidence to support the theory. There are, however, interesting and useful quotations from many Greek and Roman writers describing celestial phenomena that may be applied to an examination of ancient astronomy.

Boland, Charles M. *They All Discovered America.* New York: Pocket Books, 1961.

This book is not easily dismissed. A considerable amount of evidence is marshaled to relate Pacific and Asian cultures to artifacts and customs in the Americas. The author chafes at the somewhat mildewed but still current theory that all traffic to the Americas trudged over the Bering Land Bridge. He presents an impressive amount of evidence that there was a whole series of many voyages across both oceans from very ancient times to the period just before Columbus. Numerous strange inscriptions, rock structures, and artifacts of the most un-Indian-like character have been found in many sites on both American continents to add substance to the author's thesis. Some of the finds, of course, are under suspicion, but it seems reasonable that some day historic treatment of the discovery of America will be radically rewritten. It goes without saying that any such endeavor will be done over the dead bodies of those who write the texts today. Yet there is hope. Younger experts are actually examining the evidence.

Brasington, Virginia F. *Flying Saucers in the Bible.* Clarksburg, WV: Saucerian Books, 1963.

We can hardly dignify the thesis of the book by any discussion of it. There are several items of some interest in the book, such as a highly conjectural account relating the Bethel Stone associated with Jacob in the Old Testament with the Stone of Scone, the symbol of Scot and later British monarchy. There is also an attempt to link Irish history with names and events of the Bible.

Fox, Hugh. *Gods of the Cataclysm.* New York: Harper's Magazine Press, 1976.

Fox offers an interesting variety of evidence to support his view of an ancient, sophisticated world culture interrupted by a great catastrophic event. He cites stone monuments, carvings, and symbols, ancient literature, seafarers, myths and legends, linguistics, and artifacts of many kinds. He attempts to trace the Middle American Indians back to the Indus Valley civilization.

Fell, Barry. *America B.C. Ancient Settlers in the New World.* New York: Quadrangle, 1976; *Saga America.* New York: Quadrangle, 1980; *Bronze Age America.* Boston: Little, Brown, 1982.

Fell has written three important but highly controversial books that give some dignity to a new facet of prehistory in America. He makes a rather convincing case in showing that many ancient inscriptions, rock structures, and artifacts, almost universally written off as hoaxes in the past or merely ignored, are actually evidences of a great deal of transoceanic travel in ancient times. Fell claims that Celts, Basques, Libyans, Egyptians, Romans, Hebrews, Chinese, early Norse, and others left proof of their presence long ago in the Americas in the form of inscriptions and stone structures. In addition, there is support that the Vikings explored a greater part of the United States than previously believed. For sheer audacity in shaking conventional beliefs, these books are worthwhile reading.

Gordon, Cyrus. *Before Columbus.* New York: Crown, 1971; *Riddles in History.* New York: Crown, 1974.

Gordon was an internationally recognized scholar who disdained the conventional straitjacket of ancient history and who therefore is disdained in turn by other scholars. An extremely valuable citation of evidence of many kinds shows that the history we have learned must be radically rewritten in the future. Transoceanic contacts were common in ancient times, and the facts in support of the thesis are convincing. Some finds may be of doubtful validity. As one might expect for any unconventional book, this one has received highly critical reviews that attacked the man and the theory rather than the evidence.

In *Riddles in History* Cyrus Gordon demonstrated an ancient device of embedding messages within messages in ancient inscriptions. He anticipates that this will be a valuable tool in determining the authenticity of ancient inscriptions found, for example, in the Americas. His interpretations are strongly attacked by others. Among many interesting items in Gordon's book, he shows the possibility of locating the fabled Ophir of King Solomon's day in California and in Peru. He also cited evidence for widespread Norse explorations in America long before Columbus.

Hapgood, Charles. *Maps of the Ancient Sea Kings.* Philadelphia: Chilton, 1996.

This is an astonishing book that convincingly demonstrates that long before the age of the Greeks men had precisely mapped the entire world, including the Antarctic continent before it was ice-covered. He advocates the study of mythology throughout the world as offering historic clues to the ancient past. No reconstruction of ancient history can ignore the evidence of this book.

Hawkins, Gerald S. *Beyond Stonehenge.* New York: Harper, 1973.

This book should be read by those who have a curiosity about the distant past. Hawkins assembled a large number of curious facts from around the world to show that ancient man was indeed preoccupied with astronomical matters and that he had a fascination for numbers. Hawkins attempts to find conventional explanations for curious finds, and his book is useful for testing other theoretical viewpoints.

Heyerdahl, Thor. *Kon-Tiki.* Chicago: Rand McNally, 1959; *Aku-Aku.* Chicago: Rand McNally, 1958; *The Ra Expeditions.* New York: New American Library, 1972; *Fatu-Hiva.* Garden City: Doubleday, 1975; *Early Man and the Ocean.* New York: Vintage Books, 1978.

Besides their great interest, the works of Heyerdahl make a strong case for extensive transoceanic travel in ancient times. In addition to persistent traditions, the spread of domesticated plants and animals, curious constructions in stone, artifacts, and many other cultural similarities give support to his controversial theories. More recent research has failed to support some of his views.

Mendelssohn, Kurt. *The Riddle of the Pyramids.* New York: Praeger, 1974.

Mendelssohn accepts a conventional chronology for the history of Egypt. His book, however, includes tantalizing evidences that can be used in support of a model of history that involves catastrophes. For example, he cites periods in the history of Egypt where the record becomes hopelessly blurred. The period of making pyramids suddenly ends never to be repeated. In the 4th dynasty certain sculptures have been purposely defaced. Among the Mayas huge buildings were partly demolished in connection with cycles of Venus. The last days of the year were believed to be very unlucky and the people lived in terror until they were assured the new year had safely begun. This hints strongly of terrible things from the sky that once afflicted the earth.

Mertz, Henriette. *The Nephtali.* Chicago: Author, 1957; *The Wine Dark Sea.* Chicago: Author, 1964; *Pale Ink.* 2nd edition. Chicago: Swallow, 1972; *Atlantis: Dwelling Place of the Gods.* Chicago: Author, 1976.

Dr. Mertz is controversial. Some idea of this is the fact that three of the above titles were privately printed. She has done the unpardonable in the eyes of the conventional scholar by thinking fresh thoughts about evidence she has gathered on the ancient world.

In *The Nephtali* her thesis is that the Mandan Indians of South Dakota showed evidence of European and East Mediterranean culture, and that these people were one of the lost tribes of Israel. Of value to the study of prehistory is her discussion of transoceanic traffic in ancient times, stone carvings in America, the Old Copper culture, the blue-glass beads peculiar to the Mandans and the East

Mediterranean, and the myths and traditions of the Mandans that relate to the East Mediterranean.

In *The Wine Dark Sea* Mertz explores the age of Homer with the intriguing thesis that the Odyssey may describe an Atlantic Ocean voyage rather than one in the Mediterranean Sea. Oceanus, the circling river (currents) beyond the Pillars, was well-known to the Greeks. The ancient Greeks and Phoenicians also knew of the Sargasso Sea. Mertz describes engraved stones and other mysterious artifacts found along the Atlantic coast and in the interior of America, and she has searched old Greek writings about the Atlantic "river."

Pale Ink focuses on ancient Chinese landings, explorations, and racial traces in America. Affinities between Chinese and American culture are described, including mound building, the Folsom point and other artifacts, ancient Chinese geographical treatises, sculpture and architectural affinities, other evidences of transpacific contacts, astronomy and the calendar, engravings, effigies, and pottery. As in many other works, the conventional theory of the Alaska-Siberia land bridge as the exclusive path to the New World is shown to be untenable.

In her 1976 volume, Mertz has brought together a large amount of evidence for frequent travel between Mediterranean civilizations and North America.

Michell, John. *City of Revelation.* New York: Ballantine Books, 1972.

Michell, the author of this odd mystical book, has been described as both a genius and a madman. One useful concept developed in great detail is that earlier civilizations were more sophisticated in significant ways than civilizations that followed them. This model fits into one of a succession of catastrophes in the history of the earth.

Patten, Donald W., et al. *The Long Day of Joshua and Six Other Catastrophes.* Seattle: Pacific Meridian Publishing Co., 1973.

Patten presents the case for the planet Mars causing eight great catastrophic events on the earth, including the flood of Noah and the destruction of Sodom and Gomorrah. The author holds that Mars was not in its present orbit and made close fly-by's causing great havoc on the earth. The book is a valuable resource for the facts presented. The ancients were indeed obsessed with astronomical matters. The book includes extensive studies from the Bible, other ancient literature, and legends. A number of valuable word studies are given.

Pauwels, Louis, and Jacques Bergier. *The Morning of the Magicians.* New York: Avon, 1963.

Out of a large mass of mystical and unintelligible writing, Pauwels includes a few items of interest regarding ancient man's use of metals, a discussion of mysterious cut marks around the world, and various artifacts that show that ancient man was highly sophisticated. Pauwels fits his thinking into conventional chronology. It is not surprising to note that this book was an important source from which Erich von Daniken developed his thinking.

Pourade, Richard F. *Ancient Hunters of the Far West.* San Diego: Union-Tribune Publishing Co., 1966.

Pourade has collected an unusually fine assortment of interesting facts about ancient sites and times in the far west. While he accepts conventional dating of the past, he cites a number of incredible contradictions and problems that lend strong support to alternate views. For example, a sequence of occupations at Borax Lake believed to have covered a span of 7,000 years actually occurred within several hundred years. The same type of stone artifacts dated up to 8,000 years old in the central part of the United States are dated only 3,000 years old elsewhere. The problem is explained away by means of the expression "Arctic retardation." Artifacts of widely separated times have been found together. This is considered very baffling. Stone tools once thought to be tens of thousands of years old are now dated no older than about 2700 BC. The information Pourade has collected is very useful for studying the ancient history of America.

Riley, Carroll L., et al. *Man Across the Sea: Problems of Pre-Columbian Contacts.* Austin: University of Texas Press, 1971.

Riley has assembled an astonishing number of cultural links between America and many other cultures across both the Atlantic and the Pacific that go back to very ancient times. The details are fascinating and impressive.

Tompkins, Peter. *Secrets of the Great Pyramid,* Appendix: Livio Catullo Stecchini, *Notes on the Relation of Ancient Measures to the Great Pyramid.* New York: Harper & Row, 1971.

Tompkins treats the surprising sophistication of the ancients, particularly the early Egyptians. Undoubtedly Tompkins and Stecchini overstate the abilities of the ancients in mathematics, astronomy, and other sciences. Still they have assembled curious and impressive facts that challenge any current conventional explanation.

Toth, Max, and Greg Nielsen. *Pyramid Power.* New York: Freeway Press, Inc., 1974.

A few facts about pyramids around the world are sprinkled around in this book, but the main point of the book was to lure the public into buying cardboard pyramids that have great powers, if only you remember to align the pyramid properly. The book is a monument to public gullibility.

Temple, Robert K. G. *The Sirius Mystery.* New York: St. Martin's Press, 1976.

This is a sobering and remarkable book for the skeptical. Temple details the absolutely "impossible" astronomical knowledge of the Dogon tribe in Africa.

Somehow these people retained important knowledge of the heavens for many centuries that our modern astronomers have learned only recently, and which seems incredible without sophisticated instruments. No one has given a satisfactory explanation.

Trench, Brinsley le Poer. *Temple of the Stars.* New York: Ballantine Books, 1962.

Temple of the Stars contains little value for the study of the ancient world. Trench goes into considerable detail on the highly dubious story of huge earth works in England shaped by ancient men in the form of signs of the zodiac. One item of interest is his description of the huge chalk figures carved into the hills of England. Trench accepts catastrophic events in the past and cites considerable evidence in support of this view.

Trento, Salvatore Michael. *The Search for Lost America.* Chicago: Contemporary Books, Inc., 1978.

Trento picks up where Barry Fell left off and provides a systematic look at evidence long ignored or labeled as hoax. Large numbers of stone structures—stone piles, perched rocks, alignments of boulders, slab-roofed chambers, mysterious stone walls, and ancient carved inscriptions, particularly in northeast United States, are described and illustrated. We can be sure that other parts of the Americas will receive corresponding attention in future years. The evidence is convincing that ancient man was very much at home on the sea and that America had many significant visitors over a period of many centuries before Columbus. Mining was the lure. One can readily see from this book that modern archaeologists—not ancient men—are the ones who are afraid of travel by sea, and this has given rise to the great Bering Land Bridge humbug that attempts to explain the origin and history of all animal and human life in the Americas.

Umland, Eric, and Craig Umland. *Mystery of the Ancients.* New York: New American Library, 1974.

The Umlands have collected a modest number of presumed facts showing that ancient people were sophisticated in astronomy, in architecture, in writing, in mining, and in other technologies. Again, the question is, who were their teachers?

Velikovsky, Immanuel. *Worlds in Collision.* Garden City: Doubleday, 1958; *Ages in Chaos.* Garden City: Doubleday, 1952; *Earth in Upheaval.* New York: Dell, 1955.

Velikovsky's revolutionary theses and remarkable scholarship in reconstructing the ancient past are further developed in additional books, and in such journals *Pensee, Kronos,* and the *Journal of the Society for Interdisciplinary Studies.* We cannot endorse all the contents of these sources. Yet many of his ideas have now entered mainstream thinking, but without acknowledging their source. His writings provide refreshing insight and stimulation for understanding the past. They are close to being unique in the manner in which they speak to evidence long ignored, yet no one has ever been so viciously attacked by those holding conventional views.

Vining, Edward P. *An Inglorious Columbus.* New York: Appleton, 1885.

Vining is worth reading for two reasons. He makes a strong case for specific Asian influences in Middle American culture, particularly Chinese. These influences are shown in architecture, art work, and artifacts that have been found. He also gives many curious accounts of the existence of elephants and other animals that ought to have been long extinct in the Americas, but which apparently lived into recent times.

Vitaliano, Dorothy B. *Legends of the Earth.* Bloomington: Indiana University Press, 1973.

Vitaliano has assembled a very useful collection of legends that seem to have some kind of geological implications. Many of the legends have a catastrophic content, but loyal to conventional interpretation, Vitaliano manages a uniformitarian explanation in almost every case. It is ironic that uniformitarianism is no longer the way for explanation in evolutionary theory today. We might offer the comment that vast amounts of scholarly energy are devoted to the art of explaining away inconvenient evidence. The reader must judge how strained Vitaliano's interpretations become. Many ancients had a clear memory of the origin of a kind of glassy meteorite known as tektites, yet it is conventionally believed that the last fall of tektites was several hundred thousand years ago. The Greeks had traditions for the time when the barriers between the Mediterranean and the Black Sea burst. Vitaliano believes this event was caused by normal erosion and occurred long before the Greeks were known. (For an interesting update on this catastrophic event, see William Ryan and Walter Pitman.[5] The authors, however, confuse the mighty runoff from the ice age meltdown with Noah's flood.) Vitaliano sets aside the evidence for any kind of catastrophic extinction of elephants and other mammals in northern Siberia. She is willing to grant some connection between the events of the Exodus and the explosion of the island of Thera north of Crete. All in all, Vitaliano's book is a valuable resource in collecting legends that may have some geological significance. The material, however, is an invitation for another and better form of explanation within a biblical framework.

Allan, D. A., and J. B. Delair. *When the Earth Nearly Died.* Bath, UK: Gateway Books, 1995.

On every count this is a serious and remarkable book, documenting and treating an enormous world-wide disaster involving the whole solar system. This mighty convulsion on land, sea, and air extinguished much life violently. No chronology of the ancient world can be constructed without fitting in this traumatic event, but to date no one has attempted to do so.

von Wuthenau, Alexander. *Unexpected Faces in Ancient America, 1500 B.C.–1500 A.D.* New York: Crown, 1975.

In this remarkable and welcome book von Wuthenau demonstrates with his camera beyond contradiction that Semites, Africans, Japanese, and other peoples from across both oceans traveled to and settled in the Americas long before Columbus. The author has photographed artistic and archaeological specimens found in museums and other collections throughout the world. The result is a fascinating display of evidence that many people reached these shores centuries before the first Europeans.

Wilkins, Harold T. *Mysteries of Ancient South America.* Secaucus, NJ: Citadel Press, 1956.

Wilkins has collected a number of interesting bits of information that are relevant to the study of ancient history. He makes a strong case for catastrophic events in the past as they are evident both in traditions and in physical evidence in South America. Some background is given on the taboo against writing in civilizations where we would expect writing. Wilkins presents evidence that the Andes were suddenly uplifted in historic times, and that Lake Titicaca was once uplifted from the seashore in recent historic times. Wilkins has also collected information on giants and giantism of considerable interest.

Colby, C. B. *Strangely Enough!* Abridged. New York: Scholastic Book Service, undated.

Among many strange facts collected by Colby, several describe rapid changes on the surface of the earth. He documents a case where the ocean floor rose more than two miles within a 25-year period. He also cites the discovery of a chain of islands off the coast of South America in the seventeenth century that later disappeared completely.

Sanderson, Ivan T. *More "Things".* New York: Pyramid, 1969.

This book was not intended for more than an accounting of strange things found around the world, but there are several items of value for the study of prehistory. There are accounts of giant skeletons or bones (one of doubtful authenticity), the catastrophic and mysterious extinction of the mammoths in Alaska and Siberia, and the strange vitrified stones of forts and walls in various parts of the world.

Sprague, L., and Catherine de Camp. *Citadels of Mystery.* New York: Ballantine Books, 1964.

The de Camps cite interesting and mysterious facts, but their loyalty is to conventional history telling. Their book attempts to debunk any thought of straying from conventional explanations, though mysteries without explanation are acknowledged. The authors are particularly critical of any belief in catastrophic events in the past. Their world, mysteries and all, somehow fits into a patient and unhurried universe, a theoretical position now largely abandoned by their fellow evolutionists. Awesome stone ruins on Pacific islands, fantastic navigational ability of the Polynesians, and other mysteries are useful data that may support theses not now currently acceptable in the literature.

Edwards, Frank. *Strangest of All.* New York: Ace, 1962; *Strange World.* New York: Ace, 1964; *Stranger Than Science.* New York: Bantam, 1967.

Edwards delights in uncovering facts that jolt the reader, for they are often diametrically opposed to the usually accepted concepts of both geology and ancient history. His documentation leaves something to be desired, but useful applications are possible from his works. Edwards (1962) cites deeply buried constructions and artifacts in America that raise interesting questions about an ancient sophisticated past not treated in history books. He (1964) notes mummified seals 2500 feet above sea level in the Antarctic fifty miles from water; and pre-Indian artifacts, carvings, and stone constructions in America. Edwards (1967) cites a dinosaur carving in Arizona, skeletons of giants, support of human pygmies in strata far too old for any kind of man, and evidence of catastrophic events in historic times.

Fort, Charles. *The Book of the Damned.* New York: Ace, 1941.

No survey would be complete without mentioning the wild and wonderful writings of Charles Fort. In a whole series of similar books, Fort carries on a personal vendetta against the science establishment. Writing in machine gun bursts, he is not easy to follow. His delight is any fact that embarrasses conventional science. Gleanings of potential value are old footprints, metal artifacts found in coal or in stone, rapidly forming stones, baffling inscriptions on stone, and strange and unexpected artifacts found deeply buried.

A little antidote. Lest we get carried away on every wind of whimsy that comes along, we close this section with several authors whose viewpoints are strongly opposed to any thinking contrary to that found in standard texts commonly used today. Some of their criticism is fully justified, yet in some cases they are ignoring a great deal of compelling evidence.

Gardner, Martin. *Fads and Fallacies in the Name of Science.* New York: Dover, 1957.

Gardner's book furnishes us with a fitting close to this section. This is a useful and interesting work that exposes a number of magnificent obsessions. It is a good case study and a warning about becoming carried away by a shaky theory. It is not very useful for shedding light on prehistory, as that was not the intent of the book. His prose borders on hysteria, however, in his repeated denunciation of scholars who dare to question evolution or who advocate the possibility of catastrophic events. In his fervor he goes too far in attacking scholar's scholar Mortimer Adler and tells outright lies in order to bolster evolutionary theory, e.g., that Neanderthal man had an ape-like forehead, a head that hung forward, no chin, and non-opposable thumbs. It has been known since 1868 (yes, 1868) that Neanderthal man, while not white, middle-class Protestant, could pass unrecognized on the streets of any city.[6] We say this despite current fervent attempts to again restore Neanderthal to ape-man status.

Trefil, James. *A Consumer's Guide to Pseudoscience* 5, no. 15 (April 4, 1978): 16–21.

Trefil offered a consumer's guide to pseudoscience. We agree with much of what he says, e.g., in calling the following pseudoscience: Erich von Daniken, ancient astronauts, Loch Ness monster, and others. He is not so sure about UFO's and Bigfoot. However, he believes extraterrestrial intelligence is becoming a respectable research area. We disagree with that and his condemnation of anything to do with the Atlantis legends, and Velikovsky's works. Things are not that simple, as we have explained repeatedly in this chapter.

Williams, Stephen. *Fantastic Archaeology.* Philadelphia: University of Pennsylvania Press, 1991.

Williams teaches a course on fantastic archaeology at Harvard, with the aim of treating the wild side of North American prehistory. Where he has done his homework, the material is good in showing some of the strange and wonderful kooky frauds and humbugs about ancient America, but he often ignores this geographic limitation. Some humbugs really are humbugs, we would agree. The author's treatment of radiometric dating is a disaster, and he merely follows party-line beliefs. He accepts artifact dating by form alone, and we have shown this is a false belief. He accepts the logic of early transoceanic travel but then rejects much convincing evidence for it. He cannot resist condemning biblical creation and he expresses the hope that the truth (of evolution) will keep the public free. He holds that there is still not a single well-documented association of artifacts with extinct fauna east of the Mississippi River. This is flatly untrue.[7]

Wauchope, Robert. *Lost Tribes & Sunken Continents.* Chicago: University of Chicago Press, 1962.

Wauchope has written a book on behalf of the scientists who are weary of defending conventional points of view against such "wild" theories as Atlantis, the lost tribes of Israel settling in America many centuries ago, and Thor Heyerdahl's theories. We note in passing that Heyerdahl has become more respectable within some of the scientific establishment since this book was written. While one may agree in part and disagree in part with the views expressed by this author, it is important to see what establishment science has to say. It is even more important to see on what the arguments rest on any side of an issue.

Putting It All Together

Thinking It Through

As we ponder the works we have reviewed above, we must marvel at man's ingenuity in creating strange views of the ancient world, innocent of supporting evidence. At the same time we must conclude that there are many, many odd facts that have not yet been properly evaluated for the theoretical positions currently in vogue in our society. Some recurrent themes run through many of the books reviewed. We must somehow come to grips with some very provocative evidence about the ancient world that has not received adequate consideration up to this time. What are the alternatives? There seem to be only three ways to go in sorting out for one's self a view of prehistory, and each way necessarily includes a large proportion of conjecture. At this point and perhaps forever there are many missing elements for constructing a detailed coherent story of early man, simply because we were not there at that early time.

The Primitive-to-Modern Compulsion

The first way is the conventional primitive-to-modern view of ancient history. It follows naturally on evolutionary assumptions, and it is no surprise that almost all texts follow this pattern.

It has its problems. Attempts at a viable chronology have fallen on evil days. C14, capable of furnishing some valuable assistance, has turned out to be a false friend of history, full of contradictions and puzzles. The assumed sequence of primitive to more advanced cultures frequently is pursued in the face of contradictory evidence. The sudden inexplicable appearance of Sumer is a slap in the face to this approach.

More than a century ago a scholar observed that man had not originated from a state of barbarism, and had then risen to civilization; but wherever man has been found in a state of barbarism, it is barbarism arising from a degenerated civilization. All known peoples with a knowledge of their past have some tradition of their having been raised from barbarism by a people more civilized than themselves.[8]

Some historians have abandoned the problem and with some sense of relief have bequeathed the matter back to the anthropologists. Despite the sensationalistic manner in which new fossil finds from Africa are reported from time to time, the leading scholars in the field are in violent disagreement on the supposed sequence of fossil forms in human development. This is not to say that anyone in the field is supporting a belief in created man. Much of anthropology is devoted to safe and non-controversial analytical studies that overlap more and more with the field of sociology.

Examples of discontent with the field of ancient history should be noted. One historian observes that we know very little indeed about prehistoric man, and no two experts tend to agree on what little we do know. Although that writer accepted biological evolution, he did not want it applied to history:

> There are still many inconvenient facts which seem very difficult to explain on the basis of natural selection, and the entire theory (of evolution), if viewed dispassionately, often seems to a layman so extraordinarily unlikely, as an explanation of how the present . . . including how man evolved, as to suggest a willful perversity in present-day man.[9]

In a similar vein, Giorgio de Santillana and Hertha von Dechend question the manner in which scholars view the ages that lead from primitivism into civilization.[10] Gradually, we are told, step-by-step, men produced the arts and crafts, this and that, until they emerged into the light of history. Yet, as Alexander Marshack explains, the word *gradually* leads into all sorts of difficulties, so the word *suddenly* is used to explain what cannot be explained with this theoretical approach.[11] Almost everything we read follows this lockstep theoretical position, but we must not suppose that anything of significance is explained by it.

The Outer Space Route

The second way is one that most thoughtful students of the field abhor. This is the route that von Daniken and others have followed. Yet it is not unusual to see news reports today where serious scientists support the theory. The mysterious facts cited earlier in this book capture the imagination and are a great incentive for wildly improbable schemes. Essentially, the problem with this approach is the vast gulf fixed between the facts cited and the theories proposed. Reading such literature makes one feel very much like the victim of a hit-and-run accident. Too much "faith" is needed to follow these proposed routes. The problems of the assumptions of this approach have already been noted in some detail

Is There Another Route?

The third way, most improbable of all to many, rests on two kinds of information. Suppose one would dare to take seriously a scholar such as Nelson Glueck, who said:

The archaeologist's efforts are not directed at "proving" the correctness of the Bible, which is neither necessary nor possible, any more than belief in God can be scientifically demonstrated. It is quite the other way around. The historical clues in the Bible can lead the archaeologist to a knowledge of the civilizations of the ancient world in which the Bible developed and with whose religious concepts and practices the Bible so radically differed. It can be regarded in effect as an almost infallible divining rod, revealing to the expert the whereabouts and characteristics of lost cities and civilizations.[12]

Regarding the Table of Nations in Genesis 10, archaeologist William F. Albright noted that

• It stands absolutely alone in ancient literature without a remote parallel;

• It is an astonishingly accurate document;

• It shows such remarkable "modern" understanding of the ethnic and linguistic situation in the modern world, in spite of all its complexity.[13]

Scholars never fail to be impressed with the author's knowledge of the subject.

Among the Greeks we find the closest approach to a distribution of peoples in a genealogical framework. But this framework is mythological, and the people are all Greeks or Aegean tribes.

If a gentle bit of irony may be spoken, one might suggest that if everything else has failed, why not take the Old Testament seriously (without mangling it), as a framework for prehistory? If it is so good for history and archaeology, it might be equally good for prehistory and, of course, much more.

Explorer Vilhjalmur Stefansson made a remarkable observation more than a generation ago that has the most profound implications on our search into ancient history. His statement that man swarmed over at least three oceans long before the time of Abraham has not been challenged, but it has been ignored.[14]

In *The Evidence of Language* William F. Albright noted in a matter of fact way that the peopling of America, which began not less than 20,000 years ago (evolution time), brought groups of many racial origins and radically divergent linguistic stocks across seas and oceans to occupy every habitable part of the hemisphere, and that beginnings are now being made in locating Old World sources of elements in Paleo-American and more recent cultures.[15] We believe we have good reasons to reject his suggested date, but otherwise his statement fits well into the framework of this book.

The Way It Was

Here are several fascinating insights some scholars are suggesting to us about the way things were in the pre-flood world, fully within the framework of our young, created earth. Then, as now, there were good areas where agriculture, technology, and the arts functioned. And one of these zones was the place where family life continued, and where the ark was built.

But the Bible also tells us about the curses that were laid on Adam and Eve, on the animal and plant worlds, and on the earth itself. Only eight people knew firsthand what the world was like both before and after the greatest catastrophe that ever occurred on the earth, Noah's flood. These were the eight members of Noah's family. They all lived for centuries after the flood, and we can see traces of the stories they told around the firs to the children of the next generations. These, much altered, appear today in myths and legends around the world. The Bible has a much greater priority than to satisfy our curiosity about the pre-flood world. Scholars of the Bible, however, have described what they have concluded about the cursed world before the flood.

About 500 years ago, Martin Luther had this to say:

My opinion is that Eden was utterly destroyed and annihilated by the Flood, so no trace is visible now. . . . The Flood laid everything waste, just as it is written that all the fountains and abysses were torn open. . . . There were mountains after the Flood where previously there were fields on a lovely plain. . . . For the entire surface of the earth was changed. I have no doubt that there are remains of the Flood, because where there are now mines, there are commonly found pieces of petrified wood. In the stones themselves there appear various forms of fish and other animals . . . thus I believe the Mediterranean Sea channel was produced by the Flood (not there before); also the Red Sea was formerly a fertile plain.[16]

More than a century later John Milton described the world after the fall into sin this way: On an earth filled with storms, floods, earthquakes, violent predators, and the discomforts of changing seasons, Adam and Eve contemplated suicide. Michael comforted them and gave a vision of the future (Messiah). Man and woman followed the path leading from their paradisiacal garden to the barren and lonely world below.[17]

In their highly respected commentary on Genesis, Carl Keil and Franz Delitzsch describe the pre-flood world in this manner:

The disturbance and distortion of the original harmony (of creation) spread the whole material world (universe), so that everywhere on earth there were seen wild and rugged wastes, desolation and ruin, death and corruption . . . many things in the world and nature . . . became poisonous and destructive since his (Adam's) fall.[18]

If we ignore the conventional, commonly accepted theories, and begin all over again, we must put something like the following after the flood to the test: Long before the time of Abraham, men lived part of that time in a Golden Age on some areas of the earth, the time of the great replenishing after the flood. There was a small group of people, called the Old Ones, who for centuries were the teachers of all the most ancient cultures. They created the mysterious and baffling "explosion of innovation" in the Iranian Highlands. All the ancient technologies, miraculous then but taken for granted now, came from them.

Time is an important consideration in this matter. If we consider how the United States of America changed and developed so fantastically in not much more than two hundred years, it is easier to grasp the development of the earth and its earliest cultures after the flood in a period of time five or more times as long. For a number of reasons this writer has come to prefer the Septuagint (LXX) chronology for the ancient world.[19] The Septuagint is the Greek translation of the Old Testament made in the third century BC. Jesus and the writers of the New Testament book almost always used this translation when quoting from the Old Testament. The LXX gives 942 years for the period of the flood to Abraham, as compared with 292 years in the Hebrew Bible. The possibility of gaps in the chronology that students of the Bible can demonstrate must also be considered, and this could extend the time period further.

The Bible is almost entirely silent on the events and developments that took place between the flood and the call to Abraham. We have only Genesis 10 and 11 to give us some tantalizing glimpses into these centuries. In von Fange (1994) we have discussed these and other important aspects of this time before Abraham.[20]

The God-Kings and the Titans is an exciting example of fact and conjecture about some of the truly amazing adventures that took place during these critical centuries.[21] Despite the fact that both his orientation and chronology are non-biblical, the story he unfolds based on hundreds of myths, legends, artifacts, and other remembrances of the past, fits beautifully into the early post-flood era.

The following is a picture we may paint of the early centuries after the flood. Before and during the Golden Age, small bands of the curious and adventurous, the rebels and the ostracized, left or fled the Iranian Highlands for the great unknown world, much like the defeated Maori rebel or chief who fled with his followers to find another island home or perish in the sea. Many of the ancient ones were captivated by the lure of hunting, and they made their tools and weapons as they traveled. Archaeologists have collected the hand axes and projectile points left behind. Not until the end of the 500-year-long ice age and the great meltdown afterward did such artifacts appear in the more northerly regions. Each successive wave of tribes drove the previous ones to less and less hospitable areas. Environment and inbreeding led to gross physical degeneration and extraordinary variability, but these ancients clung invariably to some elements of their common culture, e.g., the flint tool industry.

Victor Pearce gives us this fascinating insight.[22] In the first wave of farmers migrating from the initial center in the Zagros plateau there appeared no need for weapons of war or fortifications. The same fact was reported regarding Catal Huyuk, and the earliest settlements in central Europe. Why? Everybody was a "cousin" in those first generations after the flood.

When the lowlands of the watery and unstable post-flood world became habitable, people moved down into the great valleys from the villages in the highlands. We soon read then of the Babel story and the great dispersions and migrations began, when God "splattered" the disobedient people into the far reaches of the earth.[23]

During the Golden Age deep-sea navigation extended all over the earth. Men sailed to the Antarctic and to Greenland before they were ice-covered. Men knew just where they were anywhere on the earth, and the whole earth was carefully mapped. The ancients had a profound understanding of the earth, the stars, and the calendar in a world of universally mild climate, unlike the world we know today. Sight lines, or alignments, miles long, were constructed in many parts of the world, and the stone circles were erected for astronomical purposes by which they put their universe in order. More than adventure and curiosity was involved in travel. The special lure of world-wide exploration was gold, silver, copper, precious stones, and pearls.

The ancients built elaborate irrigation systems and carried plants from one hemisphere to the other. They marked their presence by rock carvings, paintings, and effigies, and used a universal language of symbols for communication. Megaliths (huge stone structures) of many kinds were constructed involving an enormous expenditure of labor. Perhaps these ancient ones used technologies we know nothing of today. These structures most often were located near the scene of ancient mining or pearl gathering activities. In a narrow band around the earth, the great Old Kingdoms emerged. In this time of peace and great prosperity they searched out the riches of the earth. Each of the Old Kingdoms was distinctive, yet shared many cultural traits, a heritage given to each of them by the Old Ones, those who are named in Genesis 10. Satellite cities sprang up and thrived. But the earth was restless and the Golden Age ended suddenly in disaster. Somehow, precariously, many—but not all—of the old technologies survived and were passed on.

Fleeing in terror from the Arabian Gulf where disaster destroyed their civilization, a decimated people later known as the Phoenicians settled on the east coast of the Mediterranean. As they regained their strength and numbers, they became a pale reflection of what their ancestors had been in the Golden Age. Some of the old wisdom in navigation remained among another people, the Polynesians, down to our day. Knowledge of domesticated plants and animals survived among many peoples, along with some of the earlier skilled use of folk medicine.

More than one catastrophe struck the earth, and much wisdom was lost. The Middle Kingdoms were less cultured in some ways than the Old Kingdoms, and the New Kingdoms degenerated still further. Deserts such as the Gobi, Sahara, and the Arabian sprang up in an instant, destroying forever entire civilizations. Cities were destroyed, never to rise again, leveled by forces other than war or accidental fire. Forts and cities and the earth itself show signs of vitrification from blasts of intense heat from the sky. In many ways the orderly universe ended. For a time the universe no longer behaved as it had been known, and terror-driven men desperately attempted to put it together again, as at Stonehenge and at Nazca in Peru. Human sacrifice became widespread to appease the angry gods intent on destroying the earth. The obsession of early man with the planets as devastating forces of evil is evidence for the planetary origin of repeated devastations of the earth. Artifacts, deeply buried, and occasionally uncovered again by chance, testify to the violence on the earth in those times. In historic times, caves and crevices, and the Arctic muck of Alaska and Siberia, tell a tale of instant horror and extinctions at a time when men were mining metals in the far North along the Arctic coast. Not until around 600 BC did the relatively calm and predictable world we now know begin.

CONCLUSIONS

One Coherent Story

The study of ancient man and of the early history of the earth is interesting in itself, but there are better reasons to explore the evidence and the record about the ancient world. Whether we probe deeply into the earth (geology), or study early man (physical anthropology), or probe his cultural remains on the surface of the earth (archaeology), or examine the remarkable facts about the development of language and writing (historical linguistics), or look to the sky (astronomy), or read early accounts about man's history (archaeology, ancient history, biblical history, mythology), we find only one set of interwoven facts that cry for one coherent interpretation. This book explores some of these lines of evidence, fully aware of a sobering word of caution expressed a century ago:

> Ignorant people have this distinct advantage over scientific observers, that they are readily able to arrive at conclusions that are perfectly satisfactory to themselves on subjects that have been the battlefields of scholars for centuries.[24]

Yet anyone who has read even casually some of the history of science will concede that scientific observers can and do develop some very, very odd blind spots when they gather and attempt to explain data. We can't resist some comments made about researching the ancient world stated by Daniel Cohen:

I have tried to give equal exposure to the reasonable speculations of scholars and to the incredible ideas of cranks, for the cranks have often been more influential in shaping public thinking than the scholars. . .[25]

Reasonable and unreasonable, sound and foolish, they all should have their say because they are all part of the story. I have as much passion for the nice, well-constructed solution as the next one—but I have had to leave many questions unanswered, and many clues without conclusions.

The author accepts the Bible as a faithful framework for all of the past. The reader is challenged to weigh the evidence and to draw conclusions that can be defended. The threads of evidence from many sources that run through the pages of this book form a coherent pattern. They are briefly outlined as follows.

We live on a young earth.

Vast ages reported in textbooks and everywhere else are not derived from evidence, but are based on the assumption that change occurs very slowly or that such time spans are absolutely essential for evolutionary development to have taken place.

The Bible uses a special Hebrew word, *mabbul*, for the total destruction of the earth by Noah's flood. Nothing else can begin to compare with this event.

A number of other lesser catastrophic events in the past left many evidences and shaped much of the geology and the history of the earth.

Ancient man was very sophisticated; it is plausible to look at exceptions in terms of degeneration and environmental factors.

Language is a useful tool for exploring the past.

The only suitable framework for fitting the remarkable events of the past together is the biblical account of God's mighty acts.

Any reconstruction of the past is full of problems and assumptions, since we were not there to record the infinite details of the drama.

Although many questions remain and significant pieces of the mosaic are lacking, something like the above appears to be consistent both with the many kinds of evidence found in all parts of the world and with the Old Testament. This is a biblical view of a very sophisticated ancient world. No gods from outer space are needed.

For Discussion and Reflection

1. Why do you think there is such a thirst to explore some of the mysteries of the ancient past?

2. Why do you suppose there is such a wide variety of opinions about ancient world history?

3. Why is there such a lockstep in dating the universe, the world, and different periods in ancient history?

4. Are there condensed genealogies in the Bible? If so, we can never give a precise answer to the age of the world. Creationists, however, hold to a young earth, with estimates of the age of the world differing only to a minor degree. Again, we ask why evolutionists insist that we must think in terms of millions and billions of years.

5. When dealing with the history of the ancient world, the sequence for all authors in the media is to go from hunting-gathering, through assorted stone ages, to a copper age, to bronze ages, to iron ages, before getting into historic times. In this book we have shown repeatedly that this sequence has very serious faults. Why do you suppose authors continue to follow this old and faulty pattern?

6. How fitting to this chapter are these thoughts from 1 Timothy 1:3–7? . . . teachers of false doctrines and myths . . . meaningless talk . . . they want to be teachers, but they do not know what they are talking about or what they so confidentally affirm.

7. What do you suppose is behind the movement today of getting away from BC and AD and switching to Before Common Era (BCE) and Common Era (CE)?

8. Two important concepts in the study of the ancient world are diffusion and independent invention. For example, were intricate identical parts of the loom found in widely separate parts of the world invented over and over in the ancient past, or was the loom spread around the world from one original source, known as diffusion? Evolutionists strongly support independent invention. Why do creationists just as strongly support diffusion? Where do you find biblical support for your view?

PART FIVE

DISCOVERING
THE
GENESIS WORLD

CHAPTER SIXTEEN

SCIENCE AND DECEPTION

INTRODUCTION

Most of Science is Pure

As we are focusing on the exceptions to the rule in this chapter, we want to emphasize at the outset that we have the highest regard for scientific method, and for all scientific endeavors that rely on actual evidence for the conclusions drawn. We deplore deception, faked data, calling speculation fact, and confusing beliefs with science. In the following discussion, we show clearly why we part company with bad science, especially with the theory of evolution, falsely called fact. Almost all scholarly work treating the ancient world is interpreted as if the doctrine of evolution is actually true. We can learn a good deal from our study of frauds, errors, and humbugs in these attempts to explain the past.

A Notable Critique of Science

It is a pity that there is a tendency to measure the value of a book by its copyright date. A half century ago Anthony Standen's book *Science is a Sacred Cow* was highly praised by the scientific community and everywhere else.[1] There is really nothing quite like this work today, but it fits our time like a glove. The book is a most devastating attack on the arrogance of scientists insofar as they believe they are infallible and demonstrate full integrity in their callings, and by the way they are imprisoned by false theories. It is classic human nature to believe that such devastating criticism always applies to the other fellow.

Science education (not science) is a disaster, according to Standen, and, if anything, has gotten worse over the years. For several generations the great mass of college students receive one or several survey courses in the sciences that are so watered down that the only guarantee in these courses is that the student learns nothing of any value. To make doubly sure of this, students are given a broad choice of electives to help them avoid any real possibility of learning and challenge. No wonder so many of our youth hate science. Where there are exceptions to all this, we are grateful.

Science, of course, cannot touch the problem of good and evil. This was not fully recognized until the development of the atomic bomb in the 1940s. Absurdity reached new depths when a band of humanists led by journalist Bill Moyers, having written off the Book of Genesis as mere myth and superstition, attempted in a book and television series to develop a new theology derived from the theory of evolution.[2]

This effort reminds us of a huge amount of literature about a century ago by evolutionists on race. They proved by means of evolutionary principles that the white race was vastly superior to the other races, and spoke of the inferior races in the most insulting and demeaning manner. This notion has now been quietly scrubbed away. Central to the argument of evolutionists that there were superior and inferior races was the theory of recapitulation, the belief that embryos and juveniles repeat all the evolutionary adult stages of their remote ancestors. One of the earliest stages was supposedly the fish stage. Stephen Gould of Harvard states that by 1930 this notion had utterly collapsed for the best of reasons.[3] It was total nonsense. Are we surprised to find current texts still speaking of recapitulation as though it were fact?

When Anthony Standen viewed the field of biology, he found that it consisted of pompous nonsense when it attempted to do anything more than to describe life and its functions. Speaking of Darwin's *The Origin of Species*, he asks who, 150 years later, can describe the origin of any species? Of course, the answer is no one. Through slight of hand, evolutionists speaking in generalities give the impression that evolution has been proved. With millions of captive students who regurgitate nonsense on demand, evolutionists convince themselves that they are actually speaking truth instead of a large helping of nonsense.

Three Disastrous Failings Characterize Evolution

First, it is the science that consists of an unshakable faith in what this science is going to prove some day. It has nothing to show thus far, but just give it more time and more government grants.

Second, evolution consists of borrowed concepts from nineteenth century physics that physicists discarded long ago as useless. A prominent scientist observed that physicists are amused at biologists clutching onto hopelessly outmoded buggy whip concepts to support evolution.[4]

Third, evolution does not invite us into the laboratory as with other sciences. It relies on expert opinion by the brotherhood of fellow evolutionists as their source of truth.

Hoaxes are Rare?

Some years ago the Canadian Broadcasting Corporation offered a series of programs with the title of *Science and Deception.*[5] Some of the world's most prominent scientists participated. Deep problems were discussed, a number of which do not involve the creation-evolution debate. Certain aspects, however, are closely intertwined with a violent reaction against God and creation, as the reader can see with the following list of topics treated in the broadcasts. The panel discussed fraud as a fact of research life; the problem of self-deception; science for sale; faulty scientific pronouncements; and violent reactions of scientists to new ideas. In discussing how the research game is played at universities and institutes, they discussed plagiarism; inadequate peer review of research; bias against unpopular conclusions; the negative effect of prestige and personal friendship networks; and research grants given to friends who will come up with obvious results.

At about the same time we find a stark contrasting view from another scientist. Some aspects of her work were discussed in the previous chapter. In *Legends of the Earth*, Dorothy Vitaliano observed that geological and archaeological records, generally speaking, are not subject to deliberate falsification. Moreover, real scientific hoaxers are so rare that if they succeed in fooling just a few people for just a short time, they make history. Vitaliano then submits her list, that in geology there have been only two instances of such deception that came anywhere near succeeding. That statement merits a response. Anyone who has read widely in the literature of geology and archaeology, and who believes that evidence ought to accompany assertions, soon finds that a great deal of humbug is dispensed.

Let us return to the two deceptions noted above by Dorothy Vitaliano. The first one cited, the Piltdown hoax (see chapter 8), deserves a high rating for the large number of eminent scholars who accepted the fraud for two generations. There are many other skull humbugs, however, that could be described.

The second deception mentioned by Vitaliano was the bizarre joke played on Dr. Johann Beringer (1667–1740), a Wurzburg physician, who was fascinated by fossils. Two colleagues saw to it that his fossil finds became more and more unusual: worms, crabs, toads, flies, other insects, water snakes, flowering plants, and ultimately Hebrew letters. Finally Beringer found his own name as a petrified fossil. The outraged physician took the deceivers to court where they were then discredited, and the physician attempted in vain to destroy all copies of the book he had had published about those marvelous fossil discoveries.

The Donkey Man

In 1984 there was great excitement in Spain in anticipation of an international symposium to be centered about the Orce Man, identified as coming from a hominid creature that roamed near Granada about 1.3 million years ago, rivaling the claims for the earliest such fossils claimed for Africa. The symposium was called off at the last moment. While cleaning the skull fragment, paleontologists discovered that it belonged to a jackass or its ancestor.[6]

Playing with Time

The Petrona Skull, discovered in a Greek cave in 1960 was first dated at 70,000 years. Then the skull was given a date of 700,000 years, and it instantly became the oldest European known. Some time later other scientists analyzed the skull and found that it was only 160,000 to 240,000 years old.[7] Perhaps there have been other age adjustments since that time. This illustrates how arbitrary and speculative time estimates are, no matter which dating method has been used. But finding something too old for conventional dating can lead to deep trouble. When some stone tools in Ontario were dated by geologists much older than conventional limits, this is what happened:

> The sites discoverer was hounded from his Civil Service position into prolonged unemployment; publication outlets were cut off; the evidence was misrepresented by several prominent authors among the (scientific establishment there); the tons of artifacts vanished into storage bins of the National Museum of Canada; for refusing to fire the discoverer, the Director of the National Museum, who had proposed having a monograph on the site published, was himself fired and driven into exile . . .[8]

(The discovery) would have forced the rewriting of almost every book in the business. It had to be killed. It was killed.

The Standard Error Problem

All methods of radiometric dating include an estimate of the amount of error. We will avoid any technical explanation, but the concept is a very important one, yet seldom mentioned or referred to in research reports. If a Carbon 14 date shows an age of 2,000 years and an error factor of 200 years, this means we are about 68 percent sure that the true but unknown actual age falls somewhere between 1,800 and 2,200 years. If we want to be 99 percent sure of what the age is, we must add and subtract 2.5 times the standard error, that is, $200 \times 2.5 = 500$ years. We are then 99 percent sure that the true age of the sample is somewhere between 1,500 and 2,500 years old. Because of the large amount of error in attempting to date some of the Dead Sea Scrolls, we cannot answer the crucial question up to the present whether they were written before or after the time of

Christ. Error is only one of many problems involved in using radiometric dating. We must say that we were astonished to find an archaeologist state the following in his book on world prehistory: Please note that all radiocarbon and potassium argon dates should be understood to have a plus and minus factor that is omitted from this book in the interest of clarity. Stating a fixed readiocarbon date, such as 1500 BC, is not clarity. It is concealing very important information. The derived age is not a fixed amount but only a point somewhere within a range of possible dates. But as we stated, radiometric dating has many serious problems and contradictions that prevent it from rendering accurate information.[9]

Our Goal

In this chapter we shall first consider some critical statements from various sources that have to do with the manner in which prehistory is treated in the literature. We then shall consider just a few samples from an amazing display of frauds, errors, and humbugs that are notable in the literature. We quickly discover that an apology is needed to the perpetrators of many other fine humbugs that for lack of space had to be omitted from this discussion.

The Way It Ought to Be

There is no difficulty in finding statements on how things ought to be in the sciences under consideration here. Who can quarrel with the statement of T. H. Huxley? Yet these are strange words coming from Darwin's champion in the nineteenth century, as evolution is unstained by actual evidence!

> The historical student knows that his first business should be to inquire into the validity of evidence, and the nature of the record in which the evidence is contained, that he may be able to form a proper estimate of the correctness of the conclusions which may be drawn from that evidence.[10]

J. Robert Oppenheimer beautifully stated the ideal in which science blossoms:

> It is a world in which inquiry is sacred, and freedom of enquiry is sacred. It is a world in which doubt is not only a permissible thing, but in which doubt is the indispensable method of aiming at truth. It is a world in which the notion of novelty, of hitherto unexpected experience, is always with us and in which it is met by open-mindedness that comes from having known, of having seen over and over again that one had a great deal to learn.[11]

In a similar vein Dorothy Vitaliano states that once its underlying assumptions are proved untenable, a scientific theory must be rejected by anyone claiming to be a scientist.[12] At first blush, this statement sounds very good. Unfortunately, the statement is useless. If we remember our geometry, we recall that assumptions cannot be proved. They are taken on faith. Thus dialog between the creationist and the evolutionist is difficult if not impossible, not on the basis of evidence for or against a given position so much as the difference in assumptions of the two positions. Each position ultimately rests on a faith or an assumption basis. If one assumes there is no God, some form of evolution is the only view that makes any kind of sense to that person. If we assume or believe there is a God, creation is the only clear position that one may reasonably take.

The famous scientist, Descartes, showed a humility that everyone should emulate. What a privilege it would have been to work with such a man. He showed the kind of attitude that greatly stimulated scientific work:

> No sooner had I completed the course of study at the end of which one would normally be admitted to the degree of Doctor, than I changed my views completely because I found myself beset by so many doubts and errors that I felt that nothing had accrued to me from my efforts at learning but the discovery that I was more ignorant even than I had supposed.[13]

It would be difficult to better illustrate what science ought to be than the contents of an amazing book *The Encyclopedia of Ignorance*.[14] Some of the worlds leading scientists leaped at the opportunity to contribute chapters on what they did <u>not</u> know about their specialties. As the editor stated, "The more eminent they were, the more ready they were to run to us with their ignorance." Is it any surprise that this remarkable book devotes more space to the problems of evolution, including the supposed evolution of man, and mysteries of astronomy than any other pair of topics? The Big Bang and the supposed vast age of the universe, ape-to-man theory, and evolution itself are so riddled with assumptions, guesswork, and unsupported opinions that we have no actual evidence standing in the way of accepting Genesis just as it reads.

Hasty Surrender

Regarding the supposed conflict between science and religion concerning evolution, Gardner asserted that many theologians surrendered to evolution without a battle. By means of a gigantic confidence trick, by pretending that the study of man is science, by hanging on to the coattails of solid, successful, reliable physics and engineering, an army of atheists and agnostics forced many theologians to turn and flee from the Bible as God's truth.

In 1996 Pope John Paul II declared that evolution is a better explanation for the human race than the Bible. No evidence in support of evolution is cited, of course, because there isn't any. In order to keep one foot in each camp, however, he then gives God the credit for creating the soul in each person. This surrender of Scripture to false science is hardly a surprise, since Teilhard de Chardin brought evolution into the Roman church two generations earlier with the help of the Piltdown fraud (See chapter 8).

Resistance to Change

In this chapter we shall illustrate another kind of conflict. Apart from attacks on the truth of the Bible, the sciences that deal with evolution are notably resistant to change regardless of how much evidence may support a novel view. This situation is true in other fields of science as well.

A scientist from M.I.T., Giorgio de Santillana, cited Friedrich Humboldt on scholarly resistance to new discoveries or new views.[15] First, people will deny a thing; then they will belittle it; then they will decide that it had been known long ago. The great German physicist, Max Planck, reflected on how his revolutionary discoveries were received by fellow scientists. He stated that a new scientific truth does not triumph by convincing its opponents and making them see the light, but rather because its opponents die and a new generation grows up that is familiar with it. Franz Boas, a brilliant and once progressive archaeologist, stated that it would be a pity to have new evidence come to light that would overthrow all the admirable scientific work of the past that indicated the recent arrival in the New World of the American Indian. This is the kind of thinking that keeps evolution alive.

Name-calling is common. Those defending conventional views are looked on by supporters of novel theories as mentally fossilized ivory-tower isolationists. These in turn view those pushing novel solutions to complex problems as amateur, misguided mystics whose theorizing in such scholarly fields is emotional rather than intellectual. By no means are all holding novel views amateurs.

Just as bad artists invariably praise bad art, writers within the lunatic fringe do not attack other wild theorists. In a similar fashion, evolutionists rarely criticize the writings of fellow evolutionists when they are full of obvious errors and humbugs of many kinds. Perhaps the only exception of which we are aware is violent criticism of evolutionists who discover something that fits wonderfully into the young, created earth belief. And evolutionists will violently attack creationist writings even when they present good science. We must also say that sometimes creationists let good intentions rule over complex matters where they are not qualified to speak.

It is amazing that almost 200 years ago the great German philosopher, Goethe, complained that professors in his day were interested only in proving their own opinions. He charged that they concealed all experiments that would reveal the truth and would show their doctrines to be untenable. Has anything changed?

Faulty Records of the Past

Archaeologist Geoffrey Bibby emphasizes a point that has caused errors in the past.[16] Scholars forget that statistically only a very small sample of all that was written in ancient times has been preserved, and even that sample is not representative. This is an important point for scholars who study the past and attempt to reconstruct cultures. We might say that the same is true of other cultural remains. What was originally created on perishable materials—and hence not preserved for our study—may be far more vital for analysis than the stones and potsherds that were preserved. At the least, great caution and humility are called for in interpreting archaeological remains.

Can Evolutionists be Objective?

Is it possible that scientists are biased? We must say that bias does leap out at us whenever we examine the work of evolutionists. As we shall also see, human failings such as greed, envy, lust for fame and power, and the like are discovered in the scientific community much more often than we like to think. Alan and Sally Landsburg comment below on how difficult it is to be objective. When we read the works of evolutionists, we see minds imprisoned by a false view of the world:

> We tend to believe what we want to believe. It's hard to face isolated unexplained facts if they conflict with our mental picture of the world. We ourselves may be victims of mental blocks that will seem ridiculous to people a thousand years from now.[17]

In a similar vein, George Simpson, a committed evolutionist, stated the obvious that evolutionists for 150 years have studiously ignored.[18] He said that faith, dogma, and authority can make us blind to the plain evidence of our senses. What Simpson does not state is that evolution rests finally on faith or assumptions. This overpowering bias shows itself especially in attacks against the Bible. As Ian Barbour makes clear, science is full of presuppositions or unprovable assumptions, but their existence is frequently not recognized even by those most affected by them.[19] The problem has never been the evidence, but how the biased evolutionist interprets it.

Marvin Lubenow studied the literature on the supposed sequence from ape to man.[20] He concluded that any evidence for creation is automatically discarded as contamination, no matter how valid that evidence might be. Similarly, Louis Pauwels and Jacques Bergier noted that they were convinced that the study of past civilizations has been marred by numerous cases of rejected evidence, *a priori* exclusions, and inquisitorial scholarly executions.[21] Few scholars wish to risk their careers by dwelling on findings that do not fit the accepted views of their colleagues. It is not popular to rock the boat; in fact, it could be professionally fatal.

Another aspect of scholarly deception was discussed at length by Oscar Muscarella.[22] Eminent scholars will authenticate objects they know to be forgeries. They may be paid off by gifts, or they fear the powerful and wealthy owner will seek revenge. How? Deny jobs, recommendations, internships, grants, friendship.

A scholar of international note, Jacquetta Hawkes, did not like what she saw in her field of archaeology and speaks of a vast accumulation of insignificant, disparate facts, like a terrible tide of mud, quite beyond the capacity of any man to contain and mould into historical form. Further, she described archaeologists as an introverted group of specialists enjoying their often rather squalid intellectual spells and rituals at the expense of an outside world to which they will contribute nothing that is generally interesting or of historical importance. She is not amused by the large amount of bad archaeology best described as pot gathering, making poor records, bad excavating, and failure to publish or publish so poorly as to be worthless.

Evolutionists marvel that there are people today who actually believe the world was created. Philip Kitcher gives us this amazing statement, unaware that he is very accurately describing his own belief in evolution. Just substitute the word evolutionism for creationism: "I have no intention of criticizing creationism insofar as it is held as an explicitly religious belief, a belief that is recognized as running counter to the scientific evidence."[23]

Creationists in all honesty study what is claimed to be evidence and marvel that evolutionists actually believe the world and life evolved out of non-life. But only one of the above two views can be correct.

The deep bias we have described above reveals itself vividly in the scholarly journals. Here is where we expect scholarship to explore new intellectual ground and reach new heights of exploration. Then why do distinguished scholars complain that much of what the journals publish is ignorant drivel, and that at the universities there is great pressure to publish research but virtually no interest in content.[24] Researchers who come up with exciting new knowledge more often than not have their work reviewed and then rejected. The review system allows the reviewer to say unreasonable, insulting, irrelevant, and misinformed things about the research. In much of the system today, here is what the author must do in order to get published:

NOT pick an important problem
NOT challenge existing beliefs
NOT obtain surprising results
NOT use simple methods
NOT provide full disclosure, and
NOT write clearly

The above points were described in much greater detail by a distinguished scientist—not by some ignorant quack. And the above was published in a distinguished scholarly publication, *Chronicle of Higher Education*. We have probably written enough here to answer the question sometimes stated by evolutionists: Why don't creationists submit their research to scholarly journals for publication? The philosopher Kierkegaard once said that the person who is right often has to stand alone. Wherever there are exceptions to the sordid tales above, we rejoice to see that genuine scholarship is not yet dead.

Is Evolution Fact?

While today evolutionists like to speak of evolution as fact, without being able to show why it is fact, earlier evolutionists were more candid.[25] For example, some evolutionists have stated that their faith in the idea of evolution stemmed from their reluctance to accept the antagonistic doctrine of special creation. Similarly, others anger their fellow evolutionists by stating that the theory of evolution finds its support not in direct observation, but in the difficulty of forming an alternate hypothesis. In other words, anything but creation, anything but an almighty God. There are many examples of name calling to bolster a belief in evolution. Some evolutionists hold that there is no rival hypothesis to evolution except the outworn and completely refuted one of special creation, now retained only by the ignorant, the dogmatic, and the prejudiced.

Is it not strange that in all the vast areas of science, evolution is the only theory supported by ridicule, threats, and really nothing else? Why do we say nothing else? Frank Marsh pressed the prominent evolutionist Dobzhansky to give one compelling proof of macroevolution (that is one species changing into another) among living organisms.[26] The latter replied that there were no specific examples. He then referred Marsh to Simpson for examples from the fossil record. Simpson reported the same lack of any specific examples. Again, the dogma of evolution could only be accepted on a faith basis. In vivid contrast we find Gavin de Beer gushing the absurd:

> Scarcely a day passes without the appearance of new evidence confirming the truth of the theory of evolution up to the hilt, and it is now universally accepted except by those who are too ignorant or too obsessed by irrational considerations to follow scientific evidence wherever it may lead.[27]

We would much like to see a list of these daily discoveries from de Beer or any other evolutionist. We would settle for even one bit of clear evidence. Evolutionists like to say that the belief in creation is an attack on ALL of science, particularly where evolution is involved. While we speak very arbitrarily of frauds, errors, and humbugs, we are quite aware that there is much overlap. For example, the Piltdown hoax was fraud, it was error, and it was humbug, all in one. The literature is vast, and interested readers may go into a bit more depth with such eye-opening works as *Betrayers of the Truth* and *Science is a Sacred Cow*. Other titles and examples have been noted earlier in this book.

Frauds, Errors, and Humbugs

The Far Side of Pseudoscience

Alan Sokal, a physicist from New York University, achieved instant notoriety in 1996 with one journal article.[28] Who could fail to be impressed by a paper entitled *Transgressing the Boundaries: Toward a Transformative Hermeneutics of Quantum Gravity*? A few weeks later Sokal confessed that the article was a fake and the journal publishing it had been deceived by the parody. Sokal had become tired of so-called postmodern intellectuals who insisted on using mathematical and scientific theories (that is, evolution) to explain happenings in art, history, philosophy, and other fields of humanities and social sciences. Such material, he insists, is sloppy thinking, false erudition, and much nonsense. Examples of such pseudoscience are psychohistory, sociobiology, and futurology. Pseudoscholars, itching to be noticed, praise such material and pretend to understand what is said. Then they pass on the absurd ideas to captive students who also have to pretend that this is learning.

A sordid tale of twenty-first century deliberate deception and withholding the truth about archaeological fakes is described in *Odyssey*. Noted scholars authenticate objects they know to be forgeries. How could this happen? The experts do not want to offend the dealer/collector who owns the piece or the fellow scholar who had published it as genuine. They fear prominent owners of the fakes will seek revenge and deny jobs, internships, grants, recommendations—to themselves or their students, and they are right to fear reprisals.

Time—The Greatest Humbug

Evolutionists toss time about in vast quantities and give the impression that these estimates are scientific. Several examples will underline this point. In 1922 the Museum of Natural History in Toulouse, France, lost its prize fossil, the skull of a prehistoric panther thought to be tens or hundreds of thousands of years old. Then a specialist discovered that the skull was unmistakably pierced by a rifle bullet, which made it perhaps as much as 50 years old. Similarly, some camel remains from the West in the United States were believed to be at least a half million years old. Then a skull found in Utah was discovered with particles of flesh adhering to the bone. Specialists remembered that camels had been imported into the West in the 1870s.[29] Puzzling remains of a camel pelvis and human remains were excavated in Mexico in 1981. Using the latest fission-track and uranium dating of materials there, a date of about 250,000 years was obtained. Unfortunately, the stone tools in the deposit were sophisticated and obviously quite recent. So the archaeologist concluded with the wonderful phrase that such an age is essentially impossible, and left it at that.

One scholar gave this illustration about guesswork in dating. A geologist will inquire about the age of a particular layer or stratum of rock, that his geological scheme assumes to be, say, 1 billion years old. He turns to a dating laboratory for independent verification. The physicist there may then apply radioactive dating methods that were themselves developed and checked against the geological column using uniformitarian assumptions. All dates that do not fit are discarded as contaminated. To no one's surprise, then, radioactive dating methods yield appropriate dates.

The Deception of Gradualism

In his book *Ivory and the Elephant* George Kunz includes a sequence of ten drawings to show just how the elephant gradually evolved over millions of years from a dog-sized ancestor to the huge species of today.[30] The trunks and tusks gradually got longer; the ears and skull gradually became larger. All the evolutionists had to do was gather skulls from around the world and arrange them in order of size, ignoring the age and gender of the fossils, ignoring the strata in which the fossils were found, and discarding what did not fit into the scheme. The two earliest ancestors were from Egypt, the next was found in Europe, and the last seven in Nebraska. Some were mastodons, some were mammoths. This same method was used in textbooks to show supposed horse evolution, human evolution, and the evolution of other life forms. Human and horse evolution are still displayed in the textbooks today, but many prominent evolutionists have expressed their embarrassment about this fraudulent technique. There is not one bit of evidence to demonstrate how this supposed evolution actually occurred. If we did not know better, think how impressive a museum exhibit it would make if we lined up members of the cat family. We could then prove that the African lion evolved from the pussy cat over millions of years, as proved by the line of skeletons. The same could be done with the deer, the dog, and other life forms. This example is not error—it is plain fraud because no such sequences have ever been found. In fact, the sequences actually found would show many contradictions of the supposed order of development.

Another Ice Man

We no longer read about the Minnesota Ice Man hailed in the 1960s as the missing link between ape and man. Encased in a 6,000 pound chunk of ice, this hairy creature was very ape-like, but also possessed human characteristics. Reports were that it had been found by a Russian sealing ship in international waters near the Bering Sea and passed through various hands before being placed on exhibit in America. The large nose was pugged, nostrils were large, circular, and pointed straight forward. The mouth seemed to have no lips, yet the face in its entirety seemed to be human. A stench like that of decomposing flesh issued from the block of ice. Distinguished scholars in Europe, the Smithsonian Institution, Harvard, and Emory University were among the many who pronounced the remains as genuine. After all this excitement of proving evolution, the Smithsonian Institution received a call from the operator of a wax museum in California who admitted the Ice Man was the creation of five of his technicians who had fashioned him from rubber and hair.[31]

Fossil Hoaxes?

The perennial fuss over the supposed fossil feathered serpent, the Archaeopteryx, has been going back and forth for more than 130 years. Critics have demonstrated how feathers could have been pressed into a layer of cement added to the original reptile fossil. Evolutionists attach great importance to this fossil as it is one of those extremely rare supposed missing links—this time between reptiles and birds.[32] As late as 1987 the British Museum put on a special exhibit showing the arguments of scientists on both sides of the mystery. But as Ian Taylor and others have pointed out, remains of true birds have been discovered in strata supposedly of the same age and strata much older than the Archaeopteryx.

National Geographic breathlessly announced to the world a true missing link that connects dinosaurs with birds.[33] It was a primitive bird with the tail of a dinosaur. Within weeks, however, this was exposed as a fraud that had deceived some of the worlds leading paleontologists. Apparently an enterprising Chinese peasant had glued the top half of a fossil bird to the bottom half of a fossil dinosaur of about the same size. This specimen is now known as the Piltdown Chicken, in honor of an even greater fraud.

The Painted Toad

A gifted biologist, Paul Kammerer, tried to prove that acquired characteristics could become hereditary. He noted that one species of toads that spent much time in the water developed black calluses on their feet. He then arranged for some land toads to spend a similar amount of time in the water and thought they would also develop the black pigmentation. He reported that after only four generations the black areas were already showing, a marvelous proof of evolution. Another specialist, however, became suspicious and quickly proved that the acquired black spots were painted on with Chinese ink.[34]

Mistaken Discoveries

In California bones were discovered in strata underlying vast mud deposits.[35] Based on the best thinking about how the deposits were gradually laid down, the bones were dated at 75,000 years. Deeper still, an old United States Army button appeared—a good illustration of how faulty assumptions produce outrageous errors.

Another dating problem was reported in late 1990. Archaeologists examined eroded stone carvings near Grants Pass, Oregon, and they concluded that they were authentic and ancient. As the U.S. Forest Service was building an interpretive center explaining the carvings, a local artist heard of the project and sent a message to them. Some years before he had spent an afternoon carving the petroglyphs similar to ones he had seen elsewhere, just to see what was involved and how long it would take. He had photos to prove it had been his effort, and he stated that he had no intention of fooling anyone. The project was quickly dropped.

From Ape to Man?

In 1950 Dr. Robert Broom[36] became very excited about a new fossil discovery and made a prediction, as cited by a reporter from the *San Francisco Examiner*: "But science is getting so close to mans origin that Dr. Broom believes we'll find where we came from in just another ten years or so." A half century later, the situation is more confused than ever. The plain truth is that evolution has nothing to show for 150 years of very intensive and expensive work.

Le Gros Clark points out a problem that over the years has resulted in many, many humbugs in the search for the origin of man.[37] The common error, according to Clark, was to assume that any fossil was a typical specimen, so any new fossil that deviated even to a slight degree from the first was assigned to a new species or genus. Failure to recognize the great variability in humans, which any child can see, led to a multiplication of so-called human species that we now recognize as not distinguishable from Homo sapiens, that is, modern man.

Old earth—big humbug! For a very illuminating look at the fallacies of an old earth we studied the writing of the late evolutionist, Derek Ager, who was the world's leading authority on the layers or strata of the earth.[38] This slender volume is a juicy morsel for the creationist—so juicy, in fact, that Ager invoked a curse upon any creationist who used the information. One might ask how a confirmed evolutionist can shred up some of the most sacred assumptions and beliefs in conventional geology and still remain in the good graces of the profession. The answer is very simple. Ager lulled the profession with a stirring credo of his loyalty to evolution and topped this off with a scathing denunciation of the creationist view. Ager stated:

(Books are still being published by the lunatic fringe about Noah's Flood as the explanation for major geological changes in the earth.) In case this book should be read by some fundamentalist searching for straws to prop up his prejudices, let me state categorically that all my experience (such as it is) has led me to an unqualified acceptance of evolution by natural selection as a sufficient explanation for what I have seen in the fossil record. I find divine creation, or several such creations, a completely unnecessary hypothesis. Nevertheless this is not to deny that there are some very curious features about the fossil record.[39]

Nevertheless, this sharply critical statement was not enough to spare Ager from equally sharp criticisms from fellow evolutionists who told him that he had given far too much aid and comfort to creationists in this book. It is fascinating that they did not criticize his evidence and conclusions—only that creationists would make use of them to show how the world itself strongly supports the concept of a young earth. Among the points and theses Ager developed in his book are these:

- The break between Paleozoic life and Mesozoic life is in the mind of the specialists rather than in the evidence.

- Specialists do not agree on how many . . . widespread extinctions, occurred in the past.

- Extraterrestrial causes, that is, strikes from outer space, and magnetic reversals of the earth are becoming respectable explanations for great extinctions in the past.

- The most serious error of paleontologists is to measure the past by the present. [Comment: That sentence ought to be repeated ten times for its full impact!]

- Continuous sedimentation is a ridiculous illusion, full of contradictions. Remains as recent as Roman times are found at "startling" depths. Sedimentation in the past has often been very rapid indeed and very spasmodic.

- Changes do not take place gradually. We (evolutionists) are now all catastrophists!

- Dating rock by its fossils and dating the fossils by the rock is an impossible circular argument, yet we are always on dangerous ground if we accept other lines of evidence. [Comment: May we ask, Why is the hallowed science of radiometric dating not used to date the strata of the earth?]

- The history of the earth consists of long periods of boredom and short periods of terror.

The reader must excuse some of the above jargon, but suffice it to say that Ager here undermined the most basic sacred beliefs and teachings of evolutionist geology, that in turn makes hash of the notion that the strata of the earth proves a very very old earth. Ager's appeal to long periods of boredom when nothing at all happened for countless millions of years is an incredibly naive clutching at straws.

Ager goes quite far in taking an honest look at geology, but he is unable to pursue some of his points to their logical conclusion. For example, Ager in effect would say that a period like the Permian could have happened in a week or a month, but then the earth went into some kind of suspended animation for 50 million years (minus one week) until the conventional time allotted to that period was up. This is reassuring to the evolutionist who accepts the geologic column as an article of faith, but it makes no logical sense at all, nor is there supporting evidence for it. Above all, the evolutionist needs vast amounts of time for evolutionary changes to take place. Ager clings to a childish faith in the merits of radiometric dating: Radiometric dating has nearly reached the stage when we can make reasonable estimates at the stage, if not the zone level. Translated, this says that we can perhaps get within a few million years of predetermined correct dates some of the time, and it helps when we discard all the wrong dates by calling them contaminated.

The most fundamental article of faith of the evolutionist is that everything that ever happened came about very gradually. This doctrine is called uniformitarianism or gradualism. Thus evolutionists believe that enormous stretches of time produced all the geological effects we see today, such as the Grand Canyon. All in all, Ager is a curious book, illustrating the dilemma of a man faithful to his belief in evolution who attempted to face up to some devastating contradictions in the stratigraphical record.

Gradually or Suddenly?

The lifeblood of evolution is gradualism, that is, slow gradual change over infinite periods of time. From Charles Darwin on, this belief was desperately needed in order to give evolution time to happen. We have just seen how a loyal evolutionist, Derek Ager, shredded this belief on the basis of the actual evidence. Let us look at an older textbook in geology, but typical of texts for more than a century, that illustrates why the old guard sharply criticized Ager for telling the truth. Hugh Brown informs us that as the science of geology was beginning to take definite form near the start of the nineteenth century, its efforts to interpret earth history were severely handicapped by the philosophy of sudden change, of convulsions that created or destroyed mountain systems or even whole continents in a matter of hours.[40] Such a philosophy discouraged attempts to look backward, and even blinded scientists of the time to the fact that geological processes leave legible records upon the rocks for men to read. It remained for James Hutton (a retired physician) to provide the formula for a patient, unhurried universe. There had been, according to him, no convulsions of nature. Hutton, however, was an abominable writer, so it remained for Sir Charles Lyell, more than any other, who triggered an era of inspired research into geological history, that is, evolution. As we can see, evolutionists wanted to get rid of the flood, of Creation, and of God.

From Non-life to Life

Evolutionists devoted much study and research in the attempt to bridge the gap between inanimate matter and living creatures. The search continues today. Early evolutionists, such as Ernst Haeckel, were not deterred by the fact that no bridge could be found, so they "created" a group of living beings that they called the Monera. These minute organisms were supposed to cling right on the edge between life and non-life. After a flurry of papers in the literature of evolution showing this dramatic and compelling proof of evolution for the benefit of doubters, the whole scheme was exposed as a complete fiction.[41]

Alpheus Hyatt and Wurtenburger published a beautiful series of fossil ammonites in order to prove an evolutionary theory, that of recapitulation—that organisms show in their development all the evolutionary stages they had passed through. Then in 1901 A. Pavlov showed that if the shells had been arranged in that manner, the strata in which the shells had been found would have to be reversed. In other words, all the evidence had been falsified.

Another Gradualism Humbug

The authors of *Hamlet's Mill*, while quite willing to accept evolution, express astonishment that despite evidence to the contrary and admonitions from colleagues, Haeckel and others proposed to solve the world riddles by declaring that primitive man gradually evolved into modern man.[42] Even worse, the principles of evolution were unthinkingly applied to man's cultural history when, again, the evidence is just the opposite. The authors predict that future historians may declare all of us insane for these incredible blunders. With respect to mythology, which is coded history, the authors state that poets and psychologists have disfigured mythology to the point where it has become useless to most people for seeing into the past.

Conclusion

Some Observations about Science and Deception

In this chapter we are astonished, over and over again, at the workings of the human mind. We see intelligent people whose minds have been imprisoned by preconceived beliefs that have no basis in the real world. Evidence is twisted, molded, or ignored in order to put the best possible face on a theory, really a dogma. That spirit was caught in its perfect form when archaeologist Sir Flinders Petrie, who had studied the Great Pyramid in much detail, caught a pyramidologist secretly grinding down a projecting stone in the pyramid to make it fit his own theory of how this monument had recorded ancient history inch by inch in its measurements.

If one in haste thinks that this chapter is anti-science, that person has not read the message. Evolution is totally irrelevant to technology, therefore it follows that technology has thrived to an incredible level. Technology builds wonderfully on the past, so few things need to be reinvented. Where scientific method is used, the sciences prosper, subject to the kinds of evils pointed out in books such as *Betrayers of the Truth*.[43]

Now and then we read that research in the world is on such a vast scale that the amount of knowledge is believed to double every ten years (or some other similar figure). This seems most certainly correct with many of the technologies and perhaps in some of the hard sciences such as physics and chemistry. However, it is quite a different story with the sciences that treat the ancient world in some way. The soft sciences, such as sociology, psychology, archaeology, anthropology, and others undoubtedly produce almost infinite amounts of books and articles, but to say that knowledge is doubled every ten or so years is simply not true. Textbooks are revised every year or two, not because of the addition of important new knowledge, but simply as a marketing device to prevent students from reusing equally good or bad texts. Thus we are drowned in material, but seldom in new knowledge, sad to say.

Our quarrel lies with those sciences dealing with the past that are imprisoned by the sterile theory of evolution. This entire book shows clearly why we take that view.

William R. Corliss collected an immense treasure store in many volumes for the reexamination of conventional theories.[44] Corliss is not a creationist and he does not draw the violent attacks of evolutionists despite the many kinds of evidence he publishes that contradict their cherished beliefs. Here are just a few of his observations that shed some light on our examination of science and deception.

- The hard facts of science are far more tentative and provisional than we have been led to believe.

- We do not understand much of anything, from what is called the supposed Big Bang all the way down to the particles in the atoms of a cell.

- Anomalies, of which there are many thousands, tend to annoy the orthodox scientist, which prefers to view itself as a logical, virtually errorless march toward immutable truth.

- Surrounding the facts of every science is a kind of cloud of exceptions, mysteries, and contradictory facts almost always ignored by the experts most qualified to study them.

- Ancient man was smarter than he has been given credit for, and he also traveled farther and earlier than anthropologists have been willing to admit.

- Theories imprison many from taking a fresh look at the actual evidence.

The situation has deteriorated even further since 1983. William Corliss stated that his continuing comprehensive search through scientific literature has resulted in a collection of more than a staggering 6,000 anomalies. Anomalies are mysteries and discoveries that contradict current beliefs in science, chief of which is evolution. Fields examined include archaeology, astronomy, biology, chemistry, geology, geophysics, physics, and psychology. The massive campaign in our culture to force a belief in evolution ignores or trashes thousands of findings by objective scientists who have reported what they have discovered. With rare exceptions our educational system at all levels and the media teach the party line despite the evidence to the contrary.

Richard Milton, a prominent science journalist in England, is thoroughly at home with scientific research in many fields.[45] He tells about sleepless nights when he thought about his 9-year-old daughter beginning her first formal science classes. She loves fossils, dinosaurs, and science. In examining his daughter's science text, he realized that most of what she was being taught is simply not true. He saw that evolution as taught is totally an act of faith, and a dangerous one in its applications. Milton, not a Christian, not a creationist, knew that actual research provides compelling evidence that:

- There are no discovered missing links.

- Earth is much younger than previously thought.

- Radiometric dating is deeply flawed and unreliable.

- Mutations and selection cannot produce new species.

- Earth's history is one of catastrophe, not gradual change.

In the long run, evolution cannot survive this evidence. All of Milton's conclusions fit perfectly into the message and conclusions of this book because it was written within the framework of a young created earth.

Our Focus

One could track all the books used in science education from preschool through graduate school and find little or nothing of the insights described in this chapter. We have all the respect in the world for good science. It is not anti-science to point out the failings of the theory of evolution, the failure of scientists who do not or cannot distinguish between evidence and interpretation of evidence, or the blindness of scientists so deeply enmeshed in the theory of evolution that they actually believe they have no presuppositions or biases or assumptions. We look forward to the day when education can actually be the untrammeled search for truth. Creationists have nothing to fear from truth.

In *New Scientist*, Ralph Estling has written a wonderful satire about his principle of inverse irreversibility.[46] Scientists who believe they have arrived at truths objectively refuse to accept or even consider truths that other scientists believe they have arrived at objectively. The more evidence brought forward to undermine their truth, the more stubbornly they hold on to their position despite it all. This may be taken as a kind of parable about evolution and creation and shows that faith is, after all, the deciding factor. In our view, the more evidence is brought forth to undermine evolution beliefs, the more stubbornly evolutionists hold on to their beliefs despite it all. Evolutionists believe their theory has been derived scientifically, that is, objectively. Nothing could be further from the truth. Creationists believe that the ultimate truths are those in the Bible, including creation. They know of no evidence that contradicts this faith, even though there are many mysteries in nature far beyond human understanding.

Thus it all boils down to a choice between an almighty God or a belief in time and chance producing the universe and everything in it. We have examined all the evidence and our choice is God, who created what He said He did in Genesis.

For Discussion and Reflection

1. What do you suppose is the cause of deliberate deceptions that are reported from time to time in scientific fields?

2. We can't emphasize strongly enough that Christians are strongly pro-science, yet just as strongly anti-evolution. If anti-evolution is not the same as anti-science, just what is the difference?

3. What proofs or evidence do evolutionists offer in their attempt to show that evolution is no longer theory, but fact requiring no proof?

4. We must distinguish between errors and deception in fields of science. Just what is the difference?

5. Evolutionists point to erosion as proof that over many millions of years strata (layers) very gradually accumulate and show that the layers can then be dated according to the geological column. Read Job 14:18–19. Job was very much aware of erosion, but nowhere in the Bible is erosion presented as a way to date the strata of the earth. When you see layers of rock in road cuts, what do they tell you about how slowly or quickly they were formed?

6. Evolutionists do not invite creationists into the laboratory to demonstrate proofs of evolution. If that is true, why not?

7. This chapter gives an example of what may happen to a scientist when he discovers an artifact much too old to fit into the "correct" dating pattern. How

did this treatment compare with the world of science where freedom of inquiry is said to be sacred?

8. What is your view of this statement from the chapter? Belief in creation is all right as long as creationists recognize that their belief runs counter to scientific evidence. Is "evidence" the right word here?

9. Is it acceptable to display a fossil made up of the top half of a bird and the bottom half including the tail of a dinosaur? Evolutionists believe dinosaurs evolved into birds because the exhibit shows a stage that evolutionists believe happened long ago.

10. Does the creation/evolution issue reduce down to a belief in the Creator in the Bible, or belief in time and chance as the new god?

CHAPTER SEVENTEEN

THE END OF OUR SEARCH

Only One Truth

We have roamed over many fields of learning that in one way or another have treated the ancient world—the world of Genesis. We continue to have every reason to believe that there is only one story of the ancient past.

In the textbooks we are taught that evolution is the one overarching principle applied to almost every aspect of life. Moreover, students at all levels are taught that evolution is science and evolution is fact. We take strong exception to this. Our belief in the truth of the Bible is our foundation. There is the old canard that belief in creation is factless faith while evolution is faithless fact. The truth is that both creation and evolution are faiths. Just as the heavens declare the glory of God, we have discovered the same great truth in all the areas we have explored.

We do not need additional evidence so that we may come to believe the Bible. The Scriptures, however, tell us to love our God with all our minds. Not only is the study of the ancient world a deeply fascinating one in which we see the mighty acts of God, but we also see repeatedly the complete futility of evolution as explanation for past, present, or future. In our view, 2 Timothy 3:7 expresses the great contradiction of studies in evolution: ". . . always learning but never able to acknowledge the Truth."

The Bible Contradicts Evolution

In the past several decades there have been many compelling and convincing critiques of evolution that were not written by creationists. These books are the product of scholars who have weighed in on evolution and have found it wanting. Creationists are suspect when they point out the many fallacies of

Charles Darwin and his followers. It is refreshing to note again and again the works of non-creationists who have no ax to grind, except to search for truth. They have written many of the most devastating criticisms of a theory that explains nothing, predicts nothing.

Who Should Write History?

In his forward to Joseph Alsop's book *From the Silent Earth* Maurice Bowra makes an interesting statement, that history is too serious a matter to be left exclusively to professional historians.[1] Historians, for all the great work they do, are often too hampered, too cautious, too indecisive, too narrow, too much full of jargon, too fearful of criticism and ridicule from colleagues.

The world needs history written by those who are not ensnared by the crushing weight of evolutionary theory that forces all such books on the ancient world to sound exactly alike. Just look at the opening chapters of a number of world history books for mind numbing parrotlike accounts. It seems safe to say that not one new idea has penetrated the unfortunate fog for generations. No one dares to say it differently. No one dares to take a fresh, unbiased look at the actual evidence.

This book uses the framework of a young, created earth, and it is the belief of the author that the evidence presented (as opposed to speculations and interpretations) falls smoothly within this framework, except for the many mysteries where we simply to do not have enough information at hand to understand them. The Christian who believes the Bible as it is written is not surprised at how well a biblical framework fits the data.

Our objection to evolution is that students are forced to swallow and regurgitate evolution at all levels of public and much private education. The consolation, if there is one, is that national surveys continually show that the majority still does not accept the fond belief that evolution is fact. This is the case, despite more than 150 years of ceaseless propaganda in the schools and all the media.

Who Should Write the Textbooks?

There is no secret about the doctrines of evolution forced on people of all ages in our educational system and in the media. The editions of the *Humanist Manifesto* spell out this faith very clearly. Here are some samples:

> Faith in the prayer-hearing God . . . is an unproved and outmoded faith. Belief in salvation appears as harmful . . .

> Traditional religions that place revelation, God, ritual, or creed above human needs and experience is a disservice to the human species.

Any account of nature should pass the tests of scientific evidence. [**Comment**: We couldn't agree more. Evolution is by faith. Religion is by faith. We strongly support science. We strongly disagree with the doctrine of evolution.]

We can discover no divine purpose . . . for the human species . . .

Science affirms that the human species is an emergence from natural evolutionary forces.

Promises of immortal salvation or fear of eternal damnation are both illusory and harmful—no evidence that life survives the death of the body . . . [2]

The above statements are taken from the *Humanist Manifesto II*. People committed to the above beliefs dominate the halls of learning in public education, the textbook writers, all types of media, the political scene, and most other positions of power and influence. While we can rejoice at any exceptions we see, humanists described above are totally committed to destroying the Good News of the Bible. Private schools, home schooling, and Christian homes are our powerful antidote to the above.

A Creationist Stereotype

It is amusing to read a full page article in *Newsweek* about living, breathing creationists actually working in scientific settings today.[3] In a most patronizing and demeaning tone, these creatures are described as if they came from another planet. The article expresses amazement that they are intelligent beings, yet they cannot be expected to do "good science" unless they work in "disciplines far removed" from those fields most deeply involved with evolution. There is a note of alarm that "published scientists with creationist beliefs are not uncommon." The writers reassure the public in this article, however, that "evolution is the defining paradigm of biology," and that few pay attention to their journal articles and speeches. The article ends with the warning that creationism can shape some scientists' conclusions as strongly as any empirical evidence.

It is hardly necessary to say that this book shows the other side of the coin—how evolutionists deceive themselves and the public. One almost feels moved to thank *Newsweek* for allowing creationists to live. Will someone some day propose separate drinking fountains where creationists and evolutionists are found working in the same laboratory? The article illustrates, of course, the paralyzing hold evolution has today in all the media, and we need to function despite this, not in cooperation with that false belief.

A Searing Critique

How do evolutionists view studies conducted by competent scientists who also hold to the Genesis account? No one ever said it better, but in another context, than a sixteenth century Italian philosopher, Bruno:

> With a sneer, a smile, a certain discrete malice, that which they have not succeeded in proving by argument—nor indeed can it be understood by themselves—nevertheless by these tricks of courteous disdain they pretend to have proven, endeavoring not only to conceal their own patently obvious ignorance but to cast it on the back of their adversary. For they dispute not in order to find or even to seek truth, but for victory, and to appear the more learned and strenuous upholders of a contrary opinion. Such persons should be avoided by all who have not a good breastplate of patience. [4]

It is not at all difficult to find vivid examples of the above. Evidence, discussion, debate are not permitted by those in power.

Where We Have Traveled

This book has been a fascinating adventure for the author and I hope the trip has been of value to the reader. Throughout the book I have illustrated that the textbook world bears little resemblance to the real world unless it is in an area untouched by evolutionary speculation.

An examination of anthropology and archaeology found these fields wanting. The reasons are, of course, evolution, specifically the belief in the ape-to-man sequence, and gradualism, where everything happened slowly but surely over immense spans of time. Belief in creation and the world changed by catastrophic events, particularly the flood, presents an entirely different paradigm.

It is fascinating to see how the dinosaur fits into the young earth pattern. It is of interest and value to see how other non-evolutionists have treated the problem and mystery of the dinosaur.

One must be aware that almost all so-called biblical archaeology is in the hands of humanists consumed by an anti-biblical knee jerk reaction to anything that supports or seems to support the Bible. When the evidence speaks for itself, the result is wonderful illuminations of the biblical world of great benefit in the study of the Bible.

Joshua's long day occurred as Scripture tells it. It is unfortunate that well-meaning persons contrived a fiction in order to 'prove' the Bible to be correct. God does not need or want that kind of help. Just let the story of this miracle stand as written.

Bishop James Ussher believed in a young, created earth. He wrongly assumed, however, that the Bible gave a complete chronology from Adam to

Revelation. No one on earth knows the precise age of the world, but there is no doubt at all that the earth is young. Modest variations of Ussher's chronology are appropriate.

It is instructive to examine the futile attempts of the "bone peddlers," as one critic described them, to find the missing link between ape and man. In the long series of frauds and errors and humbugs on this question, the Piltdown scandal was the most outrageous. This fraud played a tragic role in bringing evolution into the Catholic Church via Teilhard de Chardin. Piltdown and another humbug, Nebraska Man, also were the key "witnesses" against the Bible in the notorious monkey trial in Tennessee in the 1920s.

There are many fictions in the literature about the supposed beginnings of animal domestication. The Bible furnishes the framework for this development. The many reports of extinctions, past and present, are evaluated.

The story of supposed horse evolution is a most interesting tale of another evolution humbug, and I examined the even stranger story of "false" horses in South America. Here I found that the science of the classification of the animal world, called taxonomy, is full of bizarre decisions. An investigation of the plant world revealed details that illuminate much about the great sophistication of the earliest times.

There are many excellent critiques of the fallacies of evolution. Charles Darwin appeared in the mid-1800s and was just what millions of itching ears wanted to hear. This faith is evaluated briefly and nothing of substance is discovered in this attempt to explain the past without God. Time, chance, and natural processes became the new god for the evolutionist. Their goal then and now is to ungod the universe.

Several pictures are painted of problems of modern astronomy. Ancient records reveal how dominant astronomy was in daily life during ancient times.

In summing up an examination of prehistory, the literature shows many attempts to deal with the past, from serious scientists who report honestly, to evolutionists whose first loyalty is to their theory despite the evidence, and to those who are often called the lunatic fringe. The purpose of this section of the book was to see on what evidence various wild theories are based. These views were then contrasted and compared with the biblical framework for all of history.

This study also salutes science as it should be and as it sometimes is. Readers must also be aware of the more than occasional expression of fraud, error, and humbug in scientific and pseudo-scientific work. After all, science is often carried out by frail humans with strong biases. The most powerful of all biases is the belief in evolution. Truth cannot rise out of fallacy.

In Praise of Real Science

As we have emphasized, we have no quarrel with science nor with scientific method. The spin-offs of science into technologies are truly awesome. But as soon as we say these things, we run headlong into the problem of good and evil. Awesome discoveries may be used for good or they may be used for evil. Moral decisions cannot arise out of science. Civilizations rise or fall on the basis of their moral fiber, not on their level of technology.

When we praise science and technology, and deservedly so, we must ask honestly to what extent is evolution the foundation and cause of the advances. Ask yourself, which major discoveries can you name that arose out of evolutionary theory? Can you name even one? There are a number of extensive lists of great Bible-believing scientists of the past. These men and women have created and pioneered the vital fields of real science.

Are our incredible space adventures dependent on evolution? The conquest of space was derived out of physics and astronomy, but no trace of evolutionary theory enters into it. What about the medical sciences? Here scientists have gone deeply into physics, chemistry, and biology, but again we can discern nothing from evolution despite claims made. Is the fantastic revolution in communications derived from evolution? Again we see there is no connection whatever.

Astronomy and physics are close partners, but only when speculation and interpretation enter into it do we see attempts to relate evolutionary beliefs into such problems as the age of the universe and how the universe was formed. What about geology? There are two fields: practical applied field geology used in engineering, oil exploration, and mining, and second, historical geology. Evolution with respect to the first is wholly irrelevant, despite futile attempts to relate the so-called geologic column to oil discoveries. We must say that nothing would change in oil exploration if there were no theory of evolution.

Historical geology is the interpretation of the strata and fossil beds of the world. Here evolution theory is applied full tilt. Nevertheless the most biting critiques of such interpretations come from the scientific community. We have this wonderful one-line critique of historical geology by Reginald Daly: "Historical geology . . . is a record of events, most of which never took place, in time, much of which never existed."[5] Much of this book illustrates the many fallacies of interpreting the earth as evolutionists do.

Archaeology and anthropology rest on assumptions derived out of evolutionary theory, such as ape-to-man development and slow gradual change over vast eons of time. These interpretations come to us despite the evidence—not out of the evidence.

One could go on. What could evolution apply to chemistry, or mathematics, or physics? The answer of course is nothing. As a final example, take a close look at biology and all its related fields. The fields are only descriptive based on careful research. Despite all the tons of literature to the contrary, evolutionists can only speculate. This book shows how disastrously wrong this guesswork can be.

CONCLUSION

I believe this has been a very honest look at the world, at prehistory, and at ancient history. We find this world is simply not as described in texts and by the media in general. Such interpretations are based on a faulty foundation full of fallacies. With the Bible as the only true foundation and framework we gain an illuminating picture of life as it was, and as it now is. A most appropriate close to this discussion is found in Acts 17:24–28:

> The God Who made the world and everything in it is the Lord of heaven and earth and does not live in temples built by hands. And He is not served by human hands, as if He needed anything, because He Himself gives all men life and breath and everything else. From one man He made every nation of men, that they should inhabit the whole earth; and He determined the times set for them and the exact places where they should live. God did this so that men would seek Him and perhaps reach out for Him and find Him, though He is not far from each one of us. For in Him we live and move and have our being.

For Discussion and Reflection

1. Science focuses on the physical world. Religion recognizes both a physical and a spiritual world. In our view there are also futile attempts to harmonize these opposite views. Should we be surprised that most evolutionists do not want to have anything to do with spiritual matters? Why? What are some invisible things you believe in?

2. Over and over in this book you have seen aspects of the two frameworks within which ancient history is organized: Bible-based or evolution-based. When you read something of the ancient world that also treats something in the Bible, can you tell which framework is in use in the article? How?

3. Can you recall where evolution was taught as truth in school? In which subject(s) and in which level(s)? Were questions and discussion permitted?

4. Would you agree or disagree that evolution theory has a paralyzing hold on all forms of the media? How can we best cope with this situation?

5. Jesus commands us to love the Lord our God with our minds. What does that mean for the study of this book?

6. What is the wisdom for you in the advice given in Colossians 4:5–6: "Be wise in the way you act toward outsiders . . . so that you may know how to answer anyone."

7. What are some of the truths you have learned in this book to assist you in your discipleship?

ABBREVIATIONS

BAR   *Biblical Archaeology Review*
CRSQ   *Creation Research Society Quarterly*
CSSHQ *Creation Social Sciences and Humanities Quarterly*
TILN   *The Illustrated London News*
JVI   *Journal of the Victorian Institute*

NOTES

Foreword

1. George Bright, ed., *The Works of the Reverend and Learned John Lightfoot, D.D.* (London: William Rawlins, 1684), 1:title page.

Chapter 1

1. Alex Haley, Roots (New York: Dell, 1976).

2. William Corliss, "Archaeological Revisionism," *Science Frontiers* 118 (July 1998).

3. Richard Rudgley, *The Lost Civilizations of the Stone Age* (New York: Free Press, 1999), 239–40, 262.

4. Marvin Lubenow, *Bones of Contention: A Creationist Assessment of the Human Fossils* (Grand Rapids: Baker, 1992), 111.

5. Harry Kaufmann, *Introduction to the Study of Human Behavior* (Philadelphia: Saunders, 1968), 27–30.

6. Dan Smith, "Scientific Method," *Kansas City Star*, March 26, 1989.

7. Charles Officer and Jake Page, *The Great Dinosaur Extinction Controversy* (Reading, MA: Addison-Wesley, 1996), 9.

8. Joseph Alsop, *From the Silent Earth* (New York: Harper & Row, 1964), 3, 12.

9. Paul Johnson, "Points to Ponder," *Reader's Digest* (October 1996): 36.

10. James Lovelock, *The Ages of Gaia: A Biography of Our Living Earth* (New York: Norton, 1988), xiv.

11. John Dayton, *Minerals, Metals, Glazing and Man* (London: Harrap, 1978), 468.

12. Peter James, quoting John Dayton, "Book Review of John Dayton, 1978," *CRSQ* 17, no. 1 (June 1980): 81.

13. Søren Løvtrup, *Darwinism: The Refutation of a Myth* (New York: Croom Helm, 1987), passim.

14. Michael Wilson, *Applied Geology and Archaeology: The Holocene History of Wyoming* (Laramie: Geological Survey of Wyoming, 1974), ii–v.

15. David Childress, *Lost Cities of Ancient Lemuria and the Pacific* (Stelle, IL: Adventures Unlimited Press, 1988), 6.

16. Lubenow, *Bones of Contention*, 44.

17. Jean-Jacques Hublin, "Rise of the Hominids," *Archaeology* 50, no. 3 (May/June 1997): 71–74.

Chapter Two

1. Cyrus Gordon, *Before Columbus* (New York: Crown, 1971), 119.

2. Grahame Clark, *Mesolithic Prelude* (Edinburgh: University Press, 1980), passim.

3. Kenneth Macgowan and Joseph Hester Jr., *Early Man in the New World*, rev. ed. (Garden City, NY: Doubleday, 1962), passim.

4. Macgowan and Hester Jr., *Early Man*, 45.

5. Will Durant, *The Story of Civilization*, 11 vols. (New York: Simon & Schuster, 1954).

6. Nelson Glueck, *The River Jordan* (New York: McGraw-Hill, 1968), 214.

7. Glueck, *River Jordan*, 214.

8. Arthur Custance, "Primitive Cultures: Their Historical Origins," in *Genesis and Early Man*, vol. 2, *Doorway Papers* (Brockville, ONT: Doorway Papers, 1960), 49.

9. Macgowan and Hester Jr., *Early Man*, 45–115.

10. Macgowan and Hester Jr., *Early Man*, 36–38.

11. James Griffin, ed., "Hopewell Chronology," *UMAP, Anthropological Papers 12* (Ann Arbor, MI: University of Michigan, 1958).

12. Michael Day, *Fossil Man* (New York: Bantam, 1970), 150–51.

13. Day, *Fossil Man*, 150–51.

14. Donovan Courville, "Evolution and Archaeological Interpretation," *CRSQ* 11, no. 1 (June 1974): 51–52.

15. Macgowan and Hester Jr., *Early Man*, 33–34.

16. Immanuel Velikovsky, "Metallurgy and Chronology," *Pensee* 3, no. 3 (Fall 1973): 5.

17. Byron Nelson, *Before Abraham: Prehistoric Man in Biblical Light* (Minneapolis: Augsburg, 1948), 18.

18. Nelson, *Before Abraham*, 18.

19. Nelson Glueck, "The Bible as Divining Rod," *Horizon* 2, no. 2 (November 1959): 9.

20. Cyrus Gordon, *Riddles in History* (New York: Crown, 1974), 146–47.

21. Geoffrey Bibby, *Looking for Dilmun* (New York: Knopf, 1970), 219, 281, 308.

22. J. Magens Brown, "The Dawn of Metallurgy," *JVI* 23 (1989): 303.

23. Immanuel Velikovsky, *Worlds in Collision* (Garden City, NY: Doubleday, 1950), 30.

24. Velikovsky, "Metallurgy and Chronology," 7.

25. William Perry, *The Children of the Sun* (London: Methuen, 1923), 391.

26. James Frazer, *The New Golden Bough*, ed. Theodor Gaster (New York: Criterion Press, 1959), 174–76.

27. Giorgio de Santillana and Hertha von Dechend, *Hamlet's Mill* (Boston: Gambit, 1969), 119.

28. Frank Edwards, *Strange World* (New York: Ace, 1964), 146–47.

29. Arthur Custance, "Longevity in Antiquity, and Its Bearing on Chronology," no. 2 (Brockville, ONT: Doorway Papers, 1957); William L. Thomas, ed., *Yearbook of Anthropology—1955* (New York: Anthropological Research Inc., 1955), 1:629; Erich von Daniken, *The Gold of the Gods* (New York: Putnam, 1973), 175; Robert Dyson, "The Archaeological Evidence of the Second Millennium B.C. on the Persian Plateau," chapter 16 in *The Cambridge Ancient History*, vol. 2, rev. ed. (Cambridge: Cambridge University Press, 1968), 20.

30. Andrew Tomas, *We Are Not the First* (New York: Bantam, 1971), 93; Arthur Custance, "Longevity in Antiquity," 40–41; Arthur Custance, "The Technology of Hamitic People," no. 43 (Brockville, ONT: Doorway Papers, 1960), 17, 27–29; Gordon, *Before Columbus*, 53–55; Cyril Aldred, *Egypt to the End of the Old Kingdom* (New York: McGraw-Hill, 1965), 93.

31. Henry Field, comp., *Contributions to the Anthropology of Saudi Arabia* (Coconut Grove, FL: Field Research Projects, 1971), 28.

32. James D. Muhly, "How Iron Technology Changed the Ancient World," *BAR* 8, no. 6 (November/December 1982): 43; Erich von Daniken, *Chariots of the Gods?* (New York: Bantam, 1969), 44.

33. Custance, "Technology of Hamitic People," 17.

34. James Mellaart, *Catal Huyuk* (New York: McGraw-Hill, 1967), 11, 131; Custance, "Primitive Cultures," 2:54.

35. Rene Noorbergen, *Treasures of the Lost Races* (Indianapolis: Bobbs-Merrill, 1982), 70; Robert Charroux, *Forgotten Worlds* (New York: Popular Library, 1971), 64; Alan and Sally Landsburg, *In Search of Ancient Mysteries* (New York: Bantham, 1974), 21.

36. *TILN*, April 15, 1973, 27–29.

37. Perry, *Children of the Sun*, 47–52, 87–93.

38. M. Rafique Mughal, "New Evidence of the Early Harappan Culture," *Archaeology* 27, no. 2 (April 1974): 110–112.

39. J. E. Howard, "On the Early Dawn of Civilisation," *JVI* 9 (1873): 258; Cyclone Covey, "Chinese Flood," *Pensee* 2, no. 3 (Fall 1972): 44.

40. Judith Treistman, *The Prehistory of China* (New York: Doubleday, 1972).

41. Perry, *Children of the Sun*, 83–85.

42. Alfred de Grazia, et al., *The Politics of Science and Dr. Velikovsky* (New York: American Behavioral Scientist, 1963), 40; Immanuel Velikovsky, *Earth in Upheaval* (New York: Dell, 1955), 279; *TILN*, April 15, 1973, 27–29.

43. Day, *Fossil Man*, 150–51; I. E. S. Edwards, "The Early Dynastic Period in Egypt," chapter 11 in *The Cambridge Ancient History*, vol. 1., rev. ed. (Cambridge: Cambridge University Press, 1964), 62; Aldred, *Egypt to the End of the Old Kingdom*, 93.

44. Aldred, *Egypt to the End of the Old Kingdom*, 57–59.

45. J. William Dawson, "Useful and Ornamental Stones of Ancient Egypt" *JVI* 26 (1892): 277.

46. Gordon, *Before Columbus*, 53–55.

47. William Corliss, comp., *Strange Artifacts*, vol. M-1 (Glen Arm, MD: Author, 1974), MMT-005; *JVI* 9 (1875): 274–75.

48. Anonymous, "Africa's Ancient Steelmakers," *Time*, September 25, 1978, 80.

49. *JVI* 9 (1875): 262; *JVI* 19 (1883): 142.

50. Corliss, *Strange Artifacts*, vol. M-1, MMT-004.

51. Adrian Bosher, "Swaziland: A Birthplace of Modern Metal, *Science Digest* 73, no. 3 (March 1973): 42–47.

52. *JVI* 13 (1877): 217.

53. Thomas, *Yearbook of Anthropology—1955*, 629.

54. Velikovsky, *Earth in Upheaval*, 179.

55. Glyn Daniel, *The Megalith Builders of Western Europe* (Baltimore: Penguin, 1962), 59–71.

56. *Ann Arbor News*, May 1, 1969.

57. Peter Kolosimo, *Not of this World* (New York: Bantam, 1971), 143.

58. James Bailey, *The God-Kings and the Titans* (London: Hodder and Stoughton, 1973), 27; James Bailey, *Sailing to Paradise* (London: Hodder and Stoughton, 1994), passim.

59. See Bailey, *Sailing to Paradise*; Henriette Mertz, *The Wine Dark Sea* (Chicago: Author, 1964); Carroll Riley, ed., *Man Across the Sea: Problems of Pre-Columbian Contacts* (Austin: University of Texas Press, 1971); Barry Fell, *America B.C. Ancient Settlers in the New World* (New York: Quadrangle, 1976); Thor Heyerdahl, *Early Man and the Ocean* (New York: Vintage, 1978);

60. Betty Sodders, *Michigan Prehistory Mysteries* (Au Train, MI: Avery, 1990), 12–45; Betty Sodders, *Michigan Prehistory Mysteries II* (Au Train, MI: Avery, 1991), 41–42.

61. Riley, *Man Across the Sea*, 285.

62. Pierre Honore, *In Quest of the White God* (New York: Putnam, 1964), passim.

63. Gordon, *Before Columbus*, 145–48.

64. Constance Irwin, *Fair Gods and Stone Faces* (New York: St. Martin's, 1963).

65. Gordon, *Before Columbus*, 204.

66. Macgowan and Hester Jr., *Early Man*, 32.

67. D. Southwall, "Pliocene Man in America," *JVI* 15 (1882): 196–210.

68. J. E. Howard, "On the Druids and Their Religion," *JVI* 14 (1881): 92; Paul Craig, "Lead, the Inexcusable Pollutant," *Saturday Review,* October 2, 1971, 68–75.

69. Southwall, "Pliocene Man," 196–199; Perry, *Children of the Sun*, 64–66.

70. Edward Lane and Robert Gentet, "The Case for the Calaveras Skull," *CRSQ* 33, no. 4 (March 1997): 248–256.

71. Riley, *Man Across the Sea*, 18–19.

72. *Popular Archeology*, January 1, 1973, 22; Edwards, *Strange World*, 110; George Fuller, ed., *Historic Michigan*, vol. 1 (National Historical Association, 1924), 36–39; *Ann Arbor News*, September 19, 1968.

73. *Saturday Review*, October 3, 1964, 53; Macgowan and Hester Jr., *Early Man*, 192; *Popular Archeology*, December 8, 1972, 20–25.

74. *Saturday Review*, October 3, 1964, 53; Poverty Point, LA., Undated. Baton Rouge LA: Tourist Commission.

75. Macgowan and Hester Jr., *Early Man*, 236.

76. H. M. Wormington, *Ancient Man in North America* (Denver: Denver Museum of Natural History, 1957), 150.

77. Perry, *Children of the Sun*, 59–60; Edward Vining, *An Inglorious Columbus* (New York: Appleton, 1885), 57.

78. Landsburg, *In Search of Ancient Mysteries*, 30; Charles Berlitz, *Mysteries From Forgotten Worlds* (New York: Dell, 1972); Joel W. Grossman, "An Ancient Gold Worker's Tool Kit," *Archaeology* 25, no. 4 (October 1972): 273–75.

79. von Daniken, *Gold of the Gods*, 103.

80. Macgowan and Hester Jr., *Early Man*, 32.

Chapter Three

1. Lorella Rouster, "Beowulf: Creationist Implications in Our Earliest English Epic," *CRSQ* 14, no. 4 (March 1978): 221–222; "The Pivotal Importance of Dinosaurs in Creationist and Evolutionist Thought," *CSSHQ* 3, no. 4 (Summer 1981): 1 ; Rouster, "The Footprints of Dragons," *CSSHQ* 8, no. 1 (Fall 1985): 25–30.

2. Gordon Lindsay, *The Dinosaur Dilemma* (Dallas: Christ For The Nations, 1982).

3. John Morris, "How do Dinosaurs Fit In?" *Institute for Creation Research: Back to Genesis* (May 1989).

4. Anonymous, "What Happened to the Dinosaurs?" *Awake* (Brooklyn), February 8, 1990.

5. Norman Hafley, "What About Dinosaurs?" *Bible-Science News* 30, no. 3 (1992): 1.

6. Erich von Fange, *Genesis and the Dinosaur*, rev. ed. (Adrian, MI: Living Word Services, 1995).

7. Associated Press, "Fossils Suggest Birds Didn't Evolve from Dinosaurs," *Ann Arbor News*, November 15, 1996.

8. Mary Lord, "The Piltdown Chicken," *U.S. News & World Report*, February 14, 2000, 53.

9. Associated Press, "Inventor Unearths Prehistoric Firsts," *The Argus*, August 4, 1996.

10. Don DeYoung and John Meyer, "Dinosaur Update," *CRSQ* 28, no. 3 (December 1991): 117–121.

11. Glennda Chui, "Jurassic Jaws," *Houston Chronicle*, August 23, 1996.

12. Carl Sagan, *The Dragons of Eden* (New York: Random House, 1977), 142.

13. William Stokes, "Creationism and the Dinosaur Boom," *Journal of Geological Education* 37, no. 1 (January 1989): 24–26.

14. Paul Taylor, *The Great Dinosaur Mystery and the Bible* (El Cajon, CA: Master Books, 1987).

15. Michael Oard, "Response to Comments on the Asteroid Hypothesis," *CRSQ* 31, no. 1 (June 1994): 12.

16. Eugene Chaffin, "Virginia Triassic Basins," *CRSQ* 31, no. 2 (September 1994): 125–26.

17. Bolton Davidheiser, "The Hip Bones of Dinosaurs and Birds," *CRSQ* 31, no. 3 (December 1994): 136.

18. Michael Oard, "Polar Dinosaurs and the Genesis Flood," *CRSQ* 32, no. 1 (June 1995): 47–56.

19. Paul Garner, et al, "Comment on Polar Dinosaurs and the Genesis Flood," *CRSQ* 32, no. 4 (March 1996): 232–34.

20. Todd Wood, "Plesiosaur of Basking Shark?" *CRSQ* 33, no. 4 (March 1997): 292–95.

21. Carl Froede Jr., "Book Review," *CRSQ* 34, no. 1 (June 1997): 61.

22. Charles Officer and Jake Page, *The Great Dinosaur Extinction Controversy* (Reading, MA: Addison-Wesley, 1996).

23. Jeremy Auldaney, et al, "Human-like Track Impressions Found," *CRSQ* 34, no. 2 (September 1997): 115–27; Jeremy Auldaney, "More Human-like Track Impressions," *CRSQ* 34, no. 3 (December 1997): 133.

24. Tom McIver, *Anti-Evolution: An Annotated Bibliography* (Jefferson, NC: McFarland, 1988).

25. Erich von Fange, *Noah to Abram* (Adrian, MI: Living Word Services, 1994).

Chapter 4

1. "Libellus of the Decian persecution," Michigan Papyri, Inv. 263 (Ann Arbor: University of Michigan), 134.

2. Mimi Mann, "Oldest Bound Book Displayed," *Houston Chronicle*, September 13, 1992.

3. Michael Coogan, "10 Great Finds," *BAR* 21, no. 3 (May/June 1995): 36–47; "Trial of the Century," *Archaeology* 58, no. 2 (March 2005): 14; Eric Meyers, "Forgery Fallout," *Archaeology* 58, no. 2 (March 2005): 16.

4. Seymour Gitin, et all, "Ekron Identity Confirmed," *Archaeology* 51, no. 1 (January 1998): 28.

5. Michael Drosnin, *The Bible Code* (New York: Simon & Schuster, 1997).

6. Keith Schoville, *Biblical Archaeology in Focus* (Grand Rapids: Baker, 1978), 97.

7. Kenneth Kitchen, *Ancient Orient and Old Testament* (Chicago: InterVarsity, 1966), 26.

8. J. Maxwell Miller, "Approaches to the Bible through History and Archaeology," *Biblical Archaeologist* 45, no. 4 (Fall 1982): 213.

9. Paul Lapp, "Palestine Known but Mostly Unknown," *Biblical Archaeologist* 26, no. 4 (December 1963): 121.

10. Schoville, *Biblical Archaeology in Focus,* 121.

11. Kenneth Kitchen, *The Bible in its World* (Downers Grove, IL: InterVarsity, 1977), 12.

12. Kitchen, *Ancient Orient*, 23.

13. George Mendenhall, Class lecture (Ann Arbor: University of Michigan, 1981).

14. Willem van Hattem, "Once Again: Sodom and Gomorrah," *Biblical Archaeologist* 44, no. 2 (Spring 1981): 87–92.

15. Larry Herr, "The Search for Biblical Heshbon," *BAR* 19, no. 6 (November/December 1993): 36.

16. Yigael Yadin, "Pottery Dating Problems," *Biblical Archaeologist* 6, no. 1 (March 1980): 1.

17. Adam Zertal, "Has Joshua's Altar Been Found on Mt. Ebal?" *BAR* 11, no. 1 (January/February 1985): 26.

18. Aharon Kempinski, "Joshua's Altar," *BAR* 12, no. 1 (January/February 1986): 42.

19. Richard Ostling, "New Grounding for the Bible?" *Time*, September 21, 1981, 76.

20. Bill Cooper, *After the Flood* (Chichester, England: New Wine Press, 1995).

21. I. Rapaport, "The Flood Story in Bible and Cuneiform Literature," *Bible and Spade* 12, no. 3–4 (Summer 1983): 57–65.

22. Charles Reed, ed., *Origins of Agriculture* (The Hague: Mouton, 1986), 57.

23. Reed, *Origins of Agriculture*, 57.

24. "Archaeologists Discover Wine Before Its Time," *Ann Arbor News*, June 6, 1986.

25. David McCrery, "Bab edh Drha," *Jordan Times*, March 1, 1978.

26. William Shea, "Two Palestinian Segments from the Eblaite Geographic Atlas," in Carol Meyers and Michael O'Connor, *The Word of the Lord Shall Go Forth* (Winona Lake, IN: Eisenbrauns, 1983), 589–612.

27. Hershel Shanks, "Three Shekels for the Lord," *BAR* 23, no. 2 (November/December 1997): 28.

28. Anonymous, "Seal of Ahaz," *BAR* 23, no. 2 (March/April 1997): 8.

29. Coogan, "10 Great Finds," 45.

30. Peter James, "A Response to Williams," *Biblical Archaeologist* 44, no. 2 (Spring 1981): 69.

31. William Albright, *The Archaeology of Palestine* (Gloucester, MA: Peter Smith, 1971), 84.

32. Anson Rainey, "Historical Geography," *Biblical Archaeologist* 45, no. 4 (1982): 217–223.

33. Adnan Hadidi, *Annual* (Amman, Jordan: Department of Antiquities, XV, 1970), 11.

34. Peter James, "A Response to Williams," *Biblical Archaeologist* 44, no. 2 (Spring 1981): 69–71.

35. Immanuel Velikovsky, *Stargazers and Gravediggers* (New York: Morrow, 1983).

36. David Rohl, *Pharaohs and Kings* (New York: Crown, 1995).

37. Peter James, *Centuries of Darkness* (New Brunswick, NJ: Rutgers, 1993).

38. Letter to the Editor, "Assessing the Evidence," *BAR* 23, no. 4 (July 1997): 8.

39. David Henige, "Comparative Chronology and the Middle East," *Bulletin of the American Schools of Oriental Research #261* (February 1986): 57–68.

40. John Garstang, *The Story of Jericho* (London, 1948).

41. Kathleen Kenyon, *Digging Up Jericho* (London: Ernest Benn, 1957), 266.

42. Bryant Wood, "Did the Israelites Conquer Jericho?" *BAR* 16, no. 2 (March/April 1990): 44–59.

43. Svi Gal, et al., "Death in Peqi'in," *Odyssey* 1, no. 4 (Fall 1998): 54.

44. J. Maxwell Miller, "Approaches to the Bible," 213.

45. Hershel Shanks, "The Sad Case of Tell Gezer," *BAR* 9, no. 4 (July 1983): 30–42.

46. Willem van Hattem, "Once Again: Sodom and Gomorrah," *Biblical Archaeologist* 44, no. 2 (1981): 87–92.

47. Grahame Clark, *Mesolithic Prelude* (Edinburgh: University Press, 1980).

48. Khair Yassine, "The Dolmens Reconsidered," *Bulletin of the American Schools of Oriental Research* no. 259 (1985): 68.

49. John Dayton, *Minerals, Metals, Glazing and Man* (London: Harrap, 1978), 238.

50. Rudolph Cohen, "The Iron Age Fortresses in the Central Negev," *Bulletin of the American Schools of Oriental Research* no. 239 (1979): 61–79.

51. Eva Danelius, "The Identification of the Biblical Queen of Sheba," *Kronos* 1, no. 4 (Winter 1976): 3.

52. J. Maxwell Miller, *Biblical Archaeologist* (1980): 133.

53. John Bimson, *Redating the Exodus and Conquest* (Sheffield, England: University of Sheffield, 1978).

54. Peter James, *Centuries of Darkness* (New Brunswick, NJ: Rutgers University Press, 1981).

55. Siegfried Horn, "What We Don't Know about Moses and the Exodus," *BAR* 3, no. 2 (March 1977): 22–331.

56. David Henige, "Comparative Chronology and the Near East," *Bulletin of the American Schools of Oriental Research* #261 (February 1986): 57.

57. Hershel Shanks, "Rx for ASOR," *BAR* 24, no. 2 (March/April 1998): 6.

58. James Moyer and Victor Matthews, "The Use and Abuse of Archaeology in Current Bible Handbooks," *Biblical Archaeologist* 48, no. 3 (September 1985): 149–59.

59. Henry Halley, *Halley's Bible Handbook* (Grand Rapids: Zondervan, 1965, 2000).

60. Edward Blair, *The Abingdon Bible Handbook* (Nashville: Abingdon, 1975).

61. James Packer and others, eds., *The Bible Almanac* (Nashville, Abingdon, 1980).

62. Moyer and Matthews, "Use and Abuse of Archaeology," 227–37.

63. J. D. Douglas, ed., *New Bible Dictionary* (Wheaton: Tyndale, 1982).

64. Kitchen, *Bible in Its World.*

65. Society of Biblical Literature, Annual Meeting program (Toronto, 1969).

66. Transcript, "Face to Face" debate between minimalists Niels Lemche and Thomas Thompson vs. humanists William Dever and P. Kyle McCarter, *BAR* 23, no. 4 (July 1997): 26.

67. John Currid, *Ancient Egypt and the Old Testament* (Grand Rapids: Baker, 1997).

68. Martin Noth, *The History of Israel* (New York: Harper, 1958).

69. "Letter to the Editor," *BAR* 7, no. 1 (January/February 1981): 18.

70. George Grinnell, "The Origins of Modern Geological Theory," *Kronos* 1, no. 4 (Winter 1976): 68–76.

71. London Observer, "British Scientists Lend Support to Noah's Ark Naysayer," *The Argus,* September 10, 1997.

72. William Corliss, *Scientific Anomalies and Other Provocative Phenomena* (Glen Arm, MD: The Sourcebook Project, 2003).

73. Anson Rainey, "Historical Geography—The Link Between Historical and Archaeological Interpretation," *Biblical Archaeologist* 45, no. 4 (1982): 217.

74. Hershel Shanks, "Clumsy Forger Fools the Scholars—But Only for a Time," *BAR* 10, no. 3 (May/June 1984): 66–72.

75. Charles Pellegrino, *Return to Sodom and Gomorrah* (New York: Avon, 1994), 156.

76. Walter Keller, *The Bible as History,* rev. ed. (London: Hodder and Stoughton, 1980).

77. Pellegrino, *Return to Sodom and Gomorrah,* 230.

78. Schoville, *Biblical Archaeology in Focus,* 97.

79. Paul Wiseman, *New Discoveries in Babylonia about Genesis* (London: Marshall, 1949).

80. William Stiebing, "Should the Exodus and the Israelite Settlement in Canaan be Redated?" *BAR* 11, no. 4 (November 1985): 58–69.

81. "Letter to the Editor" *BAR* 6, no. 6 (November/December 1980): 10.

82. George Mendenhall, *The Tenth Generation* (Baltimore: The John Hopkins University Press, 1973), 4.

83. Kenneth Kitchen, *Ancient Orient and the Old Testament* (Chicago: InterVarsity, 1966), 133.

Chapter 5

1. E. Walter Maunder, "Joshua's Long Day" *JVI* 53 (1921): 137–138.

2. Immanuel Velikovsky, *Worlds in Collision* (Garden City, NY: Doubleday, 1950), 45, 236, 237, 307.

3. C. J. Ransom, *The Age of Velikovsky* (Glassboro, NJ: Kronos, 1976), 237.

4. C. A. Totten, *Joshua's Long Day* (New York: Destiny Publishers, 1890, 1968).

5. Harry Rimmer, *The Harmony of Science and Scripture* (Grand Rapids: Eerdmans, 1936), 281–283.

6. Sidney Collett, *All About the Bible* (New York: Fleming Revell, 1933), 285–286.

7. Harold Hill, *How to Live Like a King's Kid* (Plainfield, NJ: Logos, 1974), 65–77.

8. V. L. Westberg, *The Master Architect* (Napa, CA: Author) undated.

9. Don DeYoung, "Is There an 'Empty Place' in the North?" *CRSQ* 23, no. 3 (December 1986): 129–130.

10. V. L. Westberg, *Personal Correspondence* (August 23, 1974).

11. See *Bible-Science News*, April 1970 through December 1970.

12. Robert Oden, "Joshua's Long Day" *Bible-Science News* 16, no. 5 (May 1978) 1.

13. Ransom, *Age of Velikovsky*, 237, 262–263.

14. Hill, *King's Kid*, 1970.

Chapter 6

1. Edwin Thiele, *The Mysterious Numbers of the Hebrew Kings*, 3rd edition (Grand Rapids: Zondervan, 1983).

2. Willem van Hatten, "Once Again: Sodom and Gomorrah," *Biblical Archaeologist* 44, no. 2 (Spring 1981): 70, 92.

3. James Ussher, *Annals of the Old Testment*, vol. 1 (London, 1650).

4. Homer Smith, *Man and His Gods* (Boston: Little, Brown, 1952), 324.

5. James Ussher, *Annales Veteris et Novi Testamenti*, 3 volumes (London, 1650, 1654).

6. Charles Officer and Jake Page, *The Great Dinosaur Extinction Controversy* (Reading, MA: Addison-Wesley, 1996), 48.

7. David Wallechinsky and Irvine Wallace, *The People's Almanac #2* (New York: Bantam, 1978), 634.

8. George Simpson, *Life of the Past* (New York: Bantam, 1953), 30.

9. Anonymous, "Earth Born 5,981 Years Ago?" *Ann Arbor News*, October 23, 1977.

10. F. Clark Howell, et al, *Early Man* (New York: Time-Life Books, 1973), 10; Alfred de Grazia, *Chaos and Creation* (Princeton, NJ: Metorn Publications, 1981), 2; David Watson, "Time and Ancient Records," *CRSQ* 18, no. 1 (June 1981): 30; Michael Banton, ed., *Darwinism and the Study of Society* (Chicago: Quadrangle Books, 1961), 44; George Simpson, *Life of the Past* (New York: Bantam, 1953), 30; William Jaber, *Whatever Happened to the Dinosaurs?* (New York: Julian Messner, 1978), 21; Richard Pourade, ed., *Ancient Hunters of the Far West* (San Diego: Union-Tribune), 131; V. Axel Firsoff, "Another Chronology Proposal," *S.I.S.* 5, no. 2 (1980–1981): 58.

11. James Ussher, *Annals of the Old Testament* (London: 1658), 1:23.

12. R. Buick Knox, *James Ussher, Archbishop of Armagh* (Cardiff: University of Wales, 1967), 105.

13. Andrew White, *A History of the Warfare of Science with Theology in Christendom* (New York: Appleton, 1955), 256.

14. Christopher Cerf and Victor Navasky, *The Experts Speak* (New York: Pantheon, 1984), 3.

15. Ruth Moore, *Man, Time, and Fossils* (New York: Knopf, 1953), 241.

16. Harold Gladwin, *Men Out of Asia* (New York: Whittlesey House, 1947), 65.

17. W. Raymond Drake, *Gods and Spacemen of the Ancient Past* (New York: Signet, 1974).

18. John Klotz, "Bishop Lightfoot and the Exact Hour of Creation," *CRSQ* 23, no. 4 (March 1987): 173.

19. White, *History of the Warfare of Science*, 256.

20. George Bright, ed., *The Works of the Reverend and Learned John Lightfoot D.D.* (London: William Rawlins, 1684), 1:692, 1020–1021, 1322, 1324.

21. Hudson Tuttle, *Physical Man* (Boston: Colby & Rich, 1865), 64.

22. Harold Slusher, *Critique of Radiometric Dating* (San Diego: ICR, 1973), 2.

23. Cyrus Gordon, *Riddles in History* (New York: Crown, 1974), 20.

24. Alfred de Grazia, *Chaos and Creation* (Princeton, NJ: Metron, 1981), 42. See John Woodmorappe, *The Mythology of Modern Dating Methods* (El Cajon, CA: ICR, 1999) for an extensive evaluation of radiometric dating.

25. Julius Fraser, *The Voices of Time* (New York: Braziller, 1966), 336.

26. Marvin Lubenow, *Bones of Contention: A Creationist Assessment of the Human Fossils* (Grand Rapids: Baker, 1992), 266.

27. Douglas Whitney, "The Age of the Earth" *JVI* 65 (1933): 37.

28. E. K. Victor Pearce, *Who Was Adam?* (Toronto: Paternoster Press, 1969), 33.

29. Kenneth Macgowan and Joseph Hester Jr., *Early Man in the New World*, rev. (Garden City, NY: Anchor Books, 1962), 43.

30. Andrew Tomas, *We Are Not the First* (New York: Bantam, 1971), 57.

31. Anonymous, "Earth Born 5,981 Years Ago?" *Ann Arbor News*, October 23, 1977.

32. Robert Charroux, *Masters of the World* (New York: Berkley, 1974), 49.

33. Marilyn Taylor, "Descent," *Arizona Highways* 69, no.1 (January 1993): 10–11.

34. Carl Zimmer, "How Old Is It?" *National Geographic* 200, no. 3 (September 2001): 78; Woodmorappe, *Mythology of Modern dating Methods*.

35. Bill Cooper, *After the Flood* (Chichester, England: New Wine Press, 1995).

36. Donald DeYoung and John Whitcomb, "The Origin of the Universe," *CRSQ* 18, no. 2 (September 1981): 86.

Chapter 7

1. Josh Fishman, "Tiny Survivors," *U.S. News & World Report*, November 8, 2004, 70.

2. Associated Press, "4-Million-year-old Fossil Believed to be 1st Walking Hominid," *The Toledo Blade*, March 6, 2005.

3. Johanna Rajca, "Not a Leg to Stand On," *Origins Issues, ICR* (September 2001): 10.

4. Marvin Lubenow, *Bones of Contention: A Creationist Assessment of the Human Fossils* (Grand Rapids: Baker, 1992), 206.

5. "New Debate on Lucy," *The Argus*, June 16, 1983, 35.

6. AMHS, "Biggest A. *boisei* Cranium," *Archaeology* (January 1998): 22.

7. Donald Johanson and Maitland Edey, *Lucy: The Beginnings of Humankind* (New York: Warner Books, 1981), 37.

8. Wilfred Le Gros Clark, *History of the Primates*, 5th ed. (Chicago: University of Chicago Press, 1966), passim; Philip Kitcher, *Abusing Science* (Cambridge, MA: MIT Press, 1982), 55–81.

9. William Fix, *The Bone Peddlers* (New York: Macmillan, 1984).

10. Johanson and Edey, *Lucy*, 276.

11. John Morris, "Did the 'African Eve' Leave Footprints?" *Back to Genesis*, no. 106 (October 1997): 4.

12. John Carlton, "The Great Primate Debate," *Christian News*, July 29, 1996.

13. New York Times, " 'Lucy' May Be Younger than 3.6 Million Years," *Ann Arbor News*, December 12, 1982.

14. Lubenow, *Bones of Contention*, 206.

15. Fix, *Bone Peddlers*.

16. Anomymous, "*Homo erectus* Never Existed?" *Geotimes* 37, no. 11 (October 1992): 112.

17. Fix, *Bone Peddlers*, passim.

18. Fix, *Bone Peddlers*, passim.

19. George Norrie, "New 'Missing Link' Clues Discovered in African Jungle," *San Francisco Examiner*, December 3, 1950.

20. Los Angeles Times, "Human Family Tree More Like a Brambly Bush, *Toledo Blade*, April 18, 2001.

21. Jacob Gruber, "The Neanderthal Controversy," *The Scientific Monthly* 67, no. 12 (December 1948): 436–39.

22. D. Dewar, "Theories: Origin of Organisms" *JVI* 76 (1944): 53.

23. Stewart Easton, *The Western Heritage*, 2nd ed. (New York: Holt, Rinehart, Winston, 1966).

24. Ernst Haeckel, *The Riddle of the Universe* (New York: Harper, 1900), 42–86.

25. Michael Pitman, *Adam and Evolution* (London: Rider, 1984), 89.

26. Arthur Custance, "The Place of Handicaps in Human Development," *Doorway Papers* #9 (Brockville, ONT: Doorway Papers, 1957), 5–6.

27. Alvin Josephy, ed., *The American Heritage Book of Indians* (New York: American Heritage, 1961).

28. J. Prestwich, and others, "A Possible Cause for the Origin of the Tradition of the Flood," *JVI* 27, no. 108 (1893): 296–297.

29. Michael Pitman, *Adam and Evolution* (London: Rider, 1984), 100.

30. Charles Blinderman, "The Curious Case of Nebraska Man," *Science* 85 (June 1985): 47.

31. Pitman, *Adam and Evolution*, 100.

32. Donald Johanson, "Ethiopia Yields First 'Family' of Early Man," *National Geographic* 150, no. 6 (December 1976): 798.

33. Johanson and Edey, *Lucy*, 21.

34. Rachele Kanigel, "Big Troubles for Discoverer of 'Lucy' " *The Argus*, April 9, 1995; Donald Johanson, "Lucy Turns Thirty," *Archaeology* (November 2004): 16.

35. Johanson and Edey, *Lucy*, 51.

36. Marlene Cimons, "Human Skeleton Unearthed in Egypt," *Ann Arbor News*, April 8, 1983.

37. Johanson and Edey, *Lucy*, 229.

38. Bill Dietrich, "Archaeologist Thinks Tools Date Back 2 Million Years," *Omaha World Herald*, May 1, 1994.

39. Roger Lewin, *Bones of Contention: Controversies in the Search for Human Origins* (New York: Simon & Schuster, 1987): 305, 319.

40. Fix, *Bone Peddlers*, 124.

41. Johanson and Edey, *Lucy*, 146.

42. Jack Cuozzo, *Buried Alive: The Startling Truth about Neanderthal Man* (Green Forest, AR: Master Books, 1999).

43. Brian Fagan, *The Journey from Eden: The Peopling of Our World* (New York: Thames & Hudson, 1990).

44. G. K. Chesterton, "Outlines of History," *TILN*, January 13, 1923.

45. Johanson and Edey, *Lucy*, 178.

46. Stanley L. Jaki, *The Road of Science and the Ways to God* (Edinburgh: Scottish Academic Press, 1978), 301.

47. Jerry Bergman, "Evolution, Race, and Equality of Intelligence," *CRSQ* 17, no. 2 (September 1980): 129–130.

48. Johanson and Edey, *Lucy*, 164–168, 198.

49. Lubenow, *Bones of Contention*, 15–16, 264, passim.

50. William Corliss, comp., *Ancient Man: A Handbook of Puzzling Artifacts* (Glen Arm, MD: Sourcebook Project, 1978), 636.

51. Corliss, *Ancient Man*, 636–90.

52. Michael Cremo and Richard Thompson, *Forbidden Archaeology: The Hidden Mystery of the Human Race* (San Diego: Bhaktivedanta Institute, 1993).

53. Theodore Graebner, *God and the Cosmos* (St. Louis: Concordia, 1932), 313.

54. Anonymous, "Transitions in Prehistory," *Science*, November 20, 1998, 1452.

55. G. Frederick Wright, *Origin and Antiquity of Man* (Oberlin, OH: Bibliotheca Sacra, 1912).

56. Robert Gentet "Calaveras Man" *CRSQ* 27, no 4. (March 1991): 122–27.

57. Bill Cooper, "Human Fossils from Noah's Flood," *ExNihilo* 1, no. 3 (1983): 6–9.

58. Philip Johnson, *The Wedge of Truth* (Downer's Grove, IL: InterVarsity, 2000).

59. Anonymous, "Were We Planted Here?" *Time*, September 10, 1973, 53.

Chapter 8

1. Marvin Lubenow, *Bones of Contention: A Creationist Assessment of the Human Fossils* (Grand Rapids: Baker, 1992), 44.

2. Anonymous, "News and Views: Piltdown," *Faith and Thought* 106, no. 1 (1979): 9–10.

3. John Waechter, *Prehistoric Man* (London: Octopus Books Ltd., 1977), 42.

4. Ruth Moore, *Man, Time, and Fossils* (New York: Knopf, 1953), 341–343.

5. Preson Phillips, "More Reluctance to 'Retire' Piltdown Man," *CRSQ* 13, no. 2 (September 1976): 126; Chester Reeds, *The Histomap of Evolution* (New York: Rand McNally, 1960).

6. Anonymous, "The Ape-like Progenitor of Man," *Current Opinion* 55, no. 11 (1913): 333–34.

7. Malcolm Bowden, *Ape-Men—Fact or Fallacy?* (Bromley, Kent: Sovereign Publications, 1977), 5–7, 29.

8. Rupert Furneaux, *The World's Strangest Mysteries* (New York: Ace, 1961), 117–128.

9. Frederic Brewer, "Artifacts or Geofacts," *Science* 181, no. 1202 (September 28, 1973): 1202.

10. Malcolm Browne, "Piltdown Fossils Stir New Debate," *New York Times*, March 27, 1979.

11. Furneaux, *World's Strangest*, 117–128; Ronald Millar, *The Piltdown Men*, (London: Paladin, 1972), 199; L. Harrison Matthews, "Piltdown Man," *Faith & Thought*, 108, no. 1 (1981): 24.

12. Millar, *Piltdown Men*, 96–114; Arthur Keith, "New and Views: Piltdown," *Faith & Thought* 106, no. 1 (1979): 9–11.

13. "England's most Ancient Inhabitant," *Review of Reviews* 47 (February 1913): 229–30.

14. "Piltdown," *Current Opinion* 55, no. 12 (December 1913): 421–22.

15. Adolf Rieth, *Archaeological Fakes*, trans. Diana Imber (New York: Praeger, 1967), 41.

16. Clyde Kluckhohn, *Mirror for Man* (New York: Fawcett, 1944): 63.

17. Harold Gladwin, *Men Out of Asia* (New York: Whittlesey House, 1947), 31.

18. Moore, *Man, Time, and Fossils*, 258–59; "Piltdown," *Current Opinion* 55, no. 12 (December 1913): 421–22.

19. Arthur Keith, "News and Views: Piltdown," *Faith & Thought* 106 (January 1979): 10–11.

20. Anonymous, "Teilhard in the Trenches," *Time*, April 14, 1975, 47

21. Pierre Teilhard de Chardin, *The Appearance of Man* (New York: Harper & Row, 1965), preface; Lubenow, *Bones of Contention*, 134–139; A. G. Tinley, "Reflections on Piltdown Man," *Bible-Science Newsletter* 10, no. 11 (November 1972): 4.

22. Pierre Teilhard de Chardin, *The Phenomenon of Man* (New York: Harper, 1959), 53.

23. Teilhard, *Phenomenon of Man*, 25.

24. Teilhard, *Phenomenon of Man*, 57.

25. Anonymous, "Fresh Look at the Exile Priest," *Time*, February 28, 1977, 53.

26. E. K. Victor Pearce, *Who was Adam?* (Exeter: Paternoster Press, 1969), 117–18.

27. Teilhard, *Phenomenon of Man*.

28. Bowden, *Ape-Men*, 35.

29. Millar, *Pitldown Men*, 198.

30. Moore, *Man, Time, and Fossils*, 341–403; Millar, *Piltdown Men*, 198.

31. Moore, *Man, Time, and Fossils*, 250.

32. Millar, *Piltdown Men*, 134–35, 145, 199–200.

33. Millar, *Piltdown Men*, 198; Rieth, *Archaeological Fakes*, 44.

34. Millar, *Piltdown Men*, 202–207.

35. Bowden, *Ape-Men*, 8.

36. Harry Shapiro, *Peking Man* (New York: Simon and Schuster, 1974), 47.

37. Bowden, *Ape-Men*, 8.

38. Anonymous, "Piltdown Culprit," *Time*, November 13, 1978, 82.

39. M. H. van der Veer, *Hidden Worlds* (New York: Bantam, 1972): 19–20.

40. Jerry Bergman, "The Piltdown Hoax's Influence on Evolution's Acceptance," *CRSQ* 36, no. 3 (December 1999): 145–54.

41. Bowden, *Ape-Men*, 29.

42. Anonymous, "Piltdown Culprit," *Time*, November 13, 1978, 82.

43. Millar, *Piltdown Men*, 107, 230–233.

44. John Winslow and Alfred Meyer, "*The Perpetrator at Piltdown*," *Science* 83 (September 1983): 33–43.

45. Frank Spencer, *Piltdown: A Scientific Forgery* (New York: Oxford University Press, 1990).

46. Washington Post, "Scientist Names Piltdown Man Creator," *The Argus*, May 25, 1996.

47. Teilhard, *Appearance of Man*, 32.

48. Bowden, *Ape-Men*, 32.

49. George Cornell, "Teilhard Not Hoaxer—Writer," *Ann Arbor News*, August 31, 1980.

50. Millar, *Piltdown Men*, 115–116.

51. Bowden, *Ape-Men*, 17, 26, 37.

52. Furneaux, *World's Strangest*, 119.

53. Anonymous, "Fresh Look at the Exile Priest," *Time*, February 28, 1977, 53.

54. Teilhard, *Appearance of Man*, preface.

55. Bowden, *Ape-Men*, 9–17.

56. Furneaux, *World's Strangest*, 119.

57. Theodore Graebner, *God and the Cosmos* (Grand Rapids: Eerdmans, 1932), 318.

58. Loren Eiseley, *The Immense Journey* (New York: Time, 1962), passim.

59. Anonymous, "The Piltdown Hoaxer," *Time*, July 23, 1980, 73.

Chapter 9

1. Geographica, "Decoding How a Worm Works," *National Geographic* 193, no. 5 (May 1998): unpaged.

2. Stuart Pimm, "Seeds of Our Own Destruction," *New Scientist* 146, no. 1972 (April 8, 1995): 31–35; Michael Cremo and Richard Thompson, *Forbidden Archeology* (San Diego: Bhaktivedanta Institute, 1993), 12; Anonymous, "It's a Tough World Out There," *U.S. News and World Report* 119, no. 27 (November 27, 1995): 23.

3. Hilbert Siegler, "The Magnificence of Kinds as Demonstrated by Canids," *CRSQ* 11, no. 2 (September 1974): 95.

4. Don Eicher, *Geologic Time* (Englewood Cliffs, NJ: Prentice, 1968), 98; (EB15) *Encyclopedia Britannica* 15th ed., 7 (1974), 14–16; John Whitcomb and Henry Morris, *The Genesis Flood* (Philadelphia: Presbyterian and Reformed Publishing Co., 1965), 176–178; John Moore, *Questions and Answers on Creation/Evolution* (Grand Rapids: Baker, 1976), 55; R. L. Wysong, *The Creation—Evolution Controversy* (East Lansing, MI: Inquiry Press, 1976), 287; George Price, *Common-sense Geology* (Mountain View, CA: Pacific Press, 1946), 198; Lee Gebhart, *It's Still a Mystery* (New York: Scholastic Book Service, 1970), 74.

5. Dennis Wagner, "Review of 'The Death of Species,'" *Origins Research* 14, no. 2 (1992): 10–11.

6 David Briscoe, "Mammal Extinctions," *Ann Arbor News*, November 14, 1995.

7. Philip Baffey, "Scientific Data: 50 Percent Unusable?" *CHE* 10, no. 1 (February 21, 1975): 1; Michael Satcher, "Warped School Propaganda," *U.S. News and World Report* 120, no. 23 (June 10, 1996): 63–64.

8. Carl Zimmer, "Masters of an Ancient Sky," *Discover* 15, no. 2 (February 1994): 44–45; Elso Barghoorn, "Earliest Known Form of Life," *New York Times* 61 (October 24, 1977): 1; Vincent Ettari, "The Origin of Termites" *CRSQ* 14, no.1 (June 1977): 1.

9. University of Nebraska, *Museum Notes* 56, no. 15 (January 12, 1977): 1.

10. Earth Almanac, "Mammoth Lode of Ivory from the Pleistocene," *National Geographic* 181, no. 1 (January 1992): unpaged; David Childress, *Lost Cities of Ancient Lemuria and the Pacific* (Stelle, IL: Adventures Unlimited Press, 1988), 24.

11. Don Belt, "The World's Great Lake," *National Geographic* 181, no. 6 (June 1992): 20. Herbert Wendt, *Before the Deluge* (London: Paladin, 1970), 224; Lee Gebhart, *It's Still a Mystery* (New York: Scholastic Book Service, 1970), 90. Byron Nelson, *The Deluge in Stone* (Minneapolis: Augsburg, 1931), 126; Walter Fairservis, *The Ancient Kingdoms of the Nile* (New York: NAL, 1962); William Corliss, *Scientific Anomalies and Other Provocative Phenomena* (Glen Arm, MD: The Sourcebook Project, 2003).

12. Isaac Asimov, *A Choice of Catastrophes* (New York: Fawcett Columbine, 1979), 172.

13. Eicher, *Geologic Time*, 107, 116.

14. Research News, "Traces Indicate Animal Life Existed on earth Up to 1.9 Billion Years Ago," *The Chronicle of Higher Education* (March 5, 1986): 5; Elso Barghoorn, "Earliest Known Form of Life," *New York Times* 61 (October 24, 1977): 1; Michael Pitman, *Adam and Evolution* (London: Rider, 1984), 192.

15. *Science News* (November 27, 1976).

16. Jerry MacDonald, *Earth's First Steps* (Boulder, CO: Johnson Printing, 1994), 166–170.

17. Eicher, *Geologic Time*, 110–111, 116.

18. Tom Primrose, *The Cypress Hills* (Calgary ALTA: Frontier Publisher Ltd., 1969).

19. Anonymous, "Migration of Mammalian Faunas," EB 15, vol. 19 (1997): 858.

20. Immanuel Velikovsky, *Earth in Upheaval* (New York: Dell, 1955), 93–94; Eicher, *Geologic Time*, 95; Arthur Custance, "The Influence of Environmental Pressures on the Human Skull," *Doorway Papers* no. 9 (1957b), 20–32.

21. Charles Reed, ed., *Origins of Agriculture* (The Hague: Mouton, 1977), 2.

22. Richard Milton, *Shattering the Myths of Darwinism* (Rochester, VT: Park Street Press, 1997), 35; Reiner Protsch and Rainer Berger, "Earliest Radiocarbon Dates for Domesticated Animals," *Science* 179, no. 4070 (January 19, 1973): 235–39.

23. Eicher, *Geologic Time*, 54–58; G. F. Whidborne, "Questions Involved in Evolution from a Geological Point of View," *JVI* 33 (1901); 215; Charles Warring, "Geological Exterminations," *JVI* 37 (1905): 178.

24. Eicher, *Geologic Time*, 56; Douglas Dewar, "The Supposed Fossil Links Between Man and the Lower Animals," *JVI* 67 (1935): 158; Philip Tobias, "The Tuang Skull Revisited," *Natural History* 83, no. 10 (December 1974): 38–43.

25. Reginald Daly, *Earth's Most Challenging Mysteries* (Nutley, NJ: Craig Press, 1972), 260–80.

26. Walter Fairservis, *The Origins of Oriental Civilization* (New York: New American Library, 1959): 71; James Mellaart, *Catal Huyuk* (New York: McGraw-Hill, 1967), 226–27; Robert Braidwood, *Prehistoric Men* (New York: Harper, 1967), 92, 113; Gilbert Charles-Picard, ed., "Western Asia Before Alexander," *LEOA* (1972): 191.

27. Jacquetta Hawkes, *Prehistory*, Vol. 1, Pt. 1. (New York: New American Library, 1963).

28. C. W. Ceram, *The First American* (New York: Harcourt, 1971), 179–80; Arthur Custance, *A Framework of History 29* (Brockville, ONT: Doorway Papers, 1968), 65; Fairservis, *Origins of Oriental Civilization*, 138.

29. Will Durant, *Our Oriental Heritage*, vol. 1, *The Story of Civilization* (New York: Simon & Schuster, 1954), 99.

30. E. K. Victor Pearce, *Who Was Adam?* (Exeter: Paternoster Press, 1969), 59–61; Charles Reed, *Origins of Agriculture* (The Hague: Mouton, 1977), 3; Scott Faber, "Northern Exposition," *Discover* 15, no. 2 (February 1994), 24; *The Argus* (April 12, 1999): 4.

31. Juergen Spanuth, *Atlantis of the North* (London: Book Club Associates, 1976), 121.

32. Spanuth, *Atlantis of the North*.

33. William Perry, *The Children of the Son* (London: Methuen, 1923), 52; Velikovsky, *Earth in Upheaval*, 93–94; Raoul-Jean Moulin, *Prehistoric Painting* (London: Heron, 1965), 48–51.

34. James Geikie, "The Glacial Period and the Earth Movement Hypothesis," *JVI* 26 (1890): 297–98; "Archaeology Discover Houses and Tools," *PA* 2, no. 1 (January 1973): 22.

35. Edward Vining, *An Inglorious Columbus* (New York: Appleton, 1885), 69, 100, 428, 430.

36. Arthur Custance, *Convergence and the Origin of Man .7* (Brockville ONT: Doorway Papers, 1970), 20.

37. John Cohane, *The Key* (New York: Crown, 1969), 69–72.

38. Charles Reed, "Animal Domestication in the Prehistoric Near East," *Science* 130 (December 1959): 1629–1639; Cohane, *The Key*, 68–72.

39. Sebastian Englert, *Island at the Center of the World* (New York: Scribner, 1970), 35.

40. Charles Reed, *Origins of Agriculture* (The Hague: Mouton, 1977), 89.

Chapter 10

1. Ian Taylor, *In the Minds of Men: Darwin and the New World Order* (Toronto: TFE Publishing, 1984): 152.

2. Paul Kroll, "Evolution Gets the Horse Laugh!" *PT* 10, no. 2 (April 1971): 7–10.

3. Michael Hager, *Fossils of Wyoming* Bulletin 54 (Laramie, WY: Geological Survey, 1970), 44., 252.

4. Michael Ruse, *Darwinism Defended: A Guide to the Evolution Controversies* (Reading, MA: Addison-Wesley, 1982), 311.

5. Herbert Wendt, *Before the Deluge* (London: Paladin, 1970).

6. William Fix, *The Bone Peddlers* (New York: Macmillan, 1984), 164.

7. Alan Brodrick, *Man and His Ancestry* (New York: Premier, 1964), 65–66; Frank Cousins, "Note on Unsatisfactory Nature of Horse Series of Fossils," *CRSQ* 8, no. 2 (September 1971): 99–108.

8. Leonard Radinsky, "Oldest Horse Brain," *Science* 194, no. 4265 (November 1976): 626–27; George Simpson, *Life of the Past* (New York: Bantam, 1953), 97, 196.

9. William Scheele, *The First Mammals* (Cleveland: World, 1955), 27.

10. N. J. Berrill, *Man's Emerging Mind* (New York: Fawcett World, 1955), 25.

11. E. B. Branson, *Introduction to Geology* (New York: McGraw-Hill, 1952), 451.

12. Anonymous "Horse," EB 11th edition 13 (1911), 714; *PT* (April 1971), 7; Gaius Tranquillus, *The Twelve Caesars*, trans. R. Graves (London: Folio Society, 1964), 37.

13. Brodrick, *Man and His Ancestry*, 989; Sonia Cole, *The Prehistory of East Africa* (New York: New American Library, 1963), 91–94; D. S. Allan and J. B. Delair, *When the Earth Nearly Died* (Bath: Gateway Books, 1995), 109.

14. Theodore Graebner, *God and the Cosmos* (Grand Rapids: Eerdmans, 1932), 253; Anonymous, "A Giant Horse," *Mt. Blanco Fossil News*, (October 1990): 14; Yu. Kruzhilin and V. Ovcharin, "A Horse from the Dinosaur Epoch?" *Bible-Science News* 23, no. 2 (February 1985).

15. Anonymous, "Mammals," *EB* 15th, Macropaedia 23 (1997): 445; Edward Hitchcock, *Fossil Footmarks in the Sandstone of the Connecticut Valley* (Boston: William White, 1858), 133–38.

16. Steven Stanley, "Macroevolution and the Fossil Record," *Evolution* 36, no. 3 (May 1982): 466.

17. C. Gilmore, *Fossil Footprints from the Grand Canyon* #2832 (Washington: Smithsonian, 1926), 37–39

18. William Scott, *A History of Land Mammals in the Western Hemisphere* (New York: Hafner, 1962), 431–32.

19. William Corliss, *Biological Anomalies: Mammals II* (Glen Arm, MD: Sourcebook Project, 1996), 268–70.

20. Y. Kruzhizin, "A Horse from the Dinosaur Epoch?" *Bible-Science News* 23, no. 2 (February 1985): 1; A. S. Yehuda, "Joseph in Egypt in the Light of the Monuments," *JVI* 65 (1935).

21. Henry Field, *Contributions to the Anthropology of Saudi Arabia* (Coconut Grove, FL: Field Research Projects, 1971), 27; Lady Wentworth, "The Authentic Arab Horse and His Descendants," *JVI* 82 (1950): 154; Raoul-Jean Moulin, *Prehistoric Painting* (London: Heron, 1965), 50; T. K. Callard, "The Contemporaneity of Man with Extinct Mammals," *JVI* 13 (1878): 217.

22. Evan Hadingham, *Secrets of the Ice Age* (New York: Walker, 1979), 106; Alexander Marshack, *The Roots of Civilization* (New York: McGraw-Hill, 1972), fig. 191.

23. Jacquetta Hawkes, *Prehistory* I, Part 1 (New York: New American Library, 1963), 270.

24. *TILN*, (November 1975): 87; Robert Charroux, *Forgotten Worlds* (New York: Popular Library, 1971), 117; Proinsias MacCana, *Celtic Mythology* (New York: Hamlyn, 1970), 83; Normand Hammond, "The Afghan Road to Seistan," *TILN* 250, no. 6654 (February 11, 1967): 23–25.

25. Hadingham, *Secrets of the Ice Age*, 94.

26. See Erich von Fange (1994) for a suggested chronology of this time in history.

27. Judith Treistman, *The Prehistory of China* (New York: Doubleday, 1972), 50; *SR* (September 30, 1972): 90–91; Eoin MacWhite, "On the Interpretation of Archaeological Evidence," *American Anthropologist* 58, no. 1(April 1956): 91–93; Brodrick, *Man and His Ancestry*, 203, 212.

28. Edward Vining, *An Inglorious Columbus* (New York: Appleton, 1885), 482; Henrietta Mertz, *Pale Ink*, 2nd edition (Chicago: Swallow, 1972): 62–67; Robert Marx, "Frescoes of Horses in Yucatan," *Argosy* 281, no. 4 (April 1975): 42; Anonymous, "Horse Effigy," *National Geographic* 102, no. 3 (September 1952): 389.

29. H. M. Wormington, *Ancient Man in North America* (Denver: Denver Museum of Natural History, 1957), 150, 231, 244; Gordon Smith, "E. equus: Immigrant or Emigrant?" *Science* 84, no. 5 (April 1984): 76.

30. Gavin Menzies, *1421: The Year China Discovered America* (New York: William Morrow, 2003), 42.

31. Vining, *An Inglorious Columbus*, 483; Harold Wilkins, *Mysteries of Ancient South America* (Secaucus, NJ: Citadel Press, 1956), 196; Charles Berlitz, *Mysteries From Forgotten Worlds* (New York: Dell, 1972), 133; Kenneth Macgowan and Joseph Hester Jr., *Early Man in the New World*, rev. (Garden City, NY: Anchor Books, 1962), 131.

32. Simpson, *Life of the Past*, 119.

33. Heribert Nilsson, *Synthetische Artbildung* Vol. II. (Lund, Sweden: Gleerup, 1953), 1193.

34. Anonymous, "Horse," *EB* 15th ed., Macrop 1 (1974): 425; D. Raup, "Horse Evolution Update," *Bible and Spade* (Spring 1999): 60.

Chapter 11

1. William Scott, *Reports of the Princeton University Expeditions to Patagonia 1896–1899*, Vol. VII, Part I. Litopterna of the Santa Cruz beds, (Princeton, NJ: Princeton University Press, 1910), 551.

2. William Scott, *A History of Land Mammals in the Western Hemisphere* (New York: Hafner, 1962), 397.

3. Tracy I. Storer, et al, *General Zoology*, 6th ed. (New York: McGraw-Hill, 1979), passim.

4. Scott, *Princeton University Expeditions*, 114–15.

5. R. Cifelli, *The Origin and Affinities of the South American Condylarthra and Early Tertiary Litopterna (Mammalia)*, Number 2772 (New York: American Museum Novitates, 1983), 28; Scott, *Princeton University Expeditions*, 551.

6. George Simpson, *Principles of Animal Taxonomy* (New York: Columbia University Press, 1961), 87.

7. Scott, *Princeton University Expeditions*, 551–68.

8. Anonymous, "Mammalia," *Encyclopedia Britannica* (1911): 17:526; 16:791; Bernard Peyer, *Comparative Odontology* (Chicago: University of Chicago, 1968), 317; Allen Keast, et al (eds), *Evolution, Mammals, and Southern Continents* (Albany: State University of New York Press, 1972), 283; Alfred Romer, *Vertebrate Paleontology*, 3rd ed. (Chicago: University of Chicago Press, 1966), 260; Edwin Colbert, *Evolution of the Vertebrates*, Science edition (New York: Wiley, 1967), 343, 549.

9. Peyer, *Comparative Odontology*, 274–75; Keast, *Evolution, Mammals, and Southern Continents*, 270; Romer, *Vertebrate Paleontology*, 260.

10. Scott, *Princeton University Expeditions*, 7:257, 561.

11. P. S. Martin and H. E. Wright, eds., *Pleistocene Extinctions: The Search for a Cause* (New Haven: Yale Unversity Press, 1967), passim; Peyer, *Comparative Odontology*, passim.

12. Scott, *Princeton University Expeditions*, 551.

13. Simpson, *Principles of Animal Taxonomy*, unpaged.

14. William Fix, *The Bone Peddlers: Selling Evolution* (New York: Macmillan, 1984), xxv.

15. Simpson, *Principles of Animal Taxonomy*, 85, 87, 119.

16. William Scott, *A History of Land Mammals in the Western Hemisphere*, rev. ed. (New York: Hafner, 1962), 395–97; E. Branson, et al., *Introduction to Geology*, 3rd ed. (New York: McGraw-Hill, 1952), 451.

17. Donald Johanson and Maitland Edey, *Lucy: The Beginning of Humankind* (New York: Warner Books, 1981), 176–77

18. S. Wright, "Genes and Organismic Selection," *Evolution* 34 (1980): 825–43.

19. Colbert, *Evolution of the Vertebrates*, 338.

20. Scott, *History of Land Mammals*, 114, 254–55, 406, 431.

21. Simpson, *Principles of Animal Taxonomy*, 84.

22. Peyer, *Comparative Odontology*, 317–18; Scott, *History of Land Mammals*, 397.

23. Scott, *Princeton University Expeditions*, 7:5–8.

24. Fix, *Bone Peddlers*, 169, 189.

25. Charles Hapgood, *The Path of the Pole* (Philadelphia: Chilton, 1970), 280–91.

26. David Hopkins, ed., *The Bering Land Bridge* (Palo Alto: Stanford University Press, 1967), 266–67.

27. William Scheele, *The First Mammals* (Cleveland: World, 1955), 62, 80.

28. Cifelli, *South American Condylarthra and Early Tertiary Litopterna*, 42.

29. Michael Ruse, *Darwinism Defended: A Guide to the Evolution Controversies* (Reading MA: Addison-Wesley, 1982), 311.

30. Isak Dinesen, *Out of Africa* (New York: Time, 1963), 121.

Chapter 12

1. William Corliss, "Anthropology Unbound," *Science Frontiers* 119 (September 1998).

2. "Mystery of the Stoned Pharaoh," *Science Frontiers* 116 (March 1998); Gavin Menzies, *1421: The Year China Discovered America* (New York: William Morrow, 2003), passim.

3. John Cohane, *The Key* (New York: Crown, 1969), 18, 26.

4. Jacquetta Hawkes, *Prehistory* Part l. (New York: New American Library, 1963), 1: 367–375.

5. Sebastian Englert, *Island at the Center of the World*, trans. and ed. by W. Mulloy (New York: Scribner, 1970), passim.

6. Carroll Riley, et al eds., *Man Across the Sea* (Austin: University of Texas Press, 1971), 123; Charles Miles, *Indian and Eskimo Artifacts of North America* (New York: Bonanza, 1963), 225.

7. Englert, *Island at the Center of the World*, 35, 53; Cohane, *The Key*, 209.

8. Hans Haelbaek, "Domestication of Food Plants in the Old World," *Science* 130, no. 3372 (August 14, 1959): 365; Kenneth Macgowan and Joseph Hester Jr., *Early Man in the New World*, rev. ed. (Garden City, NY: Anchor Books, 1962), 40.

9. *Time*, August 2, 1970, 66.

10. C. W. Ceram, *The First American* (New York: Harcourt, 1971): 179–186.

11. Anonymous, "The Lima Bean People of Long Ago," *Time*, August 2, 1963, 49.

12. Alan and Sally Landsburg, *In Search of Ancient Mysteries* (New York: Bantham, 1974): 36; Harold Wilkins, *Mysteries of Ancient South America* (Secaucus, NJ: Citadel Press, 1956): 186–87; Charles Berlitz, *Mysteries from Forgotten Worlds* (New York: Dell, 1972), 76–78; Andrew Tomas, *The Home of the Gods* (New York: Berkley, 1972), 15–16; Brinsley Trench, *Temple of the Stars* (New York: Ballantine Books, 1962), 22.

13. Alexander Marshack, *The Roots of Civilization* (New York: McGraw-Hill, 1972), 213.

14. *Time*, June 28, 1968, 34.

15. Ralph Solecki, *Shanidar: The First Flower People* (New York: Knopf, 1971), 248–50.

16. Peter Fleming, *Brazilian Adventure* (New York: Grosset & Dunlap, 1933), 154.

17. A. H. Finn, "The Miraculous in Holy Scripture," *JVI* 60 (1928): 187.

18. William Corliss, *Strange Artifacts* M-1 (Glen Arm, MD: The Sourcebook Project, 1974), MES-006; Russell Brubaker, *Private Correspondence* (February 1975); Edward Lain. and Robert Gentet, "The Case for the Calaveras Skull," *CRSQ* 33, no. 4 (March 1997): 248.

19. James Enterline, *Viking America: The Norse Crossings and their Legacy* (Garden City, NY: Doubleday, 1972), xvii.

20. *National Geographic* (1991); "Mission to Fossil Island," *Popular Mechanics*, December 1981; V. L. Westberg, *The Master Architect* (Napa CA: Author, undated); Hugh Brown, *Cataclysms of the Earth* (New York: Twayne Publisher, 1967).

21. I. W. Cornwall, *The World of Ancient Man* (New York: New American Library, 1964), 186.

22. Immanuel Velikovsky, *Earth in Upheaval* (New York: Dell, 1955), 197. *Anonymous,* "Noah's Park," *Time* (August 28, 1972): 64; Donovan Courville, "*Evolution and Archaeological Interpretation*," *CRSQ* 11, no. 1 (June 1974): 52.

23. Russell MacFall and Jay Wollin, *Fossils for Amateurs* (New York: Van Nostrand Reinhold, 1972), 234.

24. *Christian News*, October 18, 1971.

25. Harold Armstrong, "Determining Dates by Radioactive Carbon," *CRSQ* 2, no. 4 (January 1966): 31; R. H. Brown, "Radiocarbon Dating," *CRSQ* 5, no. 2 (September 1968): 67; Robert Whitelaw, "Radiocarbon Dating: A Reply," *CRSQ* 6, no. 2 (September 1969): 114; Franklin Folsom, *America's Ancient Treasures* (New York: Rand McNally, 1971): 160; H. M. Wormington, *Ancient Man in North America* (Denver: Museum of Natural History, 1957). 18; Velikovsky, *Earth in Upheaval.* 158; "The Pitfalls of Radiocarbon Dating," *Pensee* 3, no. 2 (Spring 1973): 13.

26. Clifford Burdick, "Microflora of the Grand Canyon," *CRSQ* 3, no. 1 (May 1966): 49; Arthur Chadwick, "Grand Canyon Palynology," *CRSQ* 9, no. 4 (March 1973): 238; George Mulfinger, "A Unique Creationist Exhibit," *CRSQ* 10, no. 1 (June 1973): 62.

27. R. L. Wysong, *The Origin of Life Controversy* (Author, undated), 268; George Mulfinger, "A Unique Creationist Exhibit," *CRSQ* 10, no. 1 (June 1973): 66; Clifford Burdick, "More Precambrian Pollen, *CRSQ* 11, no. 2 (September 1974): 122.

28. John Klotz, *Genes, Genesis, and Evolution* (St. Louis: Concordia, 1970): 188; John Whitcomb and Henry Morris, *The Genesis Flood* (Philadelphia: The Presbyterian and Reformed Publishing Company, 1965), 119; John Kirk, "Relations of Geological Science to the Sacred Scriptures," *JVI* 1 (1867): 319; MacFall and Wollin, *Fossils for Amateurs*, 246; V. L. Westberg, "The Mystery of Buried Redwood," *CRSQ* 7, no. 1 (June 1970): 80.

29. *Science* (March 26, 1934): 119.

30. Whitcomb and Morris, *Genesis Flood*, 393.

30. Herbert Sorensen, "The Ages of Bristlecone Pine," *Pensee* 3, no. 2 (Spring 1973): 13–17.

32. Darwin Lambert, "What the Ancient Pines Teach us," *Reader's Digest* 101, no. 608 (December 1972): 86–90; Michael Burton, "Europe's Cultural Pioneers," *Readers Digest* 105, no. 629 (September 1974): 15.

33. Cornwall, *World of Ancient Man*, 178–186.

34. Byron Nelson, *Before Abraham: Prehistoric Man in Biblical Light* (Minneapolis: Augsburg, 1948), 53.

35. *Ann Arbor News*, November 14, 1973.

36. Robert Heizer, *Man's Discovery of His Past* (Englewood Cliffs, NJ: Prentice-Hall, 1962), 107–108.

37. James Mellaart, *Catal Huyuk* (New York: McGraw-Hill, 1967), 63.

38. Reginald Daly, *Earth's Most Challenging Mysteries* (Nutley, NJ: The Craig Press, 1972), 189; Herbert Sorensen, "The Ages of Bristlecone Pine," *Pensee* 3, no. 2 (Spring 1973): 14; Immanuel Velikovsky, "Epilog," *Pensee* 4, no. 1 (Winter 1973): 18–19; H. B. Guppy, "Plant Distribution from an Old Standpoint," *JVI* 39 (1907): 168; David Smithers, "Hovercraft to the Heart of Africa," *Science Digest* 70, no. 6 (December 1971): 37; Thor Heyerdahl, *Kon-Tiki* (Chicago: Rand McNally, 1950), 18.

39. Macgowan and Hester, *Early Man*, 264; Riley, *Man Across the Sea*, 24;

40. Whitcomb and Morris, *Genesis Flood*, 163, 278; Daly, *Earth's Most Challenging Mysteries*, 61; Otto Stutzer, *Geology of Coal*, trans. A. Noe (Chicago: University of Chicago Press, 1940), passim; Philip LeRiche, "Scientific Proofs of a Universal Deluge," *JVI* 61, (1929): 95.

41. Robert Whitelaw, "Time, Life and History in the Light of 15,000 Radiocarbon Dates," *CRSQ* 7, no.1 (June 1970): 59–62.

42. Harold Coffin, The Classic Joggins Petrified Trees," *CRSQ* 6, no. 1 (June 1969): 35; Daly, *Earth's Most Challenging Mysteries*, 133–140.

43. John Kirk, "Relations of Geological Science to the Sacred Scriptures," *JVI* 1 (1867), 305–327; Henry Andrews, *Ancient Plants and the World They Lived In* (Ithaca: Cornell University Press, 1947), 19.

44. Whitcomb and Morris, *Genesis Flood*, 159–60.

45. Harold Armstrong, "Scientific News and Views," *CRSQ* 6, no. 4 (March 1970): 191; N. Whitley, "The Flint 'Implements' of Brixham Cavern," *JVI* 11 (1878): 27; William Corliss, *Archeological Artifacts: Wooden* (Glen Arm, MD: Sourcebook Project, 2003), 289–92.

46. Daly, *Earth's Most Challenging Mysteries*, 41–44.

47. Velikovsky, *Earth in Upheaval*, 47, 56.

48. Immanuel Velikovsky, *Ages in Chaos* (Garden City, NJ: Doubleday, 1952), 7–8; Donald Patten, "The Ice Age Phenomena," *CRSQ* 3, no. 1 (May 1966): 63–71.

49. Daly, *Earth's Most Challenging Mysteries*, 126.

50. Andrews, *Ancient Plants*, 176.

51. *The Info Journal*, 4:1: 29.

52. J. Stefansson, "Iceland: Its History and Inhabitants," *JVI* 34 (1902): 166; Velikovsky, *Earth in Upheaval*, 43–46.

53. Francis Schaeffer, *Genesis in Space and Time* (Glendale, CA: Regal, 1972), 99.

54. Daly, *Earth's Most Challenging Mysteries*, 25–26; A. P. Kelly, "Hiatuses in Plant Kingdom," *JVI* 73 (1941): 131.

55. *Ann Arbor News*, October 23,1974; *RAM, Rocks and Minerals* 49 (November 1974): 11.

56. Westberg, *Master Architect*, 6; A. E. Porsild, "Lupine Grown from Seeds of Pleistocene Age," *Science* 158, no. 3797 (October 6, 1967): 113–114.

57. Klotz, *Genes, Genesis, and Evolution*, 202: Wilbert Rusch, "The Revelation of Palynology," *CRSQ* 5, no. 3 (December 1968): 103.

58. J. Reddie, "On Geological Chronology," *JVI* 2 (1867): 325.

59. Paul Zimmerman, "Essays from the Creationist Viewpoint," *CRSQ* 7, no. 2, (September 1970): 128; Cohane, *The Key*, 58; W. P. James, "On the Relation of Fossil Botany to Theories of Evolution," *JVI* 19 (1885): 138.

60. *Popular Archaeology* 3, no. 4 (1973): 52; Riley, *Man Across the Sea*, 325; Heyerdahl, *Kon-Tiki*, 137.

61. Ivan Sanderson, *More Things* (New York: Pyramid, 1969), 103–116.

62. W. Upham, "On the Post Glacial Period," *JVI* 25 (1891): 106; Edward Lanning, *Peru Before the Incas* (Englewood Cliffs, NJ: Prentice-Hall, 1967), 50.

63. Gerald Hawkins, *Beyond Stonehenge* (New York: Harper, 1973), 189; Riley, *Man Across the Sea*, 25, 325, 403–436; Folsom, *America's Ancient Treasures*, 173.

64. Maria Ambrosini, *The Secret Archives of the Vatican* (Boston: Little, Brown, 1969), 155

65. Macgowan and Hester, *Early Man*, 236, 264

66. Cyrus Gordon, *Before Columbus* (New York: Crown, 1971), 145;

67. Lanning, *Peru Before the Incas*, 76, 81; Cohane, *The Key*, 54.

68. Anonymous, "The Lima Bean People of Long Ago," *Time*, August 2, 1963, 49.

69. Philip Schidrowitz, "Beer," *EB* 11th ed. 3 (1911), 642; Mellaart, *Catal Huyuk*, 224.

70. Anonymous, "Corn," *EB* 11, vol. 7 (1911), 162; Cohane, *The Key*, 68–72, 98–99; Jean Sendy, *Those Gods Who Made Heaven and Earth*, trans. by Lowell Bair (New York: Berkley, 1972): 147.

71. Elsie Hix, *Strange As It Seems*, 2nd ed. (New York: Bantam, 1960), 41; Anonymous, "Backward into Prehistory" *Time* (January 1,1965): 61; Grant Watson, "Facts at Variance with Organic Evolution," *JVI* 70 (1938): 55; *AAN* (May 1, 1969), 26; JudithTreistman, *The Prehistory of China* (New York: Doubleday, 1972), 63.

72. Anonymous, "Maize," *EB* 11, vol. 17 (1911), 448; *CRSQ*, 1968, 5:113; Riley, *Man Across the Sea*, 382–395; Tomas, *Home of the Gods*, 16.

73. Mellaart, *Catal Huyuk*, 224–27.

74. Henriette Mertz, *Pale Ink*, 2nd ed. (Chicago: Swallow, 1972), 61–62.

75. Riley, *Man Across the Sea*, 24.

76. Anonymous, *Pineapple* EB 11 vol 21: 625; Riley, *Man Across the Sea*, 26.

77. Englert, *Island at the Center of the World*, 43; Robert Charroux, *Forgotten Worlds* (New York: Popular Library, 1971), 138; W. Upham, "On the Post Glacial Period," *JVI* 25 (1891): 106; Adrian Boshier, "Swaziland, Birthplace of Modern Man," *Science Digest* 73, no. 3 (March 1973): 42–47.

78. Mertz, *Pale Ink*, 135; Robert Wood, "The Age of Man," *CRSQ* 2, no. 4 (January 1966): 26; Wormington, *Ancient Man in North America*, 184; Andrew Tomas, *We Are Not the First* (New York: Bantam, 1971), 24.

79. Francis Owen, *The Germanic People* (New York: Bookman Associates, 1960), 45; Cornwall, *World of Ancient Man*, 190; Mellaart, *Catal Huyuk*, 65.

80. Heyerdahl, *Kon-Tiki*; Hawkins, *Beyond Stonehenge*, 189, 216; William Corliss, *Strange Artifacts* M-1 (Glen Arm, MD: The Sourcebook Project, 1974), MLG-002; Geoffrey Bibby, *Looking for Dilmun* (New York: Knopf, 1970), 193.

81. Cohane, *The Key*, 54; Englert, *Island at the Center of the World*, 39; Riley, *Man Across the Sea*, 26.

82. *The New York Times* (July 12, 1970); Gordon, *Before Columbus*, 204.

83. Patrick McGovern, "Wine's Prehistory," *Archaeology* 51, no. 4 (July 1998): 32–34; Mellaart, *Catal Huyuk*, 224; *Ann Arbor News*, May 1,1969, 26.

84. Cohane, *The Key*, 18–22.

85. Charles Reed, Private Correspondence (1977); Cohane, *The Key*.

Chapter 13

1. Reginald Daly, *Earth's Most Challenging Mysteries* (Nutley, NJ: Craig Press, 1972), 388.

2. Randy Wysong, *The Creation-Evolution Controversy* (East Lansing, MI: Inquiry Press, 1976), 301.

3. Loren Eiseley, *The Immense Journey* (New York: Time, 1962), 7.

4. Nicholas Timasheff, *Sociological Theory: Its Nature and Growth* (Garden City, NY: Doubleday, 1955), 124–131.

5. Gould, Stephen, "The Peppered Moth Story," *Discover* 2, no. 2 (May 1981): 36; Terry Coyne, "Melanism: Evolution in Action," *Nature* 336, no. 6194 (November 3, 1988): 35–36; Jerry Adler, "Doubting Darwin," Newsweek 145, no. 6 (February 7, 2005): 48.

6. Loren Eiseley, *Darwin's Century* (Garden City, NY: Doubleday, 1961), 114.

7. Roger Lewin, "Evolutionary Theory under Fire," *Science* 210, no. 4472 (November 21, 1980): 883.

8. Tom Woodward, "Meeting Darwin's Wager," *Christianity Today* 41, no. 4 (April 28, 1997): 15.

9. Art Poettker, "Seventeen Problems for Evolutionists," *CRSQ* 14, no. 2 (September 1977): 121.

10. Luther Sunderland, *Darwin's Enigma* (El Cajon, CA: Master Books, 1985).

11. Phillip Johnson, *Darwin on Trial* (Downer's Grove, IL: InterVarsity, 1991).

12. Anthony Standen, *Science Is a Sacred Cow* (New York: Dutton, 1950), 100–104; William Howells, ed., *Ideas on Human Evolution: Selected Essays* (New York: Atheneum, 1962), v; Carl Sagan ,"Life," *EB* 15th ed., 22 (1974, 1986): 987; Ronald Duncan and Miranda Weston-Smith eds., *The Encyclopaedia of Ignorance*, (New York: Pergamon Press, 1977), 227–34.

13. Charles Darwin, *The Origin of Species*, introduction by O. H. Matthews (New York: Dutton, 1859, 1971), xi.

14. Dean Wooldridge, *Mechanical Man: The Physical Basis of Intelligent Life* (New York: McGraw-Hill, 1969).

15. George Simpson, *The Meaning of Evolution* (New Haven: Yale, 1951), 344–48.

16. James Reddie, "On Geological Chronology," *JVI* 2 (1867): 304.

17. See Raymond F. Surburg, "The Influence of Darwinism," chapter 16 in *Darwin, Evolution, and Creation*, ed. Paul Zimmerman (St. Louis: Concordia, 1959), 171–182.

18. Zimmerman, *Darwin, Evolution, and Creation*, 171–182.

19. Francis Schaeffer, *How Should We Then Live?* (Old Tappan, NJ: Fleming Revell, 1976), 150.

20. Anonymous, "Fresh Look at the Exile Priest," *Time* 109, no. 9 (February 28, 1977): 53.

21. Stewart Easton, *The Western Heritage to 1500*, 2nd ed. (New York: Holt, Rinehart, Winston, 1966), 11–12.

22. Georgio de Santillana and Hertha von Dechend, *Hamlet's Mill* (Boston: Gambit, 1969), 65–66.

23. Alexander Marshack, *The Roots of Civilization* (New York: McGraw-Hill, 1972), 11.

24. Michael Banton, ed., *Darwinism and the Study of Society* (Chicago: Quadrangle Books, 1961), 49, 63; Karl Deutsch, "Conditions Favoring Major Advances in Social Science," *Science* 171, no. 3970 (February 5, 1971): 450–59.

25. Edward Wilson, *Sociobiology: The New Synthesis* (Cambridge: Harvard University Press, 1975), passim.

26. Charles Darwin, *The Descent of Man and Selection in Relation to Sex* (New York: Collier, 1902), 180–81.

27. William James, *Principles of Psychology* (1890); Harry Kaufmann, *Introduction to the Study of Human Behavior* (Philadelphia: Saunders, 1968), passim.

28. Charles Judd, *Psychology* (1939).

29. Schaeffer, *How Should We Then Live?*, 150.

30. Ronald Gross, ed., *The Teacher and the Taught* (New York: Delta Books, 1963), 98–99.

31. John Dewey, *Democracy and Education* (1916), 1–2.

32. John Dewey, *The Influence of Darwin on Philosophy* (New York: Holt, 1910), 1–2; R. Biehler, *Psychology Applied to Teaching*, 3rd ed. (Boston: Houghton Mifflin, 1978), passim; Jerome Bruner, *Toward a Theory of Instruction* (Cambridge: Harvard University Press, 1966), chapter 4.

33. Jerome Bruner, *Toward a Theory of Instruction* (Cambridge: Harvard University Press, 1966).

34. "Darwin's Detractors," *Nature* 358, no. 6389 (August 1992): 698.

35. Jerry Bergman, "Book Review," *CRSQ* 35, no. 3 (December 1998): 170.

Chapter 14

1. Danny Faulkner and Don DeYoung, "Toward a Creationist Astronomy," *CRSQ* 28, no. 2 (December 1991): 87.

2. Andrew Tomas, *We Are Not the First* (New York: Putnam, 1971), 57.

3. Robert Bass, "Proofs of the Stability of the Solar System," *Kronos* 2, no. 2 (Summer 1976): 29–30.

4. Lee Siegel, "Big Bang Evidence Is Found," *Ann Arbor News*, April 12, 1992; Associated Press, "NASA Finds Evidence of Big Bang," *Ann Arbor News* January 8, 1993.

5. William Corliss, "Big-Bang Brouhaha," *Science Frontiers* 82 (July 1992): 2.

6. Anonymous, "Kehoutek: Comet of the Century," *Time* 70, no. 17 (December 17, 1973): 88; Anonymous, "Kehoutek," *Discover* 6, no. 7 (July 1985): 77.

7. David Levy, "Was Chicken Little Right?" *Parade* (May 10, 1998): 14; Jurgen Spanuth, *Atlantis of the North* (London: Book Club Association, 1979), 156.

8. Charles Pett, "Okay, So They Were a Little Off," *USN&WR* 124, no. 11 (March 23, 1998).

9. Cyril Aldred, *Egypt to the End of the Old Kingdom* (New York: McGraw-Hill, 1965), 57; Richard Burton, *Unexplored Syria*, vol. 1 (London: Tinsley Brothers, 1872), 14:432, 604; Colin Renfrew, "The Place of Astronomy in the Ancient World," *Archaeology* 26, no. 3 (July 1973): 222–23; Alex Haley, "My 'Search for Roots," *Readers Digest* 104, no. 625 (May 1974): 236; Agnes Clerke, "Astronomy," *EB11*, vol. 2 (1910), 809; Mrs. Walter Maunder, "History of India Reflected in the Rig-Veda," *JVI* 71: (1939), 147.

10. Joseph Seiss, *The Gospel in the Stars* (Grand Rapids: Kregel, 1882, 1972), 23; Hugh Moran and David Kelley, *The Alphabet and the Ancient Calendar Signs*, 2nd ed. (Palo Alto: Daily Press, 1969), xi, 7, 49; Richard Burton, *The Book of the Thousand Nights and a Night*, vol. 1 (Bagdad Edition #95: The Burton Club, 1885), V: 228.

11. Morris Jastrow, "Astrology," EB 11th 2 (1910), 800; Stephen Potter and Laurens Sargent, *Pedigree* (New York: Taplinger, 1973), 195; Hugh Moran and David Kelley, *The Alphabet and the Ancient Calendar Signs*, 2nd ed. (Palo Alto: Daily Press, 1969) 137; Harold Bayley, *The Lost Language of Symbolism*, volume 1 (London: Williams and Norgate, 1912), 137, 359; J. E. Howard, "On the Druids and Their Religion," *JVI* 14 (1880), 131; Joseph Murphy, "A Physical Theory of Moral Freedom," *JVI* 22 (1888), 179.

12. Harold Armstrong, "Panorama of Science," *CRSQ* 13, no. 2 (September 1976): 120; J. Reddie, "On Civilization, Moral and Material," *JVI* 6 (1872): 19.

13. Arthur Earle, *The Bible Dates Itself* (Southhampton, PA: Author, 1974) 57.

14. Anonymous, "A Few Baktuns Ago," *Time* 70, no. 17 (10/21/1957): 50; Herbert Armstrong, *Pagan Holidays* (Pasadena: Ambassador Press, 1974), 5–26; Jack Finegan, *Handbook of Biblical Chronology* (Princeton, NJ: University Press, 1964), 16; Anonymous, "Backgammon," *Jordan* 1, no. 3 (1976): 6–7; Fred Hoyle, *From Stonehenge to Modern Cosmology* (San Francisco: W. Freeman, 1972), 54.; Robert Charroux, *The Gods Unknown* (New York: Berkley, 1969), 59.

15. Thor Heyerdahl, *Kon-Tiki* (Chicago: Rand McNally, 1950), 137; Kenneth Mory, "The Coming of the Polynesians," *National Geographic* 146, no. 6 (December 1974): 732–778; Cyrus Gordon, *Before Columbus* (New York: Crown, 1971), 80; Giorgio de Santillana and Hertha von Dechend, *Hamlet's Mill* (Boston: Gambit, 1969), 62, 235.

16. John Michell, *The View Over Atlantis* (New York: Ballantine Books, 1969), 69; Evan Mackie, "Megalithic Astronomy and Catastrophe," *Pensee* 4, no. 5, Winter 1974: 5–20; Gerald Hawkins, *Beyond Stonehenge* (New York: Harper, 1973), 96–107, 171; Francis Hitching, *Earth Magic* (New York: William Morrow, 1977, 27; Andrew Tomas, *The Home of the Gods* (New York: Berkley, 1972), 68; William Corliss, *Strange Artifacts* M no.1 (Glen Arm, MD: The Sourcebook Project, (1974), MSH-015; Anonymous, "Mystery on the Mesa," *Time* 103, no. 12 (March 25, 1974): 94.

17. Hitching, *Earth Magic*, 37; Tomas, *We Are Not the First*, 58.

18. Evan Hadingham, *Circles and Standing Stones* (New York: Walker, 1975), 147.

19. Peter Stevenson, "Kohoutek, Comets, and Christianity," *CRSQ* 11, no. 1 (June 1974): 70; G. Mackinlay, "Biblical Astronomy," *JVI* 37 (1905): 142; W. Raymond Drake, *Gods and Spacemen of the Ancient Past* (New York: New American Library, 1974) 128; B. Savile, "On the Harmony Between the Chronology of Egypt and the Bible," *JVI* 9 (1875): 60; Donald Patten, et al, *The Long Day of Joshua and Six Other Catastrophes* (Seattle: Pacific Meridian, 1973), 167; William Dyrness, "Astrology: Cosmic Fatalism," *Christianity Today* 20, no. 25. (September 24, 1976): 16–17.

20. Donald Patten, et al, *The Long Day of Joshua and Six Other Catastrophes* (Seattle: Pacific Meridian, 1973), 257.

21. P. Cleator, *Lost Languages* (New York: New American Library, 1959), 108.

22. Seiss, *Gospel in the Stars*, passim; Ethelbert Bullinger, *The Witness of the Stars* (Grand Rapids: Kregel 1893, 1967), 23–27; Henry Morris, *Many Infallible Proofs* (San Diego: Creation-Life Publisher, 1974), 342–43.

23. Erich von Fange, *Noah to Abram* (Adrian, MI: Living Word Services, 1994); *Genesis and the Dinosaur* (Adrian, MI: Living Word Services, 1995).

24. Loren Steinhauer, "The Case for Global Catastrophism," *Journal of the American Scientic Affiliation* (December 1973), 25:4, p.129; E. R. Milton, ed., *Recollections of a Fallen Sky* (Lethbridge ALTA: Unileth Press, 1978), 42–86; W. R. Cooper, "Serpent Myths of Ancient Egypt," *JVI* 6 (1972), 267; de Santillana and von Dechend, *Hamlet's Mill*, cover; D. S. Allan and J. B. Delair, *When the Earth Nearly Died* (Bath, UK: Gateway Books, 1995); A. C. Crommelin, "On the Return of Halley's Comet in 1910," *JVI* 42 (1910): 18–19; Harold Armstrong, "Comets and a Young Solar System," *CRSQ* 8, no. 3 (December 1971): 192; Mel Waskin, *Mrs. O'Leary's Comet* (Chicago: Academy Chicago Publishers, 1985).

25. Milton Zysman and Clark Whelton, eds., *Catastrophism 2000* (Toronto: Heretic Press, 1972) 15; Leroy Waterman (tr), *Royal Correspondence of the Assyrian Empire*, Michigan Humanistic Series 17, Part l. (New York: Johnson Reprint Corporation, 1972); Milton, *Recollections of a Fallen Sky*, 26.

26. C. Ernest Wright, *Shechem* (New York: McGraw-Hill, 1964), 89–97.

27. Anonymous, "Billions of Tectites," *Discover* 4, no. 6 (June 1983): 75–77; Dorothy Vitaliano, *Legends of the Earth* (Bloomington, IN: Indiana University Press, 1973), 72–73; John O'Keefe, "The Tectite Problem," *Scientific American* 239, no. 2 (August 1978): 118; Stephen Talbott, "A Record of Success," *Pensee* 2, no. 2 (May 1972), 23; Peter Steveson, "Meteoric Evidence for a Young Earth," *CRSQ* 12, no. 1 (June 1975): 23; Anonymous, "Chinese Giant Meteorite," *AAN*, January 10, 1990; Research Notes, "Ores Said to Support Theory about Meteorites," *CHE* 31, no. 10 (November 6, 1985): 12.

28. Frank Edwards, *Stranger Than Science* (New York: Bantam, 1967), 70–71; William Corliss, *Strange Planet* (Glen Arm MD: The Sourcebook Project,1975).ETB1-2; AAN, 11/25/1976, p.25; Nigel Calder, *The Restless Earth* (New York: Viking Press, 1972); 91; AAN (November 25, 1977), p.B9; 1/24/1985, p.B2; Alfred deGrazia, *Chaos and Creation* (Princeton, NJ: Metron, 1981), 7; Anonymous, "Florida Bowl," *Time* 126, no. 23 (December 9, 1985), p.93; Anonymous, "Experts See Craters as Heaven Sent," *Detroit News* (October 10, 1965); Jane Forsyth, *The Geology of the Highland-Adams County Area* (Columbus, OH: Academy of Science, 1962, 18.

29. H. Nininger, "Those Earth-Grazing Asteroids," *Earth Science* 27, no. 1 (January 1974): 33.

30. Anonymous, "Were We Planted Here?" *Time*, September 10, 1974, 53.

31. Anonymous, "Delaying Doomsday," *Time*, October 15, 1973, 107.

32. Anonymous, "Clusters of Galaxies," *Ann Arbor News*, February 4, 1986.

33. Anonymous, "Fickle Universe," *Time*, January 25, 1982, 61; William Cook, "How Old Is the Universe?" *U.S. News & World Report*, May 20, 1996, 60–61; Allan Broms, *Our Emerging Universe* (New York: Dell, 1961): 57; Edwin Krupp, *Echoes of the Ancient Skies* (New York: Harper & Row, 1983), 343–46; Anonymous, "Astronomers Say Universe is Younger than it Looks," *The Argus*, January 6, 1985; Jonathan Weisman, "Doesn't Look a Day Over 14 Billion," *The Argus*, March 4, 1993; Eric Lerner, *The Big Bang Never Happened* (New York: Random House, 1991); Eric Lerner, "Bucking the Big Bang," *New Scientist* 182, no. 2448 (May 22, 2004): 20.

34. Anonymous, "Why Did Early Greeks See Star Colors Differently Than We Do?" *Ann Arbor News*, April 14, 1994; Louis Pauwels and Jacques Bergier, *The Morning of the Magicians* (New York: Avon, 1963), 308.

35. Halton Arp, *Quasars, Redshifts, and Controversies* (Berkeley: Interstellar Media, 1987), passim; C. Leroy Ellenberger, "Heretics, Dogmatics and Science's Reception of New Ideas," *Kronos* 4. no.4 (Summer 1979): 61.

36. Sharon McKern, *Exploring the Unknown* (New York: Praeger, 1972), 32; Anonymous, "Pay Dirt from the Moon," *Time* 95, no. 3 (January 19, 1970): 40; George Matzko, Moon Dust," *CRSQ* 27, no. 2 (September 1990): 77; Don DeYoung, "The Precision of Nuclear Decay Rates," *CRSQ* 13, no. 1 (June 1976): 38–39; Stephen Talbott, "A Record of Success," *Pensee* 2, no. 2 (May 1972), Anonymous, "Lunar Science: Light Amid the Heat," *Time* 100, no. 24 (December 11, 1972): 44; Arthur Custance, *Longevity in Antiquity, and Its Bearing on Chronology* 2 (Brockville ONT: Doorway Papers, 1957), 29–33.

37. Ronald Long, "The Bible, Radiocarbon Dating and Ancient Egypt," *CRSQ* 10, no. 1 (June 1973): 20; Donovan Courville, "The Use and Abuse of Astronomy in Dating Methods," *CRSQ*, 12, no. 3 (December 1976): 201–10; S. R. Glanville, ed., *The Legacy of Egypt* (Oxford: Clarendon Press, 1957), 10–11; John Dayton, *Minerals, Metals, Glazing & Man* (London: Harrap, 1978), 190.

38. Hadingham, *Circles and Standing Stones*, 14.

39. Donovan Courville, "Limitations of Astronomical Dating Methods," *Kronos* 1, no. 2 (Summer 1975): 70.

40. Michael Martin Nieto, *The Titius-Bode Law of Planetary Distances* (New York: Pergamon Press, 1972), 45; W. Mitchell, "On Falling Stones and Meteorites," *JVI* 1, no. 2 (1867): 441–42.

41. Lewis Greenberg, "Velikovsky and Venus," *Kronos* 4, no. 4 (Summer 1979): 1–15; Anonymous, "Surprise! Venus Has Craters," *Ann Arbor News* (August 5, 1973); Immanuel Velikovsky, *Worlds in Collision* (New York: Doubleday, 1950), 180–83; Arthur Sutton, "The Ruined Cities of Palestine," *JVI* 52 (1920): 52; Kurt Mendelssohn, *The Riddle of the Pyramids* (New York: Praeger, 1974): 177; William Mullen, "The Mesoamerican Record," *Pensee* 4, no. 4 (1974): 37. John Myers, "Sin and the Control Systrem," *Kronos* 2, no. 2 (November 1976): 88.

42. Donald Patten, *Catastrophism and the Old Testament* (Seattle: Pacific Meridian, 1988), 77.

43. Charles Berlitz, *Mysteries from Forgotten Worlds* (New York: Dell, 1972), 41; Peter Kolosimo, *Not of This World* (New York: Bantam, 1971), 229.

44. J. Gribben and S. Plagemann, *The Jupiter Effect* (New York: Walker. 1974); Richard Noone, *5/5/2000. Ice: the Ultimate Disaster* (New York: Harmony Books, 1982).

45. Ronald Duncan and Miriam Weston-Smith, eds., *The Encyclopaedia of Ignorance* (New York: Pergamon Press, 1977).

Chapter 15

1. Cyrus Gordon, *Before Columbus* (New York: Crown, 1971), 205.

2. Eric Powell, "Theme Park of the Gods?" *Archaeology* 57, no. 1 (January 2004): 60–67.

3. E. J. Statham, "Ancient Script in Australia," *JVI* 33 (1901): 254.

4. Anonymous, "Review: We Are Not the First," *Faith and Thought* 99:2 (1972): 156.

5. William Ryan and Walter Pitman, *Noah's Flood* (New York: Simon & Schuster, 1998).

6. James Reddie, "On the Credibility of Darwinism," *JVI* 2 (1866): 72.

7. William Corliss, *Ancient Man: A Handbook of Puzzling Artifacts* (Glen Arm MD: Sourcebook Project, 1978), 675.

8. Bishop Titcomb, "On the Antiquity of Civilisation," *JVI* 3 (1868): 21–22.

9. Stewart Easton, *The Western Heritage*, 2nd ed. (New York: Holt, Rinehart and Winston, 1966), 11.

10. Giorgio de Santillana and Hertha von Dechend, *Hamlet's Mill* (Boston: Gambit, 1969), 65–66.

11. Alexander Marshack, *The Roots of Civilization* (New York: McGraw-Hill, 1972), 11.

12. Nelson Glueck, "The Bible as Divining Rod," *Horizon* 2, no. 2 (November 1959): 6.

13. William F. Albright, *Recent Discoveries in Bible Lands*. Supplement to Robert Young, *Analytical Concordance to the Bible* (New York: Funk & Wagnalls, 1936, 1955), 25.

14. Vilhjalmur Stefansson, *Greenland* (New York: Doubleday, 1942): 25–26.

15. William Albright and T. O. Lambdin, *The Evidence of Language* (York: Cambridge University Press, 1966), 12.

16. Jaroslav Pelikan, *Genesis* (St. Louis: Concordia, 1958), 99.

17. John Milton, *Paradise Lost*, Book 1 (1608–1674).

18. Carl Keil and Franz Delitzsch, *Commentary on the Old Testament* (Grand Rapids: Eerdmans, 1975).

19. Alfred Rehwinkel, *The Age of the Earth and Chronology of the Bible* (Adelaide, South Australia: Lutheran Publishing House, 1967).

20. Erich von Fange, *Noah to Abram: the Turbulent Years* (Adrian MI, Living Word Services, 1994).

21. James Bailey, *The God-Kings and the Titans* (New York: St. Martin's Press, 1973).

22. Victor Pearce, *Who Was Adam?* (Toronto: The Paternoster Press, 1969), 71.

23. Arthur Custance, *Fossil Man and Genesis*, Doorway Paper 45 (Ottawa: Author, 1968), 39; E. O. James, *The Ancient Gods* (New York: G. P. Putnam's Sons, 1960), 19.

24. William Corliss, *Strange Artifacts* M-1 (Glen Arm MD: The Sourcebook Project, 1974), 37.

25. Daniel Cohen, *Lost Worlds* (New York: Belmont Tower Books, 1969), 8–9.

Chapter 16

1. Anthony Standen, *Science is a Sacred Cow* (New York: Dutton, 1950).

2. Bill Moyers, *Genesis: A Living Conversation* (Garden City, NY: Doubleday, 1955).

3. Stephen Gould, "This View of Life," *Natural History* 84, no. 6 (June 1975): 18.

4. Ronald Duncan, ed., *The Encyclopaedia of Ignorance* (New York: Pergamon Press, 1977): 233.

5. Canadian Broadcasting Corporation, *Science and Deception* (Toronto: CBC Transcripts, 1982); Dorothy Vitaliano, *Legends of the Earth* (Bloomington: Indiana University Press, 1973): 6–7.

6. Harry Debelius, "Scientists Make Ass of Old Skull," *London Times*, May 14, 1984.

7. Anonymous, "The Making of Mankind's History," *SIS Workshop* 4, no. 3 (December 24, 1981).

8. Anonymous, "Artifacts too Old," *Origins* (1984): 52.

9. John Woodmorappe, *The Mythology of Modern Dating Methods* (El Cajon CA: Institute for Creation Research, 1999).

10. W. G. Fearnsides and O. M. B Bulman, *Geology in the Service of Man* (Baltimore: Penguin, 1950), 9.

11. Ian Barbour, ed., *Science and Religion* (New York: Harper & Row, 1968), 68.

12. Vitaliano, *Legends of the Earth*, 7.

13. Robert Charroux, *The Gods Unknown* (New York: Berkley, 1969), v; Gardner, "Theological Surrender," *JASA* 25, no. 4 (December 12, 1973): 157; Jeffrey Sheler, "The Pope and Darwin," *USN&WR* 121, no. 12 (November 4, 1996): 12.

14. Duncan, *Encyclopaedia of Ignorance*.

15. Giorgio de Santillana and Hertha von Dechend, *Hamlet's Mill* (Boston: Gambit, 1969), xi; Stephen Talbott, "On the Need for Serious Scientific Meetings," *Pensee* 4, no. 2 (Spring 1974): 28; Kenneth Macgowan and Joseph Hester Jr., *Early Man in the New World*, rev. ed. (Garden City, NY: Anchor Books, 1962), 32; W. Raymond Drake, *Gods and Spacemen of the Ancient Past* (New York: Signet, 1974: 24.

16. Geoffrey Bibby, *Looking for Dilmun* (New York: Knopf, 1970), 181.

17. Alan and Sally Landsburg, *In Search of Ancient Mysteries* (New York: Bantham, 1974), 153.

18. George Simpson, *Life of the Past* (New York: Bantam, 1953), 5.

19. Barbour, *Science and Religion*, 62–71.

20. Marvin Lubenow, "Significant Fossil Discoveries Since 1958: Creationism Confirmed," *CRSQ* 17, no. 3 (December 1980): 152.

21. Louis Pauwels and Jacques Bergier, *The Morning of the Magicians* (New York: Avon, 1963), 161.

22. Oscar Muscarella, *The Lie Became Great* (Gronigen: Styx, 2000); David Dymond, *Archaeology and History. A Plea for Reconciliation* (London: Thames and Hudson, 1974).

23. Philip Kitcher, *Abusing Science: The Case Against Creationism* (Cambridge: MIT Press, 1982).

24. Fred Hechinger, "Scholars Fault Quality—and Quantity," *AAN* (March 15, 1987); James Lloyd, "Selling Scholarship Down the River," *CHE* 30, no. 17 (June 26, 1985): 64.

25. Loren Eiseley, *The Immense Journey* (New York: Time, 1962), passim.

26. Frank Marsh, *Fundamental Biology* (Lincoln NE: Author, 1941); William Broad, *Betrayers of the Truth* (New York: Simon & Schuster, 1982); Standen, *Science is a Sacred Cow*.

27. Alan Brodrick, *Man and His Ancestors* (New York: Previer, 1964), 62.

28. Susan Vela, "Author of Hoax Hones His Point on Intellectuals," *Ann Arbor News*, November 19, 1996; Larry Van Dyne, "The Conservative Prejudice of a Scholarly Pariah," *CHE* 25 no. 11 (November 10, 1982): 22.

29. Theodore Graebner, *God and the Cosmos* (Grand Rapids: Eerdmans, 1932), 250–51; William Corliss, "Hueyatlaco Dilemma," *Science Frontiers* 21 (May 5, 1982); RCNN, "Mystery of the Radiohalos," *Research Communications Network* (February 10, 1977): 2.

30. George Kunz, *Ivory and the Elephant* (Garden City, NY: Doubleday, 1916).

31. *Argosy*, May 1969; *Christian News*, June 30, 1969.

32. *Discover* July 1985, p.8; Ian Taylor, *In the Minds of Men* (Minneapolis: TFE Publishing, 1984), 178.

33. Christopher Sloan, "Feathers for T. Rex?" *National Geographic* 196, no. 5 (November 1999): 99; Mary Lord, "The Piltdown Chicken," *USN&WR* 128, no. 6 (February 14, 2000): 53.

34. Adolph Reith, "Archaeological Fakes," *USN&WR* 128, no. 6 (February 14, 2000): 53. Praeger, 1970), 31.

35. *Bible-Science News* (July, 1972): 7; Associated Press, "Artist Says Carvings Are His," *Ann Arbor News*, October 26, 1990.

36. George Norrie, "New 'Missing Link' Clews Discovered in African Jungle," *San Francisco Examiner* (December 3, 1950).

37. William Howells, ed., *Ideas on Human Evolution* (New York: Atheneum, 1962), 351–54.

38. Derek Ager, *The Nature of the Stratigraphical Record* (New York: Wiley, 1973), passim.

39. Hugh Brown, et al, *Introduction to Geology* (Boston: Ginn, 1958), 384–85.

40. George Howe, "Creationist Botany Today," *CRSQ* 6, no. 2 (September 1969): 85–91.

41. Arthur Custance, *The Fallacy of Anthropological Reconstructions* #33 (Brockville ONT: Doorway Papers, 1966): 18.

42. de Santillana and von Dechend, *Hamlet's Mill*, 71–74.

43. William Broad and Nicholas Wade, *Betrayers of the Truth* (New York: Simon & Schuster, 1982); W. Raymond Drake, *Gods and Spacemen of the Ancient Past* (New York: Signet, 1974), 43; Patrick Huyghe, "The Glowing Birds and 2,000 Other Mysteries that Stump Science," *Science Digest* 91, no. 2 (February 1983): 71–75.

44. William Corliss, *Scientific Anomalies and Other Provocative Phenomena* (Glen Arm, MD: Sourcebook Project, 2003), passim.

45. Richard Milton, *Shattering the Myths of Darwinism* (Rochester, VT: Park Street Press, 1997).

46. Ralph Estling, "A Sideways Look at Scientific Method," *New Scientist* 96, no. 1337 (December 23, 1982): 808–10.

Chapter 17

1. Joseph Alsop, *From the Silent Earth* (New York: Harper & Row, 1964), vii.

2. American Humanist Association, *Humanist Manifesto II* (Buffalo, NY: Prometheus Press, 1973).

3. Sharon Begley, "Heretics in the Laboratory," *Newsweek* 128, no. 12 (September 16, 1996): 82.

4. Immanuel Velikovsky, "My Challenge to Conventional Views in Science," *Pensee* 4, no. 2 (Spring, 1974): 10–14.

5. Reginald Daly, *Earth's Most Challenging Mysteries* (Nutley, NJ: Craig Press, 1972), 214.